W9-CXT-956

Thin Films—Stresses
and Mechanical Properties XI

MATERIALS RESEARCH SOCIETY
SYMPOSIUM PROCEEDINGS VOLUME 875

Thin Films—Stresses and Mechanical Properties XI

Symposium held March 28–April 1, 2005, San Francisco, California, U.S.A.

EDITORS:

Thomas E. Buchheit
Sandia National Laboratories
Albuquerque, New Mexico, U.S.A.

Andrew M. Minor
Lawrence Berkeley National Laboratory
Berkeley, California, U.S.A.

Ralph Spolenak
ETH Zurich
Zurich, Switzerland

Kazuki Takashima
Kumamoto University
Kumamoto, Japan

Materials Research Society
Warrendale, Pennsylvania

Single article reprints from this publication are available through
University Microfilms Inc., 300 North Zeeb Road, Ann Arbor, Michigan 48106

CODEN: MRSPDH

Copyright 2005 by Materials Research Society.
All rights reserved.

This book has been registered with Copyright Clearance Center, Inc. For further information, please
contact the Copyright Clearance Center, Salem, Massachusetts.

Published by:

Materials Research Society
506 Keystone Drive
Warrendale, PA 15086
Telephone (724) 779-3003
Fax (724) 779-8313
Web site: http://www.mrs.org/

Manufactured in the United States of America

CONTENTS

*Invited Paper

MECHANICAL BEHAVIOR OF
NANOSTRUCTURED FILMS

MECHANICAL PROPERTIES OF
THIN FILMS—TESTING AND ANALYSIS

THIN FILM PLASTICITY—
SIZE EFFECTS

THIN FILM PLASTICITY—
CREEP

*Invited Paper

FATIGUE AND STRESS IN
INTERCONNECT AND METALLIZATION

*Invited Paper

DEFORMATION, GROWTH AND MICROSTRUCTURE IN THIN FILMS

CHARACTERIZING AND UNDERSTANDING THIN FILM GROWTH STRESSES

THIN FILM PROCESSING

PREFACE

This proceedings volume reports on research presented at Symposium O, "Thin Films—Stresses and Mechanical Properties XI," held March 28–April 1 at the 2005 MRS Spring Meeting in San Francisco, California. This 11th symposium in the continuing series, held at 1-1/2 year intervals, provided a forum for an exchange of ideas between researchers who are interested in the mechanical behavior of thin films, broadly applied to their materials choice or methodology. This proceedings contains several papers discussing stress-related phenomena in thin films for a wide range of materials, including brittle, metallic, polymeric and biological materials. Over 100 submissions were selected for presentation at the Meeting, and 65 papers are included in these proceedings.

Thomas E. Buchheit
Andrew M. Minor
Ralph Spolenak
Kazuki Takashima

September 2005

ACKNOWLEDGMENTS

The organizers wish to thank the invited speakers, authors, and MRS staff for their contributions to the symposium. We would also like to thank the following companies that provided symposium support:

Hysitron Inc.
WITec GmbH

MATERIALS RESEARCH SOCIETY SYMPOSIUM PROCEEDINGS

Volume 828— Semiconductor Materials for Sensing, S. Seal, M-I. Baraton, N. Murayama, C. Parrish, 2005, ISBN: 1-55899-776-8

Volume 829— Progress in Compound Semiconductor Materials IV—Electronic and Optoelectronic Applications, G.J. Brown, M.O. Manasreh, C. Gmachl, R.M. Biefeld, K. Unterrainer, 2005, ISBN: 1-55899-777-6

Volume 830— Materials and Processes for Nonvolatile Memories, A. Claverie, D. Tsoukalas, T-J. King, J. Slaughter, 2005, ISBN: 1-55899-778-4

Volume 831— GaN, AlN, InN and Their Alloys, C. Wetzel, B. Gil, M. Kuzuhara, M. Manfra, 2005, ISBN: 1-55899-779-2

Volume 832— Group-IV Semiconductor Nanostructures, L. Tsybeskov, D.J. Lockwood, C. Delerue, M. Ichikawa, 2005, ISBN: 1-55899-780-6

Volume 833— Materials, Integration and Packaging Issues for High-Frequency Devices II, Y.S. Cho, D. Shiffler, C.A. Randall, H.A.C. Tilmans, T. Tsurumi, 2005, ISBN: 1-55899-781-4

Volume 834— Magneto-Optical Materials for Photonics and Recording, K. Ando, W. Challener, R. Gambino, M. Levy, 2005, ISBN: 1-55899-782-2

Volume 835— Solid-State Ionics—2004, P. Knauth, C. Masquelier, E. Traversa, E.D. Wachsman, 2005, ISBN: 1-55899-783-0

Volume 836— Materials for Photovoltaics, R. Gaudiana, D. Friedman, M. Durstock, A. Rockett, 2005, ISBN: 1-55899-784-9

Volume 837— Materials for Hydrogen Storage—2004, T. Vogt, R. Stumpf, M. Heben, I. Robertson, 2005, ISBN: 1-55899-785-7

Volume 838E—Scanning-Probe and Other Novel Microscopies of Local Phenomena in Nanostructured Materials, S.V. Kalinin, B. Goldberg, L.M. Eng, B.D. Huey, 2005, ISBN: 1-55899-786-5

Volume 839— Electron Microscopy of Molecular and Atom-Scale Mechanical Behavior, Chemistry and Structure, D. Martin, D.A. Muller, E. Stach, P. Midgley, 2005, ISBN: 1-55899-787-3

Volume 840— Neutron and X-Ray Scattering as Probes of Multiscale Phenomena, S.R. Bhatia, P.G. Khalifah, D. Pochan, P. Radaelli, 2005, ISBN: 1-55899-788-1

Volume 841— Fundamentals of Nanoindentation and Nanotribology III, D.F. Bahr, Y-T. Cheng, N. Huber, A.B. Mann, K.J. Wahl, 2005, ISBN: 1-55899-789-X

Volume 842— Integrative and Interdisciplinary Aspects of Intermetallics, M.J. Mills, H. Clemens, C-L. Fu, H. Inui, 2005, ISBN: 1-55899-790-3

Volume 843— Surface Engineering 2004—Fundamentals and Applications, J.E. Krzanowski, S.N. Basu, J. Patscheider, Y. Gogotsi, 2005, ISBN: 1-55899-791-1

Volume 844— Mechanical Properties of Bioinspired and Biological Materials, C. Viney, K. Katti, F-J. Ulm, C. Hellmich, 2005, ISBN: 1-55899-792-X

Volume 845— Nanoscale Materials Science in Biology and Medicine, C.T. Laurencin, E. Botchwey, 2005, ISBN: 1-55899-793-8

Volume 846— Organic and Nanocomposite Optical Materials, A. Cartwright, T.M. Cooper, S. Karna, H. Nakanishi, 2005, ISBN: 1-55899-794-6

Volume 847— Organic/Inorganic Hybrid Materials—2004, C. Sanchez, U. Schubert, R.M. Laine, Y. Chujo, 2005, ISBN: 1-55899-795-4

Volume 848— Solid-State Chemistry of Inorganic Materials V, J. Li, M. Jansen, N. Brese, M. Kanatzidis, 2005, ISBN: 1-55899-796-2

Volume 849— Kinetics-Driven Nanopatterning on Surfaces, E. Wang, E. Chason, H. Huang, G.H. Gilmer, 2005, ISBN: 1-55899-797-0

Volume 850— Ultrafast Lasers for Materials Science, M.J. Kelley, E.W. Kreutz, M. Li, A. Pique, 2005, ISBN: 1-55899-798-9

Volume 851— Materials for Space Applications, M. Chipara, D.L. Edwards, S. Phillips, R. Benson, 2005, ISBN: 1-55899-799-7

Volume 852— Materials Issues in Art and Archaeology VII, P. Vandiver, J. Mass, A. Murray, 2005, ISBN: 1-55899-800-4

Volume 853E—Fabrication and New Applications of Nanomagnetic Structures, J-P. Wang, P.J. Ryan, K. Nielsch, Z. Cheng, 2005, ISBN: 1-55899-805-5

Volume 854E—Stability of Thin Films and Nanostructures, R.P. Vinci, R. Schwaiger, A. Karim, V. Shenoy, 2005, ISBN: 1-55899-806-3

Volume 855E—Mechanically Active Materials, K.J. Van Vliet, R.D. James, P.T. Mather, W.C. Crone, 2005, ISBN: 1-55899-807-1

MATERIALS RESEARCH SOCIETY SYMPOSIUM PROCEEDINGS

Volume 856E— Multicomponent Polymer Systems—Phase Behavior, Dynamics and Applications, K.I. Winey, M. Dadmun, C. Leibig, R. Oliver, 2005, ISBN: 1-55899-808-X

Volume 858E— Functional Carbon Nanotubes, D.L. Carroll, B. Weisman, S. Roth, A. Rubio, 2005, ISBN: 1-55899-810-1

Volume 859E— Modeling of Morphological Evolution at Surfaces and Interfaces, J. Evans, C. Orme, M. Asta, Z. Zhang, 2005, ISBN: 1-55899-811-X

Volume 860E— Materials Issues in Solid Freeforming, S. Jayasinghe, L. Settineri, A.R. Bhatti, B-Y. Tay, 2005, ISBN: 1-55899-812-8

Volume 861E— Communicating Materials Science—Education for the 21st Century, S. Baker, F. Goodchild, W. Crone, S. Rosevear, 2005, ISBN: 1-55899-813-6

Volume 862— Amorphous and Nanocrystalline Silicon Science and Technology—2005, R. Collins, P.C. Taylor, M. Kondo, R. Carius, R. Biswas, 2005, ISBN 1-55899-815-2

Volume 863— Materials, Technology and Reliability of Advanced Interconnects—2005, P.R. Besser, A.J. McKerrow, F. Iacopi, C.P. Wong, J. Vlassak, 2005, ISBN 1-55899-816-0

Volume 864— Semiconductor Defect Engineering—Materials, Synthetic Structures and Devices, S. Ashok, J. Chevallier, B.L. Sopori, M. Tabe, P. Kiesel, 2005, ISBN 1-55899-817-9

Volume 865— Thin-Film Compound Semiconductor Photovoltaics, W. Shafarman, T. Gessert, S. Niki, S. Siebentritt, 2005, ISBN 1-55899-818-7

Volume 866— Rare-Earth Doping for Optoelectronic Applications, T. Gregorkiewicz, Y. Fujiwara, M. Lipson, J.M. Zavada, 2005, ISBN 1-55899-819-5

Volume 867— Chemical-Mechanical Planarization—Integration, Technology and Reliability, A. Kumar, J.A. Lee, Y.S. Obeng, I. Vos, E.C. Jones, 2005, ISBN 1-55899-820-9

Volume 868E— Recent Advances in Superconductivity—Materials, Synthesis, Multiscale Characterization and Functionally Layered Composite Conductors, T. Holesinger, T. Izumi, J.L. MacManus-Driscoll, D. Miller, W. Wong-Ng, 2005, ISBN 1-55899-822-5

Volume 869— Materials, Integration and Technology for Monolithic Instruments, J. Theil, T. Blalock, M. Boehm, D.S. Gardner, 2005, ISBN 1-55899-823-3

Volume 870E— Giant-Area Electronics on Nonconventional Substrates, M.S. Shur, P. Wilson, M. Stutzmann, 2005, ISBN 1-55899-824-1

Volume 871E— Organic Thin-Film Electronics, A.C. Arias, N. Tessler, L. Burgi, J.A. Emerson, 2005, ISBN 1-55899-825-X

Volume 872— Micro- and Nanosystems—Materials and Devices, D. LaVan, M. McNie, S. Prasad, C.S. Ozkan, 2005, ISBN 1-55899-826-8

Volume 873E— Biological and Bio-Inspired Materials and Devices, K.H. Sandhage, S. Yang, T. Douglas, A.R. Parker, E. DiMasi, 2005, ISBN 1-55899-827-6

Volume 874E— Structure and Mechanical Behavior of Biological Materials, P. Fratzl, W.J. Landis, R. Wang, F.H. Silver, 2005, ISBN 1-55899-828-4

Volume 875— Thin Films—Stresses and Mechanical Properties XI, T. Buchheit, R. Spolenak, K. Takashima, A. Minor, 2005, ISBN 1-55899-829-2

Volume 876E— Nanoporous and Nanostructured Materials for Catalysis, Sensor and Gas Separation Applications, S.W. Lu, H. Hahn, J. Weissmüller, J.L. Gole, 2005, ISBN 1-55899-830-6

Volume 877E— Magnetic Nanoparticles and Nanowires, D. Kumar, L. Kurihara, I.W. Boyd, G. Duscher, V. Harris, 2005, ISBN 1-55899-831-4

Volume 878E— Solvothermal Synthesis and Processing of Materials, S. Komarneni, M. Yoshimura, G. Demazeau, 2005, ISBN 1-55899-832-2

Volume 879E— Chemistry of Nanomaterial Synthesis and Processing, X. Peng, X. Feng, J. Liu, Z. Ren, J.A. Voigt, 2005, ISBN 1-55899-833-0

Volume 880E— Mechanical Properties of Nanostructured Materials—Experiments and Modeling, J.G. Swadener, E. Lilleodden, S. Asif, D. Bahr, D. Weygand, 2005, ISBN 1-55899-834-9

Volume 881E— Coupled Nonlinear Phenomena—Modeling and Simulation for Smart, Ferroic and Multiferroic Materials, R.M. McMeeking, M. Kamlah, S. Seelecke, D. Viehland, 2005, ISBN 1-55899-835-7

Volume 882E— Linking Length Scales in the Mechanical Behavior of Materials, T.J. Balk, R.E. Rudd, N. Bernstein, W. Windl, 2005, ISBN 1-55899-836-5

Volume 883— Advanced Devices and Materials for Laser Remote Sensing, F. Amzajerdian, A.A. Dyrseth, D. Killinger, L. Merhari, 2005, ISBN 1-55899-837-3

Volume 884E— Materials and Technology for Hydrogen Storage and Generation, G-A. Nazri, C. Ping, R.C. Young, M. Nazri, J. Wang, 2005, ISBN 1-55899-838-1

Prior Materials Research Society Symposium Proceedings available by contacting Materials Research Society

Elasticity in Thin Films

Mater. Res. Soc. Symp. Proc. Vol. 875 © 2005 Materials Research Society

Advanced Resonant-Ultrasound Spectroscopy
for Studying Anisotropic Elastic Constants of Thin Films

Hirotsugu Ogi, Nobutomo Nakamura, Hiroshi Tanei, and Masahiko Hirao
Graduate School of Engineering Science, Osaka University
Toyonaka, Osaka 560-8531, Japan

ABSTRACT

This paper presents two advanced acoustic methods for the determination of anisotropic elastic constants of deposited thin films. They are resonant-ultrasound spectroscopy with laser-Doppler interferometry (RUS/Laser method) and picosecond-laser ultrasound method. Deposited thin films usually exhibit elastic anisotropy between the film-growth direction and an in-plane direction, and they show five independent elastic constants denoted by C_{11}, C_{33}, C_{44}, C_{66} and C_{13} when the x_3 axis is set along the film-thickness direction. The former method determines four moduli except C_{44}, the out-of-plane shear modulus, through free-vibration resonance frequencies of the film/substrate specimen. This method is applicable to thin films thicker than about 200 nm. The latter determines C_{33}, the out-of-plane modulus, accurately by measuring the round-trip time of the longitudinal wave traveling along the film-thickness direction. This method is applicable to thin films thicker than about 20 nm. Thus, combination of these two methods allows us to discuss the elastic anisotropy of thin films. The results for Co/Pt superlattice thin film and copper thin film are presented.

INTRODUCTION

Elastic constants of thin films are required primarily for three reasons. First, they are indispensable to calculation of internal stresses in a multiphase composite caused by lattice misfit and different thermal-expansion coefficients among constituents. Second, they are needed to calculate elastic strain energy to find a minimum of the free energy for the estimation of possible microstructure. Third, they are capable of evaluating defects because defects such as voids, dislocations, and microcrackings affect the elastic constants through elastic softening. However, measurement of the elastic constants of thin films has never been straightforward due to elastic anisotropy: Thin films, even polycrystalline thin films, show different elastic properties between along the film-growth direction and along an in-plane direction. Such anisotropy originates from texture, columnar structure, oriented microcracks or precipitates, and internal stresses. The thin films then show transverse isotropy or hexagonal symmetry and possess five independent elastic constants C_{ij}:

$$[C_{ij}] = \begin{bmatrix} C_{11} & C_{12} & C_{13} & 0 & 0 & 0 \\ & C_{11} & C_{13} & 0 & 0 & 0 \\ & & C_{33} & 0 & 0 & 0 \\ & & & C_{44} & 0 & 0 \\ & sym. & & & C_{44} & 0 \\ & & & & & C_{66} \end{bmatrix}$$

when the x_3 axis is taken along the film-growth direction, where $C_{66}=(C_{11}-C_{12})/2$. Most existing methods, however, assumed thin films to be isotropic materials and deduced only one or two moduli among five. They failed to detect elastic anisotropy.

The acoustic methods we present in this paper determine four components of C_{ij} among five. They are resonant-ultrasound spectroscopy coupled with laser-Doppler interferometry and picosecond-laser ultrasound method. We demonstrate accuracy and correctness of our methods, showing the results for Co/Pt superlattice thin film and copper thin film.

KEY POINTS FOR AN ACCURATE MEASUREMENT

Previous methods fall in two groups. One is static or quasistatic methods. The other is dynamic methods. Static or quasistatic methods include the microtensile test [1-3], microbending test [4,5], and nanoindentation method. The dynamic methods include the flexural-vibration method [6,7], Brillouin-scattering method [8-10], and surface-acoustic-wave method [11,12]. These previous works involve several difficulties in obtaining the elastic constants of thin films. First of all, static methods are severely affected by errors in dimensions, especially by the film-thickness error. For example, the microbending test evaluates the mechanical behavior of thin films using a cantilever-beam specimen cut from the thin film. The deflection at the free end of the cantilever is proportional to the third power of the film thickness and second power of the width of the cantilever. Thus, the dimension errors have large influence on the measurement. Also, many previous methods are strongly affected by the gripping condition: For example, in the microbending method and flexural-vibration method, the maximum bending stress appears at the fixed end and the measurements are highly affected by the gripping condition because these methods use a simple beam-bending theory, which assumes a complete fixed (rigid) boundary. Besides in the microtensile test, it is difficult to apply a uniaxial stress in the in-plane direction and bending and torsional stresses arise, which affect significantly the stress-strain relationship. In the dynamic methods, such difficulties are less remarkable, but it is still difficult to detect the elastic anisotropy between the in-plane and out-of-plane directions.

We propose three key points for an accurate measurement of the elastic constants. First, measure frequency. Measurement accuracy of resonant frequencies is normally much higher than that for force or displacements. Also, the resonant frequency of a solid is an absolute quantity, which depends only on the mass density, elastic constants, and boundary conditions. Significant advantage of measuring resonant frequencies also involves the fact that they are much less severely affected by dimension errors than static methods, because they are

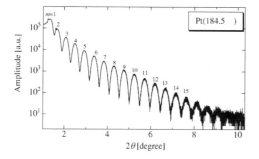
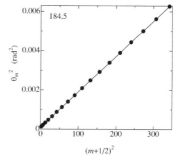

Figure 1. X-ray diffraction spectrum at low angles for platinum thin film with 184.5 Å thickness deposited on a silicon substrate (left), and the correlation between squares of the peak angle and corresponding order number (right). The slope yields the film thickness with known X-ray wavelength.

approximately expressed by $(K/M)^{0.5}$, where K and M denote the stiffness and mass of the system. If we overestimate the film thickness, the mass of the system increases. At the same time, however, the stiffness for bending or torsion increases and the dimension error is canceled.

Second, use free vibrations. Many previous methods involved ambiguous gripping condition at the fixed point. Free-vibration resonant frequencies are hardly affected by gripping conditions and the analysis can be accurate because of free boundaries.

Third, use X rays if possible. Previous works determined the thickness of the film by observing the cross-sectional area of the film with scanning electron microscopy (SEM), which caused 5% error at least. However, when we use an X-ray diffraction measurement, the error can be reduced to less than 1%. For example, Figure 1 shows the X-ray diffraction spectrum in a low-angle region observed from a 185-Å platinum film deposited on a silicon substrate. Many peaks originate from interference between the X ray reflected at the film-substrate interface and that reflected at the film surface. The mth peak angle θ_m is given by [13]

$$\theta_m^2 = \left(\frac{\lambda}{2d}\right)^2 (m+1/2)^2 + \theta_C^2, \tag{1}$$

for the platinum-silicon system. Here, d denotes the thickness of the platinum film, λ the X-ray wavelength, and θ_C the critical angle for the total reflection of X ray at platinum surface. Thus, plotting θ_m^2 versus $(m+1/2)^2$ yields a line, whose slope provides the film thickness with known X-ray wavelength (=1.548 Å for Cu-Kα). For a multilayer thin film, the X-ray diffraction spectrum show many peaks related to the bilayer thickness and we can similarly determine the bilayer thickness with the similar way and then the total thickness by multiplying the bilayer thickness by the repetition number [14].

The methodology we present here satisfies these demands. It is a combination of two advanced methods. They are (i) resonant-ultrasound spectroscopy with laser-Doppler interferometry and (ii) picosecond-laser ultrasound method. The former determines thin-film elastic constants from free-vibration resonant frequencies of the film/substrate specimen via an inverse calculation. This method is sensitive to in-plane moduli such as C_{11} and C_{66}. Its accuracy becomes worse for the out-of-plane moduli C_{33} and C_{13}. It cannot determine the C_{44}, the out-of-plane shear modulus. Thus, this technique determines four independent coefficients among five. We call this method the RUS/Laser technique. The latter method determines C_{33}, the out-of-plane modulus, with accuracy higher than that in any other methods including the RUS/Laser method. It generates the acoustic longitudinal wave propagating along the film-thickness direction by irradiating the film surface with the high-power and short-pulsed laser beam (pumping light). The pulse-echo signals of the longitudinal wave are detected by the delayed pulse light (probing light) through photoelastic phenomena.

Thus, combination of these methods allows us to determine the four moduli and to discuss the elastic anisotropy.

RUS/LASER METHOD

Resonant-ultrasound-spectroscopy (RUS) method has been applied to many anisotropic solids including monocrystals [15,16], composites [17,18], and piezoelectric materials [19-22]. It is capable of determining all the independent elastic constants from resonance frequencies; because all the elastic constants contribute to free-vibration resonant frequencies, they can be determined inversely by measuring the resonant frequencies with sufficient accuracy. This method, however, has not been successfully applied to thin films because of three difficulties. First, the resonant frequencies have to be measured with high accuracy because contributions of the elastic constants of the thin film to them are small; normalized contribution is about in the order of 10^{-2} for a 1-μm-thick film deposited on a 0.2-mm-thick substrate. Second, external forces applied to the specimen have to be eliminated to cause ideal free vibrations. Third, observed modes must be identified. At the beginning of the inverse calculation, we have to homologize individual measured frequency to calculated one. Because we exactly know vibration modes of calculated frequencies, we have to identify the observed frequencies.

The RUS/laser method overcomes these difficulties. We developed a piezoelectric tripod illustrated in Fig. 2. The tripod consists of the needle piezoelectric oscillator, needle piezoelectric detector, and a needle support. The specimen is placed on the tripod without using any coupling media between them. We apply a sinusoidal signal to the piezoelectric transducer to vibrate the specimen. The piezoelectric receiver detects the vibration of the specimen. By sweeping the frequency of the driving signal and acquiring the vibration amplitude as a function of the frequency, we obtain the resonance spectrum as shown in Fig. 3. The resonance frequencies are determined by the Lorentzian-function fitting. This setup requires no external force applied to the specimen except the specimen's weight, which allows us to carry out generating ideal free vibrations. The piezoelectric tripod we developed works for a very light-weight specimen with the mass about a few milligrams. Measurements were performed in a vacuum, eliminating damping and acoustic noise and at a constant temperature, eliminating its influence on the elastic properties of the films.

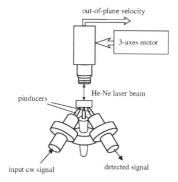

Figure 2. Measurement setup of the RUS/Laser method.

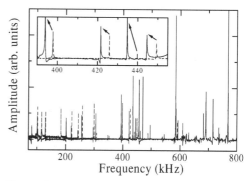

Figure 3. Resonance spectra measured by the piezoelectric tripod. Broken line shows the spectrum for a monocrystal silicon with dimensions of 6x4x0.2 mm^3 and solid line shows that after the 0.89-nm Co/Pt superlattice film was deposited on the substrate.

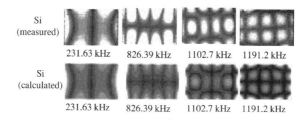

Si (measured)

231.63 kHz   826.39 kHz   1102.7 kHz   1191.2 kHz

Si (calculated)

231.63 kHz   826.39 kHz   1102.7 kHz   1191.2 kHz

Figure 4. Measured and calculated amplitude distributions of the out-of-plane displacement for a monocrystal silicon of of 6x4x0.2 mm^3.

The weak and stable acoustic coupling provides high accuracy and high reproducibility in the resonance-frequency measurement. Once a specimen is placed on the tripod, the fluctuation of measured frequencies are smaller than one part of million. The accuracy is still high among completely independent measurements, better than one part of 10,000.

Concerning mode identification, we measure distributions of the out-of-plane displacement amplitude on the vibrating specimen using laser-Doppler interferometry. They are then compared with the distributions calculated by the Rayleigh-Ritz method as shown later. Figure 4 shows examples of such comparison: Excellent agreement is always confirmed so that we identify all the observed vibration modes.

Actual procedure to determine the elastic constants are following: (i) The resonant frequencies of a silicon substrate are measured and their modes are identified. (ii) A thin film is deposited on it and resonant frequencies are measured again to record the changes of the resonant frequencies by identifying the vibration mode to find correct mode correspondence. (iii) Because the changes of the resonant frequencies can be calculated with sufficient accuracy as shown later, the elastic constants of the deposited thin film are determined by a least-squares-fitting procedure.

Analytic solutions of displacements and resonant frequencies at free vibrations are unavailable for the film/substrate specimen of rectangular-parallelepiped shape. However, they are approximately calculated by Lagrangian minimization with Rayleigh-Ritz method, which successfully satisfies accuracy required to determine reliable thin-film elastic constants through the inverse calculation [14,23]. Lagrangian for a solid subjected to free vibration is given by

$$L = \frac{1}{2} \int_V \left(S_i C_{ij} S_j - \rho \omega^2 u_i u_i \right) dV .$$

(2)

Where S_i is the engineering strain, C_{ij} the elastic constants, ρ the mass density, ω the angular resonant frequency of the system, u_i the displacement along the x_i axis, and V the volume of the film/substrate system. We make approximate estimates for displacements by linear combinations of basis functions Ψ_k

$$u_i(x_1, x_2, x_3) = \sum_k a_k^i \Psi_k^i(x_1, x_2, x_3) .$$

(3)

Here a_k denote the expansion coefficients and they provide us with the displacement distributions as show in Fig. 4. For a film/substrate rectangular parallelepiped, strains associated with the film-thickness direction (along the x_3 axis) are discontinuous across the interface because of the different moduli. Thus, we have to select basis functions that can express the broken gradients of the displacements across the interface. For this, Heyliger [24] showed that incorporation of one-dimensional Lagrangian interpolation polynomials along the x_3 direction gave good approximates for the displacements in a layered material. Thus, we use the basis functions which consist of one-dimensional Lagrangian interpolation polynomials for the x_3 direction $\xi(x_3)$ and power series for the in-plane directions (x_1 and x_2 axes):

$$\psi_k(x_1, x_2, x_3) = \left(\frac{x_1}{L_1} \right)^l \left(\frac{x_2}{L_2} \right)^m \xi_n(x_3) .$$

(4)

Here, L_1 and L_2 denote the length of the specimen along the x_1 and x_2 axes, respectively. l, m, and n are integer expressing the orders of the basis functions. Seeking the minimum of the Lagrangian, the problem is reduced to an eigenvalue problem:

$$\omega^2 [M]\{U\} = [K]\{U\}.$$

(5)

Here, [M] and [K] are matrices related to the kinetic energy and strain energy of the system, respectively. {U} is the eigenvector composed of the expansion coefficients a_k and it provides

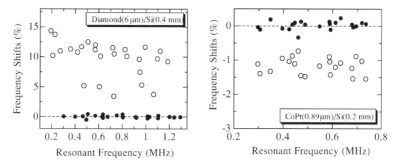

Figure 5. Changes of resonant frequencies by deposition of thin films (open marks) and differences between calculated and measured resonant frequencies (solid marks) for 6-μm-CVD-diamond film on 0.4-mm silicon substrate (left) and 0.89-μm-Co/Pt-superlattice film on 0.2-mm silicon substrate (right).

the displacement distributions through Eq. (3).

Inclusion of higher-order basis functions leads to more accurate resonance frequencies, but it takes longer calculation time. We found that use of the in-plane functions satisfying $l+m<17$ and the out-of-plane functions with $n<11$ lead to resonant frequencies with high enough accuracy and a suitable calculation time [14]. (In this case, the matrix size reaches 5000x5000, but the calculation time is drastically reduced when we consider the symmetry of vibration [14].)

Figure 5 shows resonant-frequency shifts caused by deposition of chemical-vapor-deposition (CVD) diamond film and Co/Pt superlattice film on silicon substrates. Deposition of the CVD diamond increases the resonant frequencies by 5-15% and deposition of Co/Pt superlattice film decreases them by 1-1.5 %. Resonant frequencies of the film/substrate systems are successfully calculated by the inverse calculation with the Lagrangian minimization, indicating reliable elastic constants of the thin films are determined. The rms difference between the measured and calculated resonant frequencies is typically less than 0.1%.

Figure 6 shows contributions of the thin-film elastic constants to the resonant frequencies. The RUS/Laser method is sensitive to in-plane moduli such as C_{11}, C_{12}, and C_{66} because most observed vibration modes are bending and torsional vibrations, causing larger in-plane deformation. However, it is insensitive to out-of-plane moduli such as C_{13}, C_{33}, and C_{44} because of smaller strains along the out-of-plane direction. Especially, C_{44}, the out-of-plane shear modulus, little affects the resonant frequencies and it is unavailable.

Figure 7 demonstrates tolerance of the RUS/Laser method to the error involved in film thickness. These results are for 1-μm and 3-μm copper thin films deposited on the 0.2-mm monocrystal silicon substrate. We artificially included the film-thickness error Δd in the inverse calculation. The resultant errors in the elastic constants are proportional to the film-thickness error. When the film thickness is measured by the cross-sectional observation by SEM with a 5% error, less than 5% errors are caused in the principal elastic constants.

Thus, the RUS/Laser method is capable of determining in-plane elastic constants with high

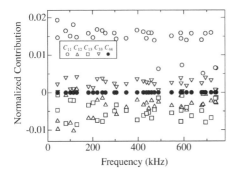

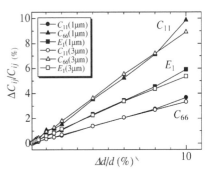

Figure 6. Contributions of the thin-film C_{ij} to resonant frequencies for 0.9-μm CoPt/0.2-mm Si system. C_{44} (solid circles) little contributes to frequencies.

Figure 7. Errors in the principal elastic constants caused by the film-thickness measurement error in the case of copper thin films deposited on 0.2-mm-thick silicon substrate.

accuracy and it is less severely affected by dimension errors. This method is applicable to thin films as thin as 200 nm, although the accuracy gets worse as the thickness decreases.

PICOSECOND-LASER ULTRASOUND METHOD

The picosecond-laser ultrasound method determines C_{33}, the out-of-plane modulus, of films thicker than 20 nm using pulse echoes of the longitudinal wave traveling along the film-thickness direction generated by the thermoelastic or photoelastic effect. There are several pioneers of this method [25-27] and following them we developed the measurement system shown in Fig. 8. We use a mode-locking titanium-sapphire pulse laser with 100-fs pulse width and 0.7-W power. The laser beam is split into two beams by a polarization beam splitter. One of the beams enters the second-harmonic-generator crystal, which outputs the light with the doubled frequency (pumping light). It is modulated by an acousto-optic modulator and then is focused on the surface of the film through the objective lens to generate the longitudinal wave. The longitudinal wave propagates along the film-thickness direction and repeats reflections between the film/substrate interface and the film surface. The other light is split into two beams; one enters the photo detector to produce the reference signal and the other irradiates the specimen surface with a time delay compared with the arrival time of the pumping light. By changing the path length of the pumping light, we can change the time delay between the pumping and probing lights and we can detect the arrival of the longitudinal wave by monitoring the modified intensity and phase of the reflected probing light. Figure 9 shows an example of such a measurement for 77-nm Co/Pt thin film. Step at t=29 ps indicates the time when the pumping beam irradiated the film surface. Clearly, pulse echoes of the longitudinal wave are observed, which yield the round-trip time and then the C_{33}.

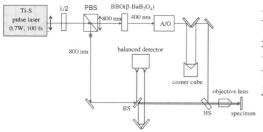

Figure 8. Optics of the picosecond-laser ultrasound method.

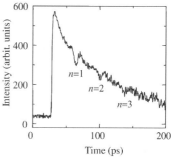

Figure 9. Pulse echoes of the longitudinal wave detected by the picosecond laser ultrasound method for 77-nm thin film of Co/Pt superlattice deposited on silicon substrate.

EXAMPLES

Co/Pt superlattice thin film

Table I shows the elastic constants of the Co/Pt multilayer thin film. We deposited 4Å Co and 16Å Pt alternately using ultrahigh-vacuum deposition method on the 0.2-mm thick monocrystal substrate [14,28]. The background pressure was of the order of 10^{-10} Torr and the deposition rate was 0.3 Å/s. The picosecond-laser (PSL) ultrasonic method deduced C_{33} smaller than that determine by the RUS/Laser method by 3%. This difference is small compared with the error limit, but such a decrease of the elastic constant determined by the picosecond-laser ultrasound method is reasonably understood considering the temperature increase due to an injection of the high-power pulse laser.

We note the elastic anisotropy between C_{11} and C_{33}. Using the elastic constants of the bulk materials, we estimated the elastic constants of the multilayer (simple rule of mixture), which predicts the in-plane longitudinal elastic constant C_{11} smaller than the out-of-plane modulus C_{33} as shown in Table I. However, their relationship is reversed; the measurement shows C_{11} larger than C_{33}. We attribute this enhancement of the in-plane elastic constant to the elastic strain caused by the lattice misfit. It is known that Co and Pt layers epitaxially bond on their closed packed planes and huge in-plane elastic strains occur because of lattice misfit [28-30]: The atomic distance in the (111) plane of Pt is larger than that in the (0001) plane of Co by 10.7%. Thus, the elastic strains reach in the order of 10^{-2}; Pt layers are compressed and Co layers are extended to a large extent. Such a large strain changes the elastic constants through lattice anharmonicity [31,32]. Because the increase of the elastic constants by contraction is more remarkable than the decrease of them in extension, the in-plane elastic constant of the multilayer is preferably affected by the enhanced elastic constants of the compressed Pt layers.

Table 1. Elastic constants (GPa) of Co/Pt superlattice thin film and copper thin film.

		C_{11}	C_{33}	C_{66}	C_{13}	C_{44}
Co/Pt superlattice	RUS/Laser (0.89 μm)	377.9 ± 7.1	367.9 ± 31.5	71.0 ± 2.0	236.9 ± 10.0	-
	PSL ultrasonics (77 nm)	-	360± 11	-	-	-
	rule of mixture (bulk)	360.1	379.4	66.4	-	-
copper	RUS/Laser (1.9 μm)	166.7 ± 14.0	122.0 ± 20.7	47.8 ± 4.1	131.8 ± 17.4	-
	polycrystal bulk	197.3	197.3	47.3	102.6	47.3
	(111) texture	213.0	237.6	50.8	86.8	35.6
	micromechanics[a]	177.6	125	47.3	65.1	40.5

[a]porosity: 6.6×10^{-4}

Copper thin film

Table I shows the elastic constants of a 1.9-μm copper thin film deposited on the silicon substrate. Significant observations are (i) C_{33} is smaller than C_{11}, and (ii) they are smaller than those of isotropic copper, which are calculated by the Hill averaging method using handbook values of monocrystal C_{ij} [33]. Because of high elastic anisotropy of copper, texture would affect the macroscopic elastic constants. The X-ray-diffraction spectrum indicated that (111) planes oriented preferentially parallel to the film surface. We then calculated the macroscopic C_{ij} of such the textured microstructure assuming that all the (111) planes of grains were aligned so as to be parallel to the film surface while their [110] directions were randomly oriented in the film plane [34]. The Hill averaging method was again used for this calculation. The results are shown in Table I. The C_{33} of the (111)-texture copper is larger than C_{11}, which is the opposite relationship to the observation. Thus, the texture cannot be a dominant factor for the observations.

We, therefore, considered the presence of the noncohesive bonded regions at grain boundaries as the dominant factor and estimated their effect by a micromechanics model. We replaced the noncohesive regions with oblate ellipsoidal microcracks. Details can be found elsewhere [35]. When major axes of the oblate ellipsoids are parallel to the film plane, the resultant elastic constants showed good agreement with the measurements with the volume fraction of the microcracks of 6.6×10^{-4} (see the last row in Table I.) Thus, presence of such an noncohesive bonded region explains the observations.

SUMMARY

The advanced ultrasound techniques, the RUL/Laser method and picosecond-laser ultrasound method overcome the difficulties that have not been solved in the past with previous methods in the determination of the elastic constants of thin films. The RUS/Laser method determines the in-plane moduli of deposited thin films but is insensitive to the out-of-plane shear modulus. It is applicable to thin films thicker than 200 nm. The picosecond-laser ultrasound method determines the out-of-plane longitudinal modulus. This method determines only this modulus, but its accuracy is the highest among the existing method. It is applicable to a very thin film, as thin as 20 nm. Thus, the combination of these two advanced acoustic methods will continue to provide us with knowledge of elasticity of functional thin films and with contributions to developments of many thin-film devices.

REFERENCES

1. W. Nix, Metall. Trans. A **20A**, 2217 (1989).
2. W. Suwito, D. Martin, S. Cunningham, and D. Read, *J. Appl. Phys.* **85,** 3519 (1999).
3. H. Huangand and F. Spaepen, *Acta Mater.* **48**, 3261 (2000).
4. K. Takashima, S. Koyama, K. Nakai and Y. Higo, 2002 MRS Fall Meeting, J3.3, (2002).
5. A. Ogura, R. Tarumi, M. Shimojo, K. Takashima, Y. Higo, *Appl. Phys. Lett.* **79**, 1042 (2001).
6. H. Mizubayashi, J. Matsuno, and H. Tanimoto, *Scripta Mater.* **41**, 443 (1999).
7. S. Sakai, H. Tanimoto, and H. Mizubayashi, *Acta Mater.* **47**, 211 (1999).
8. N. Rowell and G. Stegeman, *Phys. Rev. B* **18**, 2598 (1978).
9. A. Moretti, W. Robertson, B. Fisher, and R. Bray, *Phys. Rev. B* **31**, 3361 (1985).
10. J. Sandercock, in Light Scattering in Solids III, edited by M.Cardona and G. Güntherodt, Topics in Applied Physics Vol. 51 (Springer, New York, 1982), p. 173.
11. A. Moreau, J. Ketterson, and J. Huang, *Mater. Sci. Eng. A* **A126**, 149 (1990).
12. J. Kim, J. Achenbach, M. Shinn, and S. Barnett, *J. Mater. Res.* **7**, 2248 (1992).
13. H. Kiessig, *Annalen der Physik* **10**, 769 (1931).
14. N. Nakamura, H. Ogi, T. Ono, and M. Hirao, *J. Appl. Phys.* **97**, 013532 (2005).
15. I. Ohno, *J. Phys. Earth* **24**, 355 (1976).
16. A Migliori and J. Sarrao, *Resonant Ultrasound Spectroscopy* (Wiley-Interscience, New York, 1997).
17. H. Ledbetter, C. Fortunko, and P. Heyliger, *J. Appl. Phys.* **78**, 1542 (1995).
18. H. Ogi, M. Dunn, K. Takashima, and H. Ledbetter, J. Appl. Phys. 87, 2769 (2000).
19. I. Ohno, *Phys. Chem. Minerals*, **17**, 371 (1990).
20. H. Ogi, H. Ledbetter, Y. Kawasaki, and K. Sato, *J. Appl. Phys.* **92**, 2451 (2002).
21. H. Ogi, N. Nakamura, K. Sato, M. Hirao, and S. Uda, *IEEE Trans. Ultrason. Ferroelectr. Freq. Ctrl.* **50**, 553 (2003).
22. H. Ogi, M. Fukunaga, M. Hirao, and H. Ledbetter, *Phys. Rev. B* **69**, 024104 (2004).
23. H. Ogi, G. Shimoike, M. Hirao, K. Takashima, and Y. Higo, *J. Appl. Phys.* **91**, 4857 (2002).
24. P. Heyliger, *J. Acous. Soc. Am.* **107**, 1235 (2000).
25. D. Hurley and O. Wright, *Opt. Lett.* **24**, 1305 (1999).
26. T. Saito, O. Matsuda, and O. B. Wright, *Phys. Rev. B* **67**, 205421 (2003).
27. O. Matsuda, O. B. Wright, D. H. Hurley, V. E. Gusev, and K. Shimizu, *Phys. Rev. Lett.* **93**, 095501 (2004).
28. N. Nakamura, H. Ogi, T. Ono, and M. Hirao, *Appl. Phys. Lett.* **86** (2005), in press.
29. T. Kingetsu, Y. Kamada, and M. Yamamoto, *Sci. Tech. Adv. Mater.* **2**, 331 (2001).
30. Y. Kamada, Y. Hitomi, T. Kingetsu, and M. Yamamoto, *J. Appl. Phys.* **90**, 5104 (2001).
31. Y. Hiki and A. Granato, *Phys. Rev.* **144**, 411 (1966).

32. H. Ogi, N. Suzuki and M. Hirao, *Metall. Mater. Trans. A*, **29A**, 2987 (1998).
33. H. Ogi, S. Kai, H. Ledbetter, R. Tarumi, M. Hirao, and K. Takashima, *Acta Mater.* **52**, 2075 (2004).
34. N. Nakamura, H. Ogi, and M. Hirao, *Acta Mater.* **52**, 765 (2004).
35. M. Hirao and H. Ogi, *EMATs for Science and Industry* (Kluwer-Academic, Boston, 2003).

Mater. Res. Soc. Symp. Proc. Vol. 875 © 2005 Materials Research Society   O1.2

Elastic Constants and Graphitic Grain Boundaries of Nanocrystalline CVD-Diamond Thin Films: Resonant Ultrasound Spectroscopy and Micromechanics Calculation

Hirotsugu Ogi[1], Nobutomo Nakamura[1], Hiroshi Tanei[1], Ryuji Ikeda[2,3], Masahiko Hirao[1], and Mikio Takemoto[3]
[1]Graduate School of Engineering Science, Osaka University, Toyonaka, Osaka 560-8531, Japan
[2]Asahi Diamond Ind Co Ltd, Res & Dev, Chiba 290-0515, Japan
[3]Faculty of Science and Engineering, Aoyama Gakuin University, Kanagawa 229-8558, Japan

ABSTRACT

Using resonant-ultrasound spectroscopy coupled with laser-Doppler interferometry, we determine the independent elastic constants of nanocrystalline CVD-diamond thin films with thickness between 2-12 μm. They are deposited on oriented monocrystal silicon substrates by the hot-filament methane/nitrogen CVD method. The diagonal components of the elastic constants are smaller than those of microcrystalline CVD diamond films and bulk diamond. However, the off-diagonal component is larger. We attribute these observations to the presence of sp^2-bonded graphitic phase at grain boundaries. A micromechanics model assuming inclusions of thin graphitic plates consistently explains the observations.

INTRODUCTION

Nanocrystalline diamond (NCD) originates an emerging materials-science field because of their remarkable properties such as high stiffness, high hardness, enhanced electrical conductivity, and flat surface, which greatly improve surface-acoustic-wave (SAW) devices, field-emission transistor, and machine tools for high-precision processing. Recently their microstructure and bond configuration are studied by transmission-electron microscopy [1-4], Raman-spectroscopy [1-5], and electron energy-loss spectroscopy [3]. These efforts reveal that NCD films consist of sp^3-bonded diamond grains and sp^2-bonded grain boundaries. Therefore, it is expected that strongly-bonded diamond grains in NCD films are less strongly connected via sp^2 bonds. Such a specific microstructure can show anomalous elastic properties. However, few reports appear concerning the elastic constants of NCD films because of the difficulty of the measurement.

The purpose of this study is to measure the independent components of the elastic-constant matrix of NCD thin films and analyze the results with micromechanics modeling. For this, we use resonant-ultrasound spectroscopy with laser-Doppler interferometry (RUS/Laser method) [6-8], which determines independent elastic constants of anisotropic thin films. Several methods are reported for evaluating the elastic constants of thin films, including microtensile tests, microbending tests, and nanoindentation method. However, these previous methods involve difficulties for determining the thin-film elastic constants. First, they are severely affected by dimension errors. Second, some of them involve ambiguous gripping conditions. Third, they cannot determine independent elastic-constant component simultaneously: Polycrystalline thin films usually show elastic anisotropy between the film-growth direction and

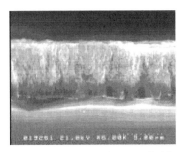

Figure 1. Cross-sectional area of 6.3-μm thick NCD film observed by a scanning-electron microscope.

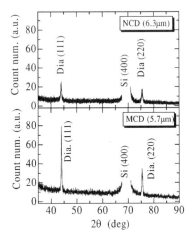

Figure 2. X-ray diffraction spectra for 6.3-μm NCD film and 5.76.3-μm MCD film.

an in-plane direction because of texture, columnar structure, oriented microcracks and internal stresses. The thin films then show transverse symmetry and have five independent elastic constants. However, most previous works assumed thin films to be isotropic and evaluated only one or two moduli. The RUS/Laser method we developed is capable of determining the independent elastic constants of thin films with much higher accuracy than in previous methods.

Here, we show that the off-diagonal component of the elastic constants of NCD films is significantly larger than that of bulk diamond. Such an enhancement of the off-diagonal stiffness is not observed for microcrystalline diamond films.

SPECIMENS

Five NCD thin films were deposited on the rectangular-parallelepiped monocrystal silicon substrate by the hot-filament chemical-vapor-deposition (CVD) method with various thicknesses between 2-12 μm. The substrate were 6-mm long, 4-mm wide, and 0.4-mm thick. Source gasses were 96.5%H_2, 3%CH_4, and 0.5%N_2. Deposition rate was 0.33 μm/h and substrate's temperature was 620 °C. The grain size was between 5 and 30 nm. Figure 1 shows the cross-sectional view of an NCD film. For comparison, we prepared three microcrystalline-diamond (MCD) thin films on the silicon substrates; source gasses were 98.5 H_2 and 1.5%CH_4. The deposition rate was 0.64 μm/h and substrate's temperature was 720 °C. The film thickness varied from 3.8 to 17.3 μm. The grain size was of the order of the film thickness. The film thickness and grain size were measured by a field-emission scanning-electron microscope. We performed visible Raman spectroscopy. The Raman spectra for MCD films showed a sharp and high peak at 1330 cm^{-1}, indicating a large volume fraction of the sp^3-bonded region. Those of NCD films, however, showed a much lower

sp^3-bonded peak and higher and broad peak centered near 1500 cm^{-1}, indicating larger volume fraction of distorted sp^2-bonded region [3-5]. Figure 2 compares the X-ray diffraction spectrum of NCD film to that of MCD film. Diffraction-peak intensities from NCD films were smaller than those of MCD films, indicating larger volume fraction of distorted and amorphous phases in NCD films.

The elastic constants of polycrystalline thin films are generally given by

$$[C_{ij}] = \begin{bmatrix} C_{11} & C_{12} & C_{13} & 0 & 0 & 0 \\ & C_{11} & C_{13} & 0 & 0 & 0 \\ & & C_{33} & 0 & 0 & 0 \\ & & & C_{44} & 0 & 0 \\ sym. & & & & C_{44} & 0 \\ & & & & & C_{66} \end{bmatrix}, \tag{1}$$

with the coordinate system where the x_3 axis is along the film-growth direction.

RUS/LASER METHOD

Mechanical resonant frequencies of free vibrations of a solid are determined by the density, dimensions, and all the independent elastic constants of the solid. Because the density and dimensions are measurable, measuring many resonant frequencies allows us to determine the independent elastic constants. This method is called resonant-ultrasound spectroscopy (or RUS) and has been applied to many anisotropic solids to determine their all independent elastic constants [9-11]. In order to apply this method to thin films, however, we have to measure the changes of the resonant frequencies caused by the deposition of the film with high enough accuracy because of small contributions of film's elastic constants to the resonant frequencies of the film/substrate system. In addition, we must identify the observed resonant frequencies because mode misidentification causes physically meaningless elastic constants. To achieve these demands, we developed the RUS/Laser method [6-8], which detects the free-vibration resonant frequencies with the piezoelectric tripod without using any coupling media and without applying any external force to the specimen. Thus, ideal free vibrations are caused and the measurement accuracy for frequencies is very high, of the order of 0.0001%. This method also measures distributions of the out-of-plane displacement amplitude on the vibrating specimen. By comparing these distributions with those calculated, we can identify the vibrations modes unambiguously [6].

We first measured the resonance frequencies of silicon substrate alone and then measured them after the deposition of the diamond film. All measurements were performed in vacuum (~1 Pa) to avoid the acoustic noise and at a constant temperature. Figure 3 shows examples of resonant spectra observed from the silicon substrate and from 6.34-μm NCD/Si specimen. It is never straightforward to identify the vibration modes only by looking at the resonant spectra. However, because we can identify all the vibration modes by measuring the vibration-amplitude distributions, the mode correspondence between before and after the deposition is unambiguous.

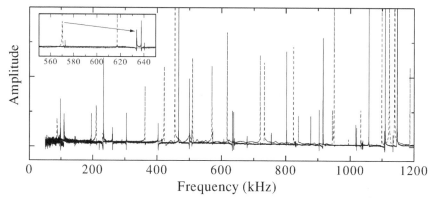

Figure 3. Resonant spectrum of the monocrystal silicon substrate (6x4x0.4 mm^3) (dashed line) and that of 6.3-μm NCD/silicon (solid line). All resonant frequencies increased 5-15% by the deposition of the diamond film.

The resonant frequencies increased by 5-20% after the deposition of the diamond thin films. Using 25-30 resonant modes showing good spectrum shape, we inversely determined the set of the elastic constants that yield the best fits between measured and calculated resonant-frequency shifts [7,8]. After the convergence of the inverse calculation, the measured and calculated resonant frequencies agreed within 0.1% error. The elastic constants C_{11}, C_{12}, and C_{66}, which are associated with the in-plane displacements, can be determined accurately under the favor of their large contributions to frequencies. But, we failed to determine the elastic constants C_{33}, C_{13}, and C_{44}, related to the out-of-plane displacement, because of their too small contributions. Thus, we here present the three elastic constants as shown in Fig. 4.

RESULTS AND DISCUSSION

Horizontal broken lines in Fig. 4 are the elastic constants of ideal polycrystalline (isotropic) diamond calculated by the Hill-averaging method [12]. There are thee significant observations: (i) The diagonal elastic constants C_{11} and C_{66} of the NCD films are smaller than those of bulk diamond and MCD films. (ii) The diagonal elastic constants of the NCD films decrease as the film thickness decreases. (iii) The off-diagonal elastic constant C_{12} of the NCD films is significantly larger than that of bulk diamond, while that of the MCD films is smaller. One may attribute (i) and (ii) to incohesive bonds at grain boundaries because the volume fraction of the grain boundary in the NCD films is larger. Indeed, such an incohesive bonded region at grain boundary successfully explained the decrease of the elastic constants of the MCD film [7]. However, the incohesive bonds cannot explain the enhanced off-diagonal elastic constant of the NCD films.

To explain observations (i)-(iii) consistently, we perform micromechanics calculations, modeling the diamond thin film with a two-phase composite consisting of polycrystalline (isotropic) diamond and randomly oriented thin pancake-shape inclusions. For example, we

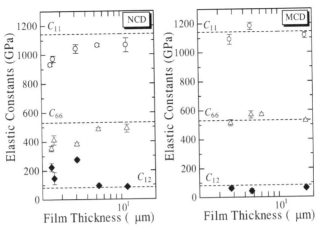

Figure 4. Elastic constants of NCD (left) and MCD (right) thin films. Horizontal broken lines are the elastic constants of polycrystalline diamond.

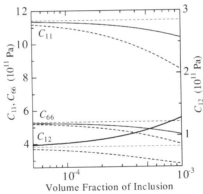

Figure 5. Elastic constants of the composite consisting of isotropic diamond matrix and randomly oriented thin pancake-shape inclusions calculated by micromechanics modeling. Horizontal broken lines are the elastic constants of isotropic diamond. Broken and solid curves are calculations for microcrack inclusions and graphite-plate inclusions, respectively. The aspect ratio of the inclusions is assumed to be 1000.

can estimate the softened elastic constants due to the incohesive bonds at grain boundaries by assuming microcracks as inclusions [13]. The calculation takes two steps [13]. First, we calculate the hexagonal elastic constants of the composite with the inclusions, whose minor axes are parallel to one direction. Eshelby's equivalent-inclusion model and Mori-Tanaka's mean-field theory are used in this calculation. Second, we average the hexagonal elastic constants to yield the isotropic elastic constants by the Hill-averaging method. The micromechanics calculations were done for the microcrack inclusions and for graphite inclusions whose c-axes are along the minor axes. Figure 5 shows the results.

Both the diagonal and off-diagonal elastic constants decrease with the increase of the volume fraction when the inclusions are microcracks. This model then explains the decrease of the elastic constants of the MCD films, but fails to clarify the enhanced off-diagonal elastic constant of the NCD films. When we assume the graphite-phase inclusions at grain boundaries, the diagonal elastic constants again decrease with the increase of the volume fraction, whereas the off-diagonal elastic constant remarkably increases, being consistent with the measurements. Therefore, the enhanced off-diagonal elastic constant of NCD films indicates the presence of thin graphitic plates between grains. This is possible because grain boundaries of NCD films contain larger fraction of the sp^2 bond compared with those of MCD [2-4].

CONCLUSION

Resonant-ultrasound spectroscopy with the laser-Doppler interferometry successfully determined the independent elastic constants of NCD and MCD thin films and an important relationship was found between the off-diagonal elastic constant and the sp^2-bonded regions at grain boundaries. The diagonal elastic constants of the NCD films are smaller than those of bulk diamond, but the off-diagonal modulus is much larger. The micromechanics calculation consistently explained this unusual trend by taking account of randomly oriented thin pancake-shape graphitic inclusions.

REFERENCES

1. D. Zhou, A. R. Krauss, L. C. Qin, T. G. McCauley, D. M. Gruen, T. D. Corrigan, R. P. H. Chang, and H. Gnaser, *J. Appl. Phys.* **82**, 4546 (1997).
2. S. Bhattacharyya, O. Auciello, J. Birrell, J. A. Carlisle, L. A. Curtiss, A. N. Goyette, D. M. Gruen, A. R. Krauss, J. Schlueter, A. Sumant, and P. Zapol, *Appl. Phys. Lett.* **79**, 1441 (2001).
3. James Birrell, J. A. Carlisle, O. Auciello, and J. M. Gibson, *Appl. Phys. Lett.* **81**, 2235 (2002).
4. James Birrell, J. E. Gerbi, O. Auciello, J. M. Gibson, D. M. Gruen, and J. A. Carlisle, *J. Appl. Phys.* **93**, 5606 (2003).
5. A. C. Ferrari and J. Robertson, *Phys. Rev. B* **61**, 14095 (2000).
6. H. Ogi, K. Sato, T. Asada, and M. Hirao, *J. Acoust. Soc. Am.* **112**, 2553 (2002).
7. N. Nakamura, H. Ogi, and M. Hirao, *Acta Mater.* **52**, 765 (2004).
8. N. Nakamura, H. Ogi, T. Ono, and M. Hirao, *Appl. Phys. Lett.* **86** (2005), in press.
9. I. Ohno, *J. Phys. Earth* **24**, 355 (1976).
10. A Migliori and J. Sarrao, *Resonant Ultrasound Spectroscopy* (Wiley-Interscience, New York, 1997).
11. H. Ledbetter, C. Fortunko, and P. Heyliger, *J. Appl. Phys.* **78**, 1542 (1995).
12. O. L. Anderson, in Physical Acoustics, Vol. IIIB, ed. W. P. Mason (Academic, New York, 1965), p. 43.
13. H. Ogi, S. Kai, H. Ledbetter, R. Tarumi, M. Hirao, and K. Takashima, *Acta Mater.* **52**, 2075 (2004).

Mater. Res. Soc. Symp. Proc. Vol. 875 © 2005 Materials Research Society   O1.3

Mechanical properties and size effect in nanometric W/Cu multilayers

P. Villain, D. Faurie, P.-O. Renault, E. Le Bourhis, P. Goudeau, K.-F. Badawi

Laboratoire de Métallurgie Physique, UMR 6630 CNRS - Université de Poitiers, SP2MI, Bd Marie et Pierre Curie, BP 30179, 86962 Futuroscope Chasseneuil Cedex, France.

ABSTRACT

The mechanical behavior of W/Cu multilayers with periods ranging from 24 down to 3 nm prepared by ion beam sputtering was analyzed using a method combining X-ray diffraction and tensile testing, and instrumented indentation. Cracks perpendicular to the tensile axis observed by optical microscopy were generated in the films under the largest applied tensile stresses. These cracks may appear in the multilayer while W layers are still in a compressive stress state. Elastic modulus and hardness values were extracted from nano-indentation data. Crack initiation and elastic constants were observed to depend on the period of these multilayers.

Keywords: *Multilayers, X-ray diffraction, mechanical properties, cracks, size effects*

INTRODUCTION

In nanocrystalline materials or thin films with nanometer scale thickness, surface contribution becomes preponderant yielding deviations from the average elastic behavior of the material. Moreover, the mechanical properties are known to differ from those expected from the bulk state. Multilayers have attracted much attention since one dimension can be tailored down to the nano-scale, leading then to novel electronic, magnetic, optical and mechanical applications. A lot of studies [1-6] have been devoted for experimentally characterizing the length scale dependence of strength and also the theory and modelling of deformation mechanisms in nano-scale multilayers. Deviations from Hall–Petch extrapolation at nanoscale layer thickness are observed as well as very high yield strengths. These behaviors cannot be then explained by a simple extrapolation of scaling laws such as the Hall–Petch relationship. This indicates also the crucial role played by interfaces on the deformation behavior of nano-scale materials.

Here, W/Cu multilayers with periods ranging from 24 down to 3 nm have been prepared by ion beam sputtering and characterized using X-ray reflectometry, X-ray diffraction, instrumented indentation and energy dispersive analysis in a scanning electron microscope. The elastic behavior of W sub-layers has been analyzed using a new method combining X-ray diffraction and tensile testing [7]. Young's modulus softening has already been observed when reducing the period down to 3 nm. A similar effect is also observed for the stress free lattice parameter. Extended X-ray Absorption Fine Structure (EXAFS) measurements have revealed that surface alloying may occur in tungsten sub-layers for the smallest periods leading then to lattice parameter decrease [8].

In the present paper, we will focus on the crack initiation under tensile testing and indentation, and the correlation between film damage and microstructural and elastic features.

EXPERIMENTAL DETAILS

Specimen preparation and characteristics: W/Cu multilayers were deposited by ion beam sputtering in a NORDIKO-3000 sputtering chamber at room temperature (the substrate-holder was water-cooled), tungsten being the first deposited layer. Three types of substrates were used: 200 μm and 600 μm thick natural oxidized Si (001) wafers, and 127.5 μm thick polyimide (Kapton®) dogbone foils. The nominal values of the multilayer period Λ were 3, 6 and 24 nm with equal thickness of W and Cu, and the total film thickness ranged from 220 to 240 nm.

The as-deposited layers were examined by small and high angle X-ray diffraction (XRD) to determine the multilayer period Λ and to characterize the initial microstructural and mechanical state respectively. Tungsten and copper sub-layers were polycrystalline and presented a (110) and (111) fiber texture respectively.

The residual stresses in tungsten layers were evaluated by means of XRD ($\sin^2\psi$ method), using the elastic constants of bulk tungsten (E = 400 GPa and ν = 0.28) in a first approximation; tungsten is clearly submitted to strong compressive residual stresses. The global residual stresses in the multilayers were analyzed using the curvature method with the multilayers deposited on 200 μm Si cantilevers (E/(1-ν) = 180.5 GPa for Si (001) wafers). These stresses are compressive and decrease with Λ. Neglecting the contribution of the interfaces, the residual stresses in copper sub-layers can then be deduced; it clearly appears that copper is under tensile stress (table I).

Table I: Residual stress in the as-deposited W/Cu multilayers: the mean stress was determined by the curvature method on 200 μm Si substrate, the stress in W sub-layers was analyzed by XRD, and the stress in Cu sub-layers was deduced from the previous experimental data.

Multilayer period Λ (nm)	3	6	24
Mean stress (GPa)	- 0.2	-1.2	-1.9
Stress in W (GPa)	-2.0	-2.5	-5.1
Stress in Cu (GPa)	+2.8	+1.1	+2.1

Tensile testing and X-ray diffraction: Combined tensile tests and XRD measurements were realized both at the French synchrotron radiation facility L.U.R.E. (Orsay, France) on the H10 beam line and in our laboratory on a Seifert four-circle diffractometer. A Deben™ mini-tensile testing device allows performing in situ tensile tests on most XRD goniometers [9].

The technique used here is based on the well known "$\sin^2\psi$ method" [10,11]; it consists in applying a uniaxial tensile force to the multilayer/substrate set and monitoring the shift of a {hkl} peak position ({211} W peak here) as a function of $\sin^2\psi$ (ψ being the angle between the diffracting planes and the normal to the sample surface) and the applied force. The global applied load is recorded via a load cell. The X-ray measurements on W (211) reflection allow us to determine the mean elastic strains in tungsten layers: in the case of a polycrystalline elastically isotropic material such as tungsten, a linear relation is expected between $\ln(a_\psi)$ and $\sin^2\psi$, where a_ψ is the lattice parameter of W for each value of the ψ angle (Figure 1); the slope and the intercept of the linear regressions of these experimental data allow calculating the elastic strain ε_{11}^A applied to the tungsten layers for each applied load (Figure 2a).

Nanoindentation tests: The samples were deformed at room temperature by a Berkovich diamond pyramid using a nanohardness tester machine from CSEM (Switzerland). The tests were performed in air with the force-control mode of the machine. The loading-unloading procedure was as follows: loading to maximum load in 30 s, holding at maximum load for 30 s, unloading in 30 s. The maximum load was varied between 0.5 and 20 mN. The calibration procedure suggested by Oliver and Pharr [12] was used to correct the data for the load-frame compliance of the apparatus and for the imperfect shape of the indenter tip. The unloading curves were used to determine the contact projected area between the sample and the indenter and hence to determine both the hardness H and the indentation effective modulus E_{eff} ($1/E_{eff} = 1/E_r + (1-v_D^2)/E_D$ with E_D and v_D the elastic modulus and Poisson's ratio of Diamond). We plot below $E_r = E/(1-v^2)$, where E and v denote the specimen Young's modulus and Poisson's ratio, respectively.

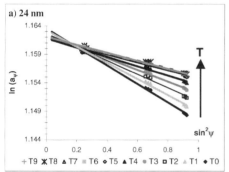 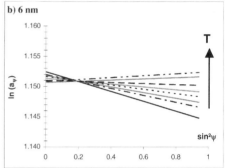

Figure 1: $\sin^2\psi$ curves relative to the {211} planes of the tungsten sublayers (a) in the $\Lambda = 24$ nm W/Cu multilayer and (b) in the $\Lambda = 6$ nm W/Cu multilayer for several increasing values of the applied load. The straight lines represent the linear regressions of the experimental data.

RESULTS AND DISCUSSION

Tensile testing and X-ray diffraction: Figure 1 clearly shows that W layers are initially under a compressive stress state (negative slope for T0). Then the slopes increase as the applied load increases (from T1 up to T6). For applied stresses higher than a critical value, the slopes saturate (from T6 to T9). The X-ray elastic strain ε_{11}^A in W sublayers remains constant (as shown on Figure 2a). The maximum value of ε_{11}^A increases with the multilayer period Λ, from 0.4% ($\Lambda = 3$ nm) to 1.1 % ($\Lambda = 24$ nm).

Ex-situ post-mortem optical microscopy observations have evidenced the presence of straight cracks perpendicular to the tensile axis (Figure 3). These cracks induce a stress relaxation near their edges, which can account for saturating global elastic strain: as the X-ray projected beam has an area of about 1 mm^2, the strain measured here is a mean value on stressed adhesive parts of the film and relaxed cracked regions.

It is also worth noting that the slope of the sin$^2\psi$ curve becomes positive before crack formation in the case of the Λ = 6 nm W/Cu multilayer whereas the Λ = 24 nm multilayer is damaged when tungsten sublayers are still under compression. It means that tungsten layers crack while they are in a compressive stress state. This indicates that cracks should be initiated in Cu sublayers. Indeed, these layers are already under tensile stresses in the as-deposited state (Table I). Using the stress values in W sub-layers and assuming equal strains in all layers (and the substrate), we have estimated the stress in Cu sub-layers when the multilayer is cracked. The values are plotted in Figure 2b as a function of the period Λ.

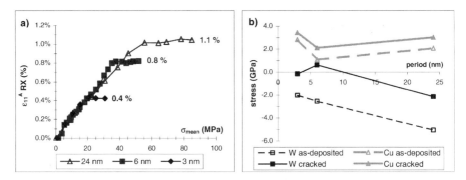

Figure 2: a) Elastic strain $\varepsilon_{11}{}^A$ in tungsten layers deduced from the sin$^2\psi$ curves as a function of the mean applied stress, for the 3 values of the multilayer period
b) Evolution of the stress in W (measured by XRD) and Cu (deduced from W measurements) sub-layers between the as-deposited state and the crack formation.

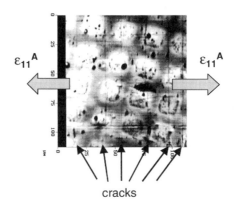

cracks

Figure 3: Optical microscope image of cracks observed after the tensile test on the surface of the Λ = 3nm specimen. Gold dots (diameter ~ 20µm) were deposited for macrostrain measurements using an optical method (not commented here, [13]).

Nano-indentation : When thin films are tested under increasing load, the substrate contribution to the mechanical response becomes prominent. In Figure 4 the hardness and indentation modulus of the films are plotted as a function of the h_c/t ratio, where h_c denotes contact depth and t the total film thickness. In the studied range of contact depth to thickness ratio, the indentation modulus increases by 25 % while the hardness increases by 18% due to substrate contribution to the mechanical response. Work is in progress to produce thicker films (~700 nm) using magnetron sputtering in order to get a response at light loads that is more representative of the films. For the load range used here observations of the indent sites revealed no radial cracking.

Using the data obtained under the lightest loads, one observes that hardness H values are approximately given by a rule of mixture that indicates a multilayer value about the average of W and Cu values (equal thickness of W and Cu was used). Only small influence of the period is observed (no Hall-Petch like strengthening).

The indentation modulus E_r slightly decreases when the period is reduced. In all cases, it does not follow a rule of mixture between W and Cu ($E_{r(mixture)} \approx 250$ GPa). In contrast, the response is very close to that of Cu indicating that the softer layers rule out the multilayer behavior. This multilayer system is analogue to the case where two strings are considered to act in a parallel mode instead as in a series mode. The equivalent modulus then obtained using the following relation

$$Er_{//} = 2 \left[(Er_{Cu} \times Er_W) / (Er_{Cu} + Er_W) \right] \tag{1}$$

is equal to the experimental value (~160 GPa) plotted on figure 4.

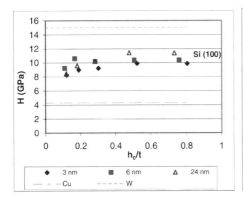
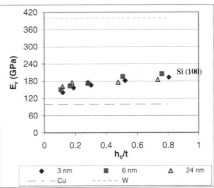

Figure 4: Hardness and indentation modulus of the films for the values of multilayer period plotted as a function of indentation depth normalized by the total film thickness t. Dotted lines are the corresponding values of sputtered Cu and W [14] thin films.

CONCLUSIONS

Size effects on the resistance to crack formation have been evidenced. Cracks can appear in the multilayer while tungsten layers are still in a compressive stress state. This means that cracks should be initiated in copper sub-layers due to the tensile stress state of copper. Interfacial regions probably also play an important role: as the roughness level increases when the period decreases, crack initiation is favored (lower value of K_c). Under that condition, increasing the number of interfaces (decreasing the period) increases the probability of early damage process.

In the case of the $\Lambda = 3nm$ multilayer, copper incorporation in tungsten layers may also facilitate the propagation of the cracks initiated in copper through the whole thickness of the film. Further experiments will aim at directly monitoring the stress state evolution during mechanical testing of copper by means of XRD.

ACKNOWLEDGEMENTS

The authors are gratefully indebted to Dominique Thiaudière and Marc Gailhanou for assistance during XRD experiments at LURE, and Philippe Guérin at LMP for preparing the multilayered thin films.

REFERENCES

1. A. Misra and H. Kung, *View point set "deformation and stability of nanoscale metallic multilayers"*, *Scripta Mat.* **50** (2004).
2. Y.Y. Tse, G. Abadias, A. Michel, C. Tromas and M. Jaouen, *Mater. Res. Soc. Symp. Proc.* **778**, U6.8 (2003)
3. J. Musil, *Surf. Coat. Techn.* **125**, 322 (2000)
4. M.A. Haque and M.T.A. Saif, *Thin Solid Films* **484**, 364 (2005)
5. H.D. Espinosa, B.C. Probok and B. Peng, *J. mechanics and physics of solids* **52**, 667 (2004)
6. R.C. Hugo, H. Kung, J.R. Weertman, R. Mitra, J.A. Knapp, D.M. Follstaedt, *Acta Mater.* **51**, 1937 (2003)
7. P. Villain, P. Goudeau, P.-O. Renault and K.F. Badawi, *Appl. Phys. Lett.* **81**, 4365 (2002)
8. P. Goudeau, P. Villain, T. Girardeau, P.-O. Renault and K.F. Badawi, *Scripta Mat.* **50**, 723 (2004)
9. K.F. Badawi, P. Villain, P. Goudeau and P.-O. Renault, *Appl. Phys. Lett.* **80**, 4705 (2002)
10. C. Noyan, J.B. Cohen, *Residual stress measurement by diffraction and interpretation.* New York, Springer (1987)
11. V. Hauk, *Structural and residual stress analysis by non destructive methods: evaluation, application, assessment.* Amsterdam, Elsevier (1997)
12. W.C. Oliver, G.M. Pharr, *J. Mater. Res.* **7**, 1564 (1992)
13. P. Villain, PhD thesis, Université de Poitiers, France (2002)
14. R. Saha, W.D. Nix, *Acta Mater.* **50**, 23 (2002).

Mater. Res. Soc. Symp. Proc. Vol. 875 © 2005 Materials Research Society O1.5

Improvement of the Elastic Modulus of Micromachined Structures using Carbon Nanotubes

Prasoon Joshi[1], Nicolás B. Duarte[1], Abhijat Goyal[1], Awnish Gupta[2], Srinivas A. Tadigadapa[1] and Peter C. Eklund[2]
Departments of [1]Electrical Engineering, and [2]Physics, Pennsylvania State University, University Park, PA 16802.

ABSTRACT

Microelectromechanical flexural structures have been fabricated using sandwiched multi-layers consisting of bundled singled walled carbon nanotubes(SWNTs) incorporated into silicon nitride (Si_3N_4) films. The Si_3N_4-SWNT composite layer was patterned by reactive ion etching followed by release in XeF_2 to create freestanding bridge structures. The mechanical stiffness of the micromechanical bridges was monitored via force-displacement (F-D) curves obtained using an Atomic Force Microscope (AFM). Inclusion of SWNTs resulted in an increase in the spring constant of the bridge by as much as 64%, with an average increase of 25%. In a second experiment, micromachined bridges fabricated using dissolved wafer process were coated with debundled SWNTs. The SWNTs suspended in N-methyl-2-pyrrolidinone (NMP) solvent were sprayed locally on each bridge using a piezoelectric print head. Resonance frequency measurements were done in vacuum ($\sim 10^{-4}$ Torr) on the bridges after successive SWNT depositions. A 20% increase in the resonance frequency of the bridges was observed. The observed increase in stiffness in the first set of experiments as well as the observed increase in the frequency in the second set of experiments can be attributed to the high axial modulus of elasticity (~ 1 TPa) of the carbon nanotubes.

INTRODUCTION

There is considerable interest in high frequency mechanical resonators and mechanical structures for RF applications [1]. However, high frequency mechanical resonators can be achieved only via miniaturization of the structures and for frequencies in the 100MHz-1GHz range the size of these resonators start approaching the nanometer scale. This makes the practical implementation of such high frequency resonators quite cumbersome if not impossible. Even using the torsional mode of operation, the micron scaled resonators are typically limited to a maximum frequency of ~ 1 GHz [2, 3]. At the micrometer dimensions one way to achieve further improvements in the resonator characteristics is to use higher stiffness materials. Carbon nanotubes (CNT) have been measured to have very high axial modulus of elasticity ~ 1 TPa [4] and the incorporation of these high elasticity nanotubes into typical thin film materials used in micromechanical structures is expected to improve the elastic properties of these thin films. Motivated by this possibility, we incorporated CNTs into micromechanical structures in two different ways and independently observed an increase in the stiffness of these structures. In this paper we report an observation of the improvement of the stiffness of microelectromechanical structures by the addition of carbon nanotubes.

EXPERIMENTAL DETAILS

PECVD silicon nitride (Si_3N_4) bridges were fabricated on a Si wafer. Both, test samples as well as control samples were prepared. The test samples had a five layer structure with alternating layers of silicon nitride and CNTs as shown in Figure 1. The control sample had a three layer structure of nitride alone without any CNT. CNTs suspended in isopropyl alcohol (IPA) were spray deposited using an air brush. This resulted in a random orientation of the CNTs. This method of deposition did not allow for any uniformity control of the aerial density of deposited CNTs and often the CNTs agglomerated into clusters during the drying of the solvent. The wafer with the sandwich layers was patterned using standard photolithography techniques and etched in SF_6 plasma. Finally, the wafer was released in XeF_2 to define the bridge structure as shown in Fig. 1. The microfabricated bridges were then evaluated for mechanical stiffness by use of an AFM system. Force-Displacement (F-D) curves were hence generated, the slope of which determined the spring constant (and hence the stiffness) of the bridges.

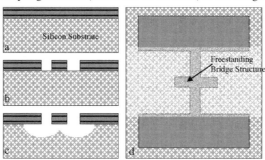

Figure 1: *(a) Deposit 150nm of PECVD Si_3N_4 ▬, spray coat CNT ▬, deposit 300nm of PECVD Si_3N_4, spray coat CNT, and deposit 150nm of PECVD Si_3N_4 on top of Si, (b) Ppattern and etch through the silicon nitride and CNT layer, . (c) Rrelease in XeF_2, and (d) Top view of the fabricated structure..*

In the second part of the experiment, silicon micromachined bridges and cantilevers were fabricated using dissolved wafer process [5]. 4" Silicon wafer, with a resistivity of 1-10 ohm-cm was doped with boron (~10^{19} cm^{-3}) on single side of the wafer to achieve a diffusion depth of ~4μm. The wafer was subsequently annealed in nitrogen atmosphere (1175^0C) for 45mins to minimize any stress gradients in the boron doped layer. The wafer was then diced into 1"x1" squares, patterned using photolithography and reactive ion etched using SF_6 gas. A glass wafer was used to serve as the eventual substrate for the free standing structures. It was cleaned in H_2SO_4:H_2O_2::1:1 solution, diced into 1"x1" pieces and the electrode patterns were defined using photolithography. A thin film of Cr and Au measuring 200Å and 2000 Å, respectively, was evaporated using e-beam evaporation and defined by lifting off the metal. The 1"x1" pieces of Si and glass were aligned and anodically bonded [6, 7]. The bonded samples were subsequently diced into individual dies, and the substrate silicon was etched away in ethylene diamine pyrocatechol (EDP) leaving behind the boron doped silicon bridges on the glass substrates. The released structures were dried using critical point drying and wire bonded. See Fig. 2 for fabrication details.

Flat bridges were selected for experiments after observing them under a white light interferometer. Single walled carbon nanotubes suspended in a solution of NMP (N-methyl-2-pyrrolidinone) were dispensed on the surface of the bridges using a piezoelectric print head. The NMP was allowed to evaporate from the surface, leaving the SWNTs behind. Resonance frequency measurements using an impedance analyzer were carried out in vacuum ($\sim 10^{-4}$ Torr) for an extended period of time ranging from 12 to 48 hours and the resonance frequency was sampled every 4 minutes. These measurements were done both before and after SWNT deposition. In all, up to three rounds of SWNT depositions were done allowing study of progressive change in resonance frequency with additional deposition of carbon nanotubes.

RESULTS AND DISCUSSION

The stiffness of the bridge was measured using an AFM. Figure 3 shows the typical force-displacement curves obtained using this technique. As shown in Fig. 1, the center of the bridges was not lithographically marked and since this central area of the bridges in the present case was much larger in dimensions (80 µm by 80 µm) as compared to the tip of the AFM which is comparatively small, it was important to generate the F-D curves over a large number of points around the center of the bridge. F-D curves at points not at the geometric center of the bridge were non-linear because of the complex deformation of the bridge in several planes. Hence, the F-D curves were generated for 50 points

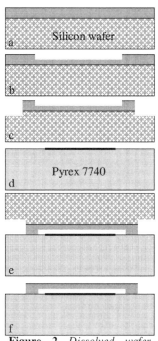

Figure 2 *Dissolved wafer process fabrication sequence. (a) 4µm deep boron doped silicon wafer, (b) 2 µm gap etch, (c) 6µm feature etch, (d) Electrode pattern on a Pyrex 7740 wafer, (e) Alignment and anodic bonding, and (f) Dissolution of undoped bulk silicon substrate in EDP.*

over an area of 25µm x 25µm around the visually determined center point of the bridge. For unidirectional deformation of the bridge, at points in the vicinity of the geometric center of the bridge, linear F-D curves were obtained. The value of the stiffness was calculated from the slope of the curves. The measured bending stiffness includes the stiffness of the AFM cantilever as well however, since the AFM cantilever is unchanged it is

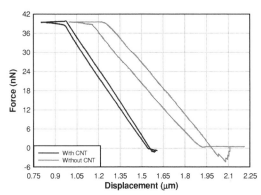

Figure 3 *Typical force-displacement curves obtained using an AFM in silicon nitride bridges with and without embedded carbon nanotubes.*

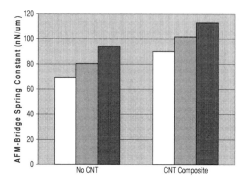

Figure 4: *Spring constant measurements using AFM of three different Nitride samples with and without inclusion of CNTs.*

assumed its stiffness remains the same and thus the measured results directly show the change in the stiffness of the measured micromechanical structures. Figure 4 shows the average values of the spring constants for three separate bridges with and without the incorporation of CNTs. Spring constants of three representative bridges without CNTs were measured to be 68.9, 80.2 and 93.9 nN/µm. The same measurements on bridges with CNTs incorporated in them resulted in a spring constant of 89.8, 101.5 and 113.1 nN/µm. This amounts to an average increase in the spring constant approximately 25% which is significant improvement in the overall bending stiffness of the micromechanical structures.

From these observations, it can be concluded that incorporation of CNTs in the bulk or on the surface of the microfabricated bridges showed an increase in the stiffness of the microstructures. In the design of the experiment, care was taken that the CNT layer was incorporated into the silicon nitride layer as far away from the bending neutral plane as possible to obtain the maximum impact on the spring constant of the micromechanical structure upon the inclusion of such high axial stiffness material. Thus a technique has been developed whereby traditional silicon micromachining materials with well developed fabrication processes can be used in the design of high stiffness micromechanical structures with the CNTs included into specific planes for tuning the overall stiffness of these devices. In this current set of experiments, CNTs suspended in isopropyl alcohol were sprayed onto the silicon nitride layer using an air brush and allowing the solvent to evaporate. This resulted in CNT films which were not particularly uniform. Some areas of large agglomeration could be seen optically. Further, any variations in the stress of the deposited silicon nitride film can easily show as spread in the stiffness of the micromechanical structures and can explain the spread in the results on the same kind of structures. However, we expect this spread to be smaller than the 25% change observed when the carbon nanotubes are incorporated.

Microfabricated structures owing to their large surface to volume ratio are extremely sensitive to variations in their surface properties. Hence, the next step in this investigation was to deposit nanotubes on the top surface of a bridge and investigate the effect of these on the mechanical properties of these structures. Upon deposition of the carbon nanotubes, an increase in the resonance frequency of the bridges was observed which implies an increase in their stiffness as well. Figure 5 shows the observed increase in the resonance frequency of two micromachined bridges fabricated using dissolved wafer process as a function of the number of the deposited carbon nanotube layers. The deposition was carried out sequentially in steps in order to observe the effect of increasing quantity of carbon nanotubes on the surface of the resonator. As shown in Figure 5 (a), the base frequency of an uncoated resonator was 73.0 kHz. Upon stepwise successive deposition of carbon nanotubes on the resonator surface, the resonance increased to 82.0, 86.0 and 86.5 kHz after deposition of one layer, two layers and three layers

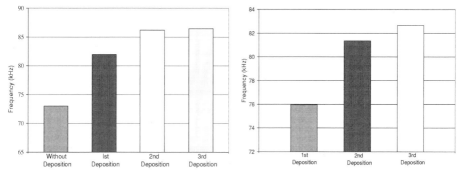

Figure 5: *Maximum resonance frequency of two different resonators plotted as a function of the number of SWNT depositions.*

respectively. This amounts to an increase of ~20% in the resonance frequency of the resonators. These increases in resonance frequency were corroborated by repeating the experiment on another similar bridge which yielded similar results. The deposition of the high stiffness CNT layer on the surface of the mechanical resonators once again increases the bending stiffness of the bridges and explains the observed increase in the resonance frequency. Implicit in this explanation is the assumption that the carbon nanotubes adhere strongly to the surface of the silicon micromachined structure without slipping.

Additionally, the resonance frequency was found to increase logarithmically when the coated resonators were kept in vacuum for 24 hours. This logarithimic increase in the frequency can be attributed to the decrease in mass stored on the resonator due to desorption of gases from the carbon nanotubes. Figure 6 shows a representative curve for the increase in the resonance frequency of the nanotubes coated resonator when kept in vacuum over a period of 24 hours. This sensitivity of the resonating bridge to desorption of gases from the carbon nanotubes on top of it opens possible avenues for configuring such an arrangement as a gas sensor, since the carbon nanotubes are now known to be storehouse of gases [8-11]. The bridges and even cantilevers, functionalized as above using a coating of carbon nanotubes, thus promise to be an excellent platform for sensing applications.

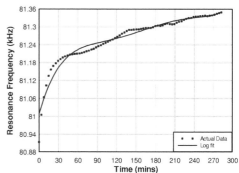

Figure 6: *Logarithmic change in the resonance frequency of the bridge loaded with carbon nanotubes indicating stochastic desorption of physisorbed, chemisorbed and trapped gases from carbon nanotubes as a function of time in a vacuum of ~10^{-4} Torr. The line is a logarithmic fit to the data and is a guide to the eye.*

CONCLUSIONS AND FUTURE WORK

In summary, we have successfully demonstrated that incorporation of carbon nanotubes on the surface and in the bulk of a microfabricated bridges can result in the increase in the stiffness of these structures. Dynamic testing of the bridges under resonance conditions showed an average increase of 20% in resonance frequency. These observations were made in spite of relatively poor control over the aerial density uniformity, and lack of control over orientation of the deposited nanotubes. The authors believe that better control over these variables would result in much higher increases in the stiffness of the micromachined structures. The observation of increase in the resonance frequency of the bridges with desorption of gases opens up the possibility of configuring the present structures as biochemical sensors through functionalization of carbon nanotubes.

ACKNOWLEDGEMENTS

The authors would like to thank Paul D. Sunal for fabricating the devices using dissolved wafer process. The project is supported by an NSF-NIRT grant MCE 0304178.

REFERENCES

1.   Lakin, K.M., G.R. Kline, and K.T. McCarron, *Development of miniature filters for wireless applications.* IEEE Transactions on Microwave Theory and Techniques 1995. **43**(12, pt.2): p. 2933-9.

2.   Wang, K., A.-C. Wong, and C.T.-C. Nguyen, *VHF free-free beam high-Q micromechanical resonators.* Journal of Microelectromechanical Systems. 2000, **9**(3): p. 347-60.

3.   Wang, J., Z. Ren, and C.T.-C. Nguyen. *1.14-GHz self-aligned vibrating micromechanical disk resonator.* in *IEEE Radio Frequency Integrated Circuits Symposium (RFIC), 8-10 June 2003.* 2003. Philadelphia, PA, USA: IEEE.

4.   Popov, V.N., *Carbon nanotubes: properties and application.* Materials Science & Engineering R: Reports, 2004. **R43**(3): p. 61-102.

5.   Tadigadapa, S. and S.M. Ansari. *Applications of high-performance MEMS pressure sensors based on dissolved wafer process.* in *Proceedings SENSORS EXPO Baltimore.* 1999. Baltimore, MD, USA: Helmers Publishing.

6.   Wei, J., et al., *Low temperature wafer anodic bonding.* Journal of Micromechanics and Microengineering, 2003. **13**(2): p. 217-222.

7.   Goyal, A., J. Cheong, and S. Tadigadapa, *Tin-based solder bonding for MEMS fabrication and packaging applications.* Journal of Micromechanics and Microengineering, 2004. **14**(6): p. 819-825.

8.   Li, J., et al, *Carbon nanotube sensors for gas and organic vapor detection.* Nano Letters, 2003. **3**(7): p. 929-933.

9.   Sumanasekera, G.U., et al. *Thermoelectric study of hydrogen storage in carbon nanotubes.* in *Making Functional Materials with Nanotubes. Symposium, 26-29 Nov. 2001.* 2002. Boston, MA, USA: Mater. Res. Soc.

10.   Dillon, A.C., et al., *Storage of hydrogen in single-walled carbon nanotubes.* Nature, 1997. **386**(6623): p. 377-9.

11.   Dujardin, E., et al., *Capillarity and wetting of carbon nanotubes.* Science, 1994. **265**(5180): p. 1850-2.

Characterizing Thin Films
by Nanoindentation

Micro/Nano Indentation and Micro-FTIR Spectroscopy Study of Weathering of Coated Engineering Thermoplastics

Samik Gupta, Jan Lohmeijer[1], Savio Sebastian, Nisha Preschilla and Amit Biswas
GE India Technology Center, Hoodi Village, Whitefield Road, Bangalore-560066, India
[1]GE Advanced Materials-Europe, Plasticlaan-1, Bergen op Zoom, The Netherlands

ABSTRACT

A novel combination of depth-sensing nano-indentation, micro-indentation and micro-FTIR techniques is employed towards understanding the durability of coating layers used on engineering thermoplastics upon exposure to harsh weathering environments. This combination of techniques enables study of changes in surface-to-bulk properties in the clearcoat-substrate system upon weathering; typically observed as a degradation starting from the surface and then proceeding inwards to the bulk of the material. Nano-indentation measurements carried out to understand the mechanical properties of the coating layer provide insights into the changes in hardness and modulus upon prolonged weathering exposure. Depth-sensing micro-indentation and micro-FTIR spectroscopy studies performed to evaluate mechanical performance and chemical changes, respectively, explain the influence of the substrate on the coating layer, especially at the interface upon weathering. This unique combination of depth-sensing indentation and micro-FTIR spectroscopy has led to an understanding of the properties of the coating layer and the substrate individually as well as an integral system as a function of weathering exposure time. Finally, the physico-chemical properties of the coating and substrate are linked to performance prediction, enabling optimization of coating-substrate combinations.

INTRODUCTION

High performance Engineering Thermoplastics (ETP's) have widespread outdoor applications and are consequently susceptible to deterioration in physical properties upon exposure to different environmental stresses such as ultraviolet radiation, changes in humidity and ambient temperature etc., though these changes are maximal in the near surface [1,2]. Therefore, a clearcoat layer is usually applied on an ETP, primarily for protecting the bulk of the material from this weathering induced degradation. ETP's generally possess poor abrasion resistance and inferior gloss, hence clearcoat layers also function as a scratch resistant shielding layer with superior aesthetics thereby offering an overall enhanced visual appeal. However, exposure to ultraviolet radiation may cause photo oxidation of the clearcoat and result in changes in the physical properties affecting the synergistic performance of the coated ETP, typically seen as a catastrophic failure of the coating substrate system or coating delamination [3]. It is thus important to explore and comprehend these weathering induced changes in the mechanical properties and chemical composition of both the coating layer as well as the substrate while developing a coated ETP.

Numerous surface analysis and depth sensing techniques have been recently reported towards studying the large changes seen in the surface properties of coated ETP's upon natural exposure [4-10]. Depth-sensing indentation, calorimetry, spectroscopy and microscopy techniques, mostly in combination with each other, has developed into a widely adopted analysis methodology in studying the ageing consequences, especially at a nano or micro scale

[1,7,11,12]. These techniques together explain changes in chemical composition of the material leading to changes in mechanical properties and hence have been particularly applied to develop coated ETP's with superior resistance to weatherability and also to predict their lifetime. Depth sensing indentation is a powerful tool towards characterizing mechanical properties of coatings as well as substrate surface on weathering [9,13]. A simple load-penetration curve obtained from a depth-sensing curve in a nanoindentor not only provides the hardness and modulus values, but is also capable of providing other valuable information such as in-plane elastic modulus, or residual tensile strength in the film [14,15]. A combination of depth sensing nano and micro indentation techniques is an ideal choice to understand the surface mechanical performance of the substrate coated with a thin film. If the mechanical properties of the coating layer alone must be studied on an integral coating- substrate system; care should be taken to restrict to the critical penetration depth beyond which substrate effects become considerable. The ratio of the critical penetration depth to coating thickness is independent of coating thickness, but is sensitively dependent on the differences in elastic properties of the coating –substrate system [16]. Though this ratio of upto 40% has been reported for many coating substrate systems, the conventional "rule of thumb" value of 10% can be used to isolate substrate effects [17,18], less than which effect of substrate is not expected to be significant.

Exposure to weathering also induces chemical changes in both the organic coating as well as the ETP substrate, and it is important to monitor these properties along with gauging changes in mechanical properties. These chemical changes normally triggered off by the influence of many environmental factors, are often very complex in nature. Among the wide range of effective techniques capable of depth-sensing composition analysis, FTIR spectromicroscopy technique qualifies as a straightforward choice due to ease of sample preparation and fast data acquisition [19]. Changes in absorbance in the infrared spectra highlight many possible changes in chemical composition as a result of degradation, hydrolysis, photo-oxidation etc. Depth-sensing Attenuated Total Reflectance FTIR (ATR-FTIR) on sections prepared laterally at different depths, employing different crystals and reflection angles so as to perform depth-sensing ATR-FTIR analysis or transmission FTIR of microtomed cross-sections in conjunction with indentation techniques have been reported for coating evaluation [8,9]. However, analytical methods exploring the possible interdependence of chemical and mechanical properties of the coating and substrate thereby affecting the synchronous performance of a coated ETP upon weathering have not been reported to our knowledge. In this report, we present a combination of micro/nano indentation techniques and micro-FTIR as a proficient approach for monitoring individual as well as combined properties of the coating and substrate of a coated ETP upon weathering.

EXPERIMENTAL DETAILS

ETP's used in this study were a set of polycarbonate-based formulations containing saturated or unsaturated impact modifiers and various additives such as antioxidants, fillers, flow promoters, hydro-stabilizers and photo-stabilizers. Substrates of dimensions 3 cm x 3 cm x 3 mm were prepared for weathering studies. An acrylic clearcoat of thickness 40 microns was applied on these substrates. In order to understand the mechanical properties of the coating layer and changes in its properties upon weathering, a free standing clearcoat film of equal thickness, i.e., 45 microns was also prepared.

Accelerated weathering of uncoated and clearcoated samples as well as the clearcoat film was performed according to ISO 4892-2A test protocols in an Atlas Ci 4000 weatherometer. This protocol simulates Florida noon environmental conditions with a Xenon lamp source of 6000W power and irradiance of 0.5 W/m^2 at 340nm, with two cycles consisting of 102 minutes of incident beam at 65°C, relative humidity of 50% and 18 minutes of deionized water spraying at 50°C and 80% relative humidity.

Two different instruments were used to cover a large depth range of indentation. A Zwick micro-hardness tester with a Vickers tip as the probe, suited to study indentations in the load range of 200 N with a resolution of 1 mN was used for micro-indentation studies. Since the clearcoat thickness for the samples is 45 microns, its properties were studied by indenting to depths less than 4 microns and that of the substrate properties were studied by indenting much larger depths, say larger than 100 microns. An MTS Nanoindenter XP capable of indentation in the load range of 500 mN with a resolution of 50 nN was used for nanoindentation so as to evaluate coating properties. A Berkovitch diamond tip was used as the nanoindentor probe. Data obtained from depth-sensing indentation is a force (P) $vs.$ depth (h) plot, from which the hardness and elastic modulus were computed using well-known relationships of contact mechanics [4]. The tip-area function, which represents the cross-sectional area of the indenter as a function of the axial distance, is calibrated with fused silica according to the standard procedure. A series of indentations were made on each sample and the properties reported are the averages of 10 indentation measurements.

Depth dependent composition of samples was studied using a Perkin Elmer Spectrum GX FTIR spectrometer coupled with an Auto-IMAGE FTIR microscope. Sections of 10 micron thickness, required for depth-sensing micro-FTIR studies, were prepared on a microtome from cross-sections of uncoated and coated samples. Absorbance spectra were recorded at specific wave numbers at 30μm intervals from the surface to the core of the substrate, with the spectra normalized at the absorbance with respect to methylene peak at 2968 cm^{-1}. Absorbance peaks, especially the O-H stretching at 3400-3500cm^{-1} and C=C stretching at 966 cm^{-1}, respectively showing effect of hydrolysis and the photo-oxidation of unsaturated components [20] were monitored to capture the changes in chemical composition.

RESULTS AND DISCUSSION

Micro/Nano indentation studies as a function of weathering

Microindentation measurements carried out to understand the effect of weathering on the top layer of coated ETP comprising of the clearcoat and the substrate top surface is shown in fig. 1a and 1b. Substrate was a polycarbonate blend having unsaturated impact modifiers but no photo / hydro stabilizers. Micro-indentation measurements providing hardness and modulus for the integral coating-substrate system showed that the hardness of around 81 MPa (fig. 1a) and modulus of 1.7GPa did not change appreciably upon weathering exposure. Interestingly, nanoindentation measurements on the clearcoat of the same sample indicated a steep increase in hardness and modulus upon weathering exposure. After about 1000 hours of exposure, the clearcoat hardness increased from 180 MPa to 235 MPa (fig. 1c), while the modulus increased from 2.9GPa to 3.6GPa (fig. 1d). Nanoindentation measurements on a freestanding clearcoat also showed similar trend in hardness and modulus sharply increasing upon 1000 hours of weathering exposure. Detailed nano-indentation and micro-FTIR studies of the clearcoat were performed

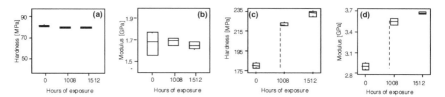

Figure 1. Hardness and modulus plots for the (a)&(b) top surface of coated ETP estimated from micro-indentation measurements and (c)&(d) free standing clear coat film from nanoindentation.

to investigate this weathering behavior, especially on substrates with varying formulations. The results are discussed below.

The dependence of hardness and modulus on weathering, obtained from nanoindentation measurements for the clearcoat applied on a substrate not containing any photo or hydro stabilizers, is shown in fig. 2a and 2b. Very similar to the freestanding clearcoat behavior discussed earlier, this clearcoat also shows increase in hardness and modulus after weathering exposure of 1000 hours. However, addition of photo and hydro stabilizers in the substrate formulation showed a large difference in weathering exposure time dependence of hardness and modulus of the clearcoat. Fig. 2c and 2d shows nanoindentation results for a clearcoat when applied on a substrate containing photo and hydro stabilizers. Durability of the clearcoat for this particular substrate formulation was found to be considerably larger; the increase in hardness and modulus is seen only after about 2000 hours of accelerated weathering exposure. Hardness for the clearcoat showed a sudden large increase from 175 MPa to 215 MPa, while the modulus also showed a jump from 3.0GPa to 3.45GPa upon 2000 hours of weathering.

Weathering time-dependent micro and nanoindentation experiments were also carried out on a saturated impact modifier substrate-clearcoat system. In this sample, the unsaturated impact modifiers in the polycarbonate blend were substituted with saturated impact modifiers, but no hydro or photo stabilizers were added to the system. Micro-indentation analysis on the integral coating-substrate system as well as nanoindentation analysis for isolated clearcoat showed highly comparable trend in hardness and modulus as for the substrate containing unsaturated impact modifiers, as shown in Fig 1. Thus, it was concluded that addition of photo and hydro stabilizers to the substrate formulations almost doubled the coating lifetime. This could be due to their gradual leaching into the coating. Subsequently, chemical stability of substrate surface seemed to play a vital role in the durability of clearcoat and was explored through depth-sensing micro-FTIR analysis as a function of weathering exposure. Micro-FTIR results are discussed below.

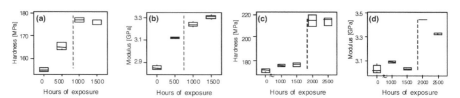

Figure 2. (a) Hardness and (b) modulus from nanoindentation for substrate with no photo / hydro stabilizers. (c) and (d) shows respective plots for a blend containing these components.

Chemical response of the substrate

A wide range of substrate formulations containing saturated or unsaturated impact modifiers as well as hydro and / or photo stabilizers were chosen for micro-FTIR studies, similar to those chosen for indentation studies. Depth-sensing micro-FTIR spectra monitoring the O-H vibration peak for a polycarbonate blend containing unsaturated impact modifiers but no photo and hydro stabilizers, prior to and after 1000 hours of weathering exposure, respectively, are shown in Fig. 3a and 3b. While the unexposed sample showed only a low intensity O-H peak prominent within the first 30 microns of the sample surface; weathering exposed sample showed a large extent of photo-oxidation on the surface progressively decreasing in intensity until disappearance at a depth of 100 microns into the substrate. Fig. 4a and 4b shows depth dependence of O-H stretching vibrations (3500 cm^{-1}) and C=C stretching (966 cm^{-1}) vibrations respectively correlating to surface oxidation and effect of photo-oxidation of unsaturated impact modifiers in the blend for the sample after weathering exposure of 1000 hours. At both wavelengths, oxidation was observed only on the surface to a depth of top 100 microns into the substrate from the clearcoat-substrate interface.

A formulation containing hydro and photo stabilizers also showed surface hydrolysis within 100 microns of the substrate surface (fig 4d), but notably lesser compared to formulations without these components (fig. 4c). While sharp increase in absorption was seen for the polycarbonate formulation without any photo or hydro stabilizers after an exposure time of 1000 hours, the hydro and photo-stabilized formulation appears to possess excellent chemical stability (fig 4d). Comparing micro-FTIR and nanoindentation results, it can be concluded that chemical changes in substrate directly influence weathering performance of clearcoat.

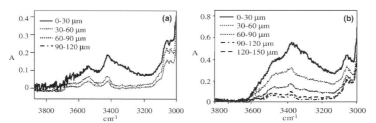

Figure 3. Depth sensing micro-FTIR spectra showing decrease of O-H absorption intensity as a function of depth for (a) unexposed blend and (b) after 1000 hours weathering exposure.

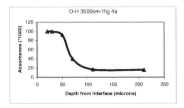
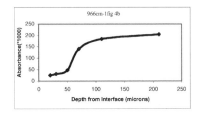

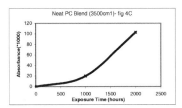
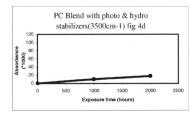

Figure 4. Influence of accelerated weathering showing depth dependence of (a) O-H absorption and (b) C=C stretching (966 cm^{-1}). Absorption at 30μm depth from the interface as a function of exposure time at 3500cm^{-1} for (c) neat polycarbonate blend and (d) polycarbonate blend with photo and hydro stabilizer.

CONCLUSION

In this paper we have reported investigation of the weathering performance of a clearcoated polycarbonate-based blend with depth-sensing indentation and micro-FTIR techniques. Nano-indentation measurements showed a substantial increase in hardness and modulus of the clear coat after approximately 1000 hrs of accelerated weathering exposure for substrate formulations not containing any hydro or photo stabilizers. Clearcoat durability was considerably larger for the substrate formulations with photo and hydro stabilizers, therefore indicating direct influence of chemical composition of substrate on clearcoat performance. Micro-FTIR analysis provided insights into chemical changes on the substrate surface with respect to formulation differences that impact clearcoat performance. From these experiments, it may be concluded that depth sensing indentation and micro-FTIR techniques can be together used as an efficient approach for coating-substrate performance evaluation and optimization. Future studies will be targeted towards correlating the mechanical property and chemical structure changes observed during weathering to the impact performance of the coated ETP.

REFERENCES

1. B. Claudè, L. Gonon, V. Verney and J.L. Gardette, Polymer Testing, **20**, 771 (2001)
2. M.E. Nichols, J.L. Gerlock, C.A. Smith and C.A. Darr, Progress in Organic Coatings, **35**, 153 (1999)
3. N. Tahmassebi and S. Moradian, Polymer Degradation and Stability, **83**, 405 (2004).
4. W.C. Oliver and G.M. Pharr, J. Mat. Res., **19**, 3 (2004).
5. S. Etienne-Calas, A. Duri and P. Etienne, J. Non-Crystalline Solids, **344**, 60-65 (2004)
6. C.M. Chan, G.C. Cao, H. Fong, T. Robinson and N. Nelson, J. Mat. Res., **15**, 14 (1999)
7. S. Roche, S. Pavan, J.L. Loubet, P. Barbeau and B. Magny, Progress in Organic Coatings, **47**, 37 (2003).
8. K. Adamsons, Progress in Organic Coatings, **45**, 69 (2002)
9. A.C. Tavares, J.V. Gulmine, C.M. Lepienski and L. Akcelrud, Polymer Degradation and Stability, **81**, 367 (2003)
10. M. Osterhold and P. Glöckner, Progress in Organic Coatings, **41**, 177 (2001)
11. W.C. Oliver and G.M. Pharr, J. Mat. Res., **4**, 1564 (1992)
12. J. Menčík, D. Munz, E.R. Weppelmann and M.V. Swan, J. Mater. Res., **12**, 2475 (1997)
13. S. Brunner, P. Richner, U. Müller and O. Guseva, Polymer Testing, **24**, 25 (2005).
14. S. Suresh and A. Giannakopoulos, Acta Mater, **46**, 5755 (1998)
15. C.A. Taylor, M.F. Wayne and W.K.S. Chiu, Thin Solid Films, **429**, 190 (2003)
16. N. Panich and Y. Sun, Surface and Coatings Technology, **182**, 342 (2004)
17. G.N. Peggs and I.C. Leigh, Recommended Procedure for Micro-indentation Vicker's hardness test, Report MOM 62, UK, National Physical Laboratory (1982)
18. Z.H. Hu, " *Mechanical characterization of coatings and composites – depth sensing indentation and finite element modeling*" (Doctoral dissertation, Royal Institute of Technology, Sweden, 2004).
19. B. Mailhot, P. Bussière, A. Rivaton, S. Morlat-Thérias and J. Gardette, Macromolecular Rapid Communications, **25**, 436 (2004)
20. R.M. Silverstein, G.C. Bassler and T.C. Morril, "*Spectrometric identification of organic compounds*", John Wiley & Sons (1981).

Mater. Res. Soc. Symp. Proc. Vol. 875 © 2005 Materials Research Society

Mechanical Characterization of Multilayer Thin Film Stacks Containing Porous Silica Using Nanoindentation and the Finite Element Method

Ke Li [a], Subrahmanya Mudhivarthi [b,c], Sunil Saigal [a], and Ashok Kumar [b,c]
[a] Department of Civil Engineering,
[b] Department of Mechanical Engineering, and
[c] Nanomaterials and Nanomanufacturing Research Center,
University of South Florida, Tampa, FL 33613, U.S.A.

ABSTRACT

Novel metal/dielectric material combinations are becoming increasingly important for reducing the resistance-capacitance (RC) interconnection delay within integrated circuits (ICs) as the device dimensions shrink to the sub-micron scale. Copper (Cu) is the material of choice for metal interconnects and SiO_2 (with a dielectric constant k = ~ 3.9) has been used as an interlevel dielectric material in the industry. To meet the demands of the international road map for semiconductors, materials with a significantly lower dielectric constant are needed. In this study, the effects of porosity and layer thicknesses on the mechanical properties of a multilayer thin film (Cu, Ta and SiO_2)-substrate (Si) system are examined using nanoindentation and finite element (FE) simulations. A micromechanics model is first developed to predict the stress-strain relation of the porous silica based on the homogenization method for composite materials. An FE model is then generated and validated to perform a parametric study on nanoindentation of the $Cu/Ta/SiO_2/Si$ system aiming to predict the mechanical properties of the multilayer film stack.

INTRODUCTION

A number of new organic and inorganic materials are being investigated for their applicability as interlayer dielectric materials [1]. One way to obtain dielectric materials with a low dielectric constant (k) is to introduce air voids into silica (SiO_2). However, the introduction of porosity deteriorates the mechanical properties of the dielectric material and thus the mechanical reliability of the entire IC device when undergoing chemical-mechanical planarization processes. Thus, in-depth understanding of the mechanical behavior of this class of low-k materials (i.e., porous silica) and the overall mechanical performance of thin film stacks containing porous silica layers is essential for the reliable application of these dielectric materials.

An extensively used technique for characterizing the mechanical behavior of thin films is nanoindentation, whereby properties such as hardness and elastic modulus can be obtained [2]. The applicability of nanoindentation to porous low-k materials, nevertheless, needs further justifications for lack of understanding of pore crushing involved in the indentation process [1]. In this work, a parametric study is performed to investigate the mechanical performance of thin film stacks containing Cu, tantalum (Ta) and SiO_2 (solid or porous) as the interconnect layer, the barrier layer, and the dielectric layer, respectively. Porous silica produced by Iskandar et al. [3] is taken as the porous dielectric material. A micromechanics model is first developed to determine the effective mechanical properties of the porous silica. A finite element (FE) model is then generated to simulate the nanoindentation process of thin film stacks. Controlling parameters include the thicknesses of the Cu layer and the SiO_2 layer.

MICROMECHANICAL MODELING OF POROUS SILICA

Iskandar et al. produced silica films containing three-dimensional (3-D) ordered macropores using a spray drying method [3]. Scanning electronic microscopy (SEM) observations indicated that the porous silica contains well-ordered pores with an approximately regular hexagonal close-packed (HCP) microstructure (see Figure 1). A unit cell of the HCP structure is illustrated in Figure 2. Taking advantage of the symmetry in geometry, one twelfth of the HCP cell can be adopted as the minimum repeating unit, as shown in Figure 2 (in green). A micromechanics model is developed based on this repeating unit. The overall mechanical properties of the repeating unit and thus of the porous silica are obtained by employing the homogenization method for composite materials.

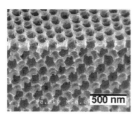

Figure 1. SEM images of the porous silica [3].

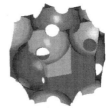

Figure 2. An HCP unit cell.

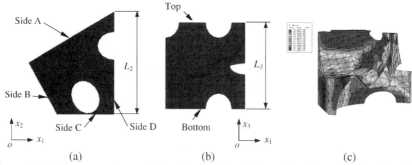

(a)   (b)   (c)

Figure 3. Minimum repeating unit ($d = 79$ nm): (a) undeformed configuration (top view), (b) undeformed configuration (front view), and (c) deformed configuration.

A finite element study is conducted to obtain the elastic properties and the stress-strain relationship of the porous silica using the commercial FE package ABAQUS 6.4. The FE model shown in Figure 3 is generated based on the minimum repeating unit (see Figure 2). A total of 1846 tetrahedral elements are employed in the model. The Young's modulus and yield strength of solid silica are, respectively, taken to be 73 GPa and 7.1 MPa [4], and the Poisson's ratio is 0.28 [5]. A uniaxial compressive test is simulated by applying a prescribed displacement to the nodes on the top plane of the repeating unit along the vertical direction (see Figure 3). Symmetry boundary conditions are imposed on the nodes on planes A and D. The nodes on the bottom plane are restrained to move in the vertical direction. In order to ensure that the effective properties of the repeating unit represent those of the porous material, periodic boundary conditions are applied at the nodes on planes B and C [6].

To obtain the effective elastic properties of the porous silica, a small compressive strain in the amount of -0.001 is applied in the loading direction to ensure elastic deformations needed. The Young's modulus E_3 and Poisson's ratio v_{32} are then obtained, respectively, as

$$E_3 = -\frac{F}{\varepsilon_3 A}, \quad v_{32} = -\frac{u_2}{\varepsilon_3 L_2}, \tag{1a,b}$$

where F is the total reaction force along the x_2 direction; $\varepsilon_3 = -0.001$ is the applied compressive strain, $A = \frac{\sqrt{3}}{4}L^2$ is the cross-sectional area of the minimum repeating unit perpendicular to the x_3 direction; u_2 is the lateral displacement along the x_2 direction; and L_2 is the dimension of the repeating unit along the x_2 direction (see Figure 3(a)). The magnitudes of the true stress, σ_T, and true strain, ε_T, are determined by first calculating the magnitudes of the engineering stress, σ_E, and engineering strain, ε_E, using

$$\sigma_E = \frac{F}{A}, \quad \varepsilon_E = \frac{u_3}{L_3}, \tag{2a,b}$$

where u_3 is the applied displacement and L_3 denotes the height of the repeating unit (see Figure 3(b)), and then employing

$$\sigma_T = \sigma_E(1 + \varepsilon_E), \quad \varepsilon_T = \ln(1 + \varepsilon_E). \tag{3a,b}$$

The predicted effective Young's modulus and Poisson's ratio are found to be the same for the porous silica material with three different pore sizes (i.e., $d = 79$ nm, 178 nm or 254 nm). They are, respectively, 6.3 GPa and 0.025. From Figure 4 it is noted that the stress-strain curves are identical for varying pore sizes. This is because the stress-strain response is dependent on porosity of the layer, which was kept constant. Figure 4 also gives the yield strength of the porous silica being 0.65 MPa.

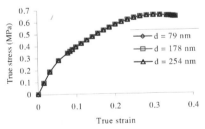

Figure 4. True stress vs. true strain relations of porous silica samples with different pore diameters.

Figure 5. Thin film stack under complete deformation (maximum depth of penetration).

FE ANALYSIS OF MULTILAYER THIN FILM STACKS

In order to capture the microstructural details of the porous silica, continuum elements with sufficiently small sizes are needed to mesh the repeating unit (see Figure 3). This requires that a huge number of FE elements be used if a porous silica layer comprised of hundreds of HCP cells is modeled for simulation of the nanoindentation process. This computationally prohibitive modeling method is unsuitable for a parametric study. An alternative approach is to homogenize the microstructural features of a heterogeneous material such that the material can be treated as an equivalent homogeneous continuum, thereby leading to a significantly less computationally demanding nanoindentation problem [7]. The heterogeneous material and its homogenized counterpart have the same overall stress-strain relation, which can be determined by uniaxial compressive tests. It should be pointed out that the use of a homogenized material to replace its heterogeneous composite may not be able to describe all of the deformational phenomena that would occur with the heterogeneous composite involved in the nanoindentation process. Nevertheless, since the indenter is not in direct contact with the silica layer, this homogenization approach is believed to hold good validity for comparing the effective mechanical reliability of thin film stacks.

An axi-symmetric FE model is developed to simulate the nanoindentation of the multilayer thin film stacks. The thickness of the barrier layer (Ta) is taken to be fixed at 40 nm [1], and that of the Cu layer varies as 1000 nm, 1200 nm, 1400 nm and 1600 nm, and that of the silica layer changes as 500 nm, 600 nm, 700 nm and 800 nm. The mechanical properties (Young's modulus E, Poisson's ratio v, yield strength σ_y, and tangent modulus E_t) of the constituent films are obtained from micromechanical modeling as well as literature, as listed in Table I. The tip of the diamond indenter has a radius of 125 nm. A total of 4050 CAX4 elements are employed to mesh the thin film stack and the substrate, and 460 CAX3 elements are used to model the indenter. Symmetry boundary conditions are applied at the nodes lying on the left side of the nanoindentation system, and displacement constraints are imposed on the bottom nodes of the substrate. Perfect bonding is assumed at the interfaces between the three layers and between silica layer and the substrate. A coefficient of friction of 0.2 is used at the interface between the Cu layer and the indenter. A preliminary analysis is performed on the nanoindentation of commercially pure aluminum. The predicted loading-unloading curve agrees well with the experimentally obtained results [4], thereby validating the model. This validated FE model allows that simulations be performed for the nanoindentation of multilayer thin film stacks with varying thicknesses of the Cu layer (t_c) and the silica layer (t_s).

Table I. Material properties used in FE analyses

Material	E (GPa)	σ_y (MPa)	E_t (GPa)	v
Cu	145 [8]	275 [8]	14.5 [8]	0.24 [5]
Ta	185 [9]	354 [10]	1.1 [10]	0.34 [5]
Porous SiO$_2$	6.43	12.86		0.025
Si	127 [11]	4410 [11]	0	0.278 [5]
Solid SiO$_2$	73 [4]	7100 [4]	0	0.28 [5]
Diamond	1140 [4]			0.07 [4]

RESULTS AND DISCUSSION

The loading-unloading curves for thin film stacks containing porous or solid silica layers are shown in Figures 6-9. It follows from Figures 6 and 7 that stack with a thicker porous silica layer have smaller magnitudes of the peak indentation load, while stacks with varying thicknesses of the copper layer have insignificant changes in their loading-unloading curves. The reason for this lies in that the overall load-carrying capability of a thin film stack depends on that of the porous silica layer, which has the lowest yield strength (see Table I). As t_s increases, there is a larger volume of pores present in the stack, thereby deteriorating the overall mechanical properties. When solid silica is used as the dielectric material, the copper layer is the most deformable layer in the stack (see Table I), and hence plays a dominant role in determining the peak indentation load of the stack. Since the thickness of the copper layer is fixed, the pattern of the loading-unloading curve remains almost unchanged as t_s varies from 500 nm to 800 nm (see Figure 8). However, with a constant thickness of the silica layer, the peak indentation load decreases in a monotonic fashion as t_c increases (see Figure 9). In addition, it is clearly shown in Figures 6-9 that the peak indentation loads are significantly smaller for stacks containing porous silica layers than for stacks including solid silica layers.

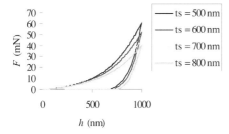

<table>
<tr><td>

Figure 6. Load vs. displacement for stacks containing porous silica (t_c = 1000 nm).

</td><td>

Figure 7. Load vs. displacement for stacks containing porous silica (t_s = 500 nm).

</td></tr>
</table>

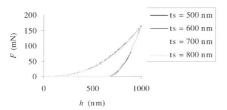 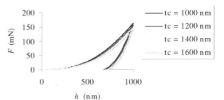

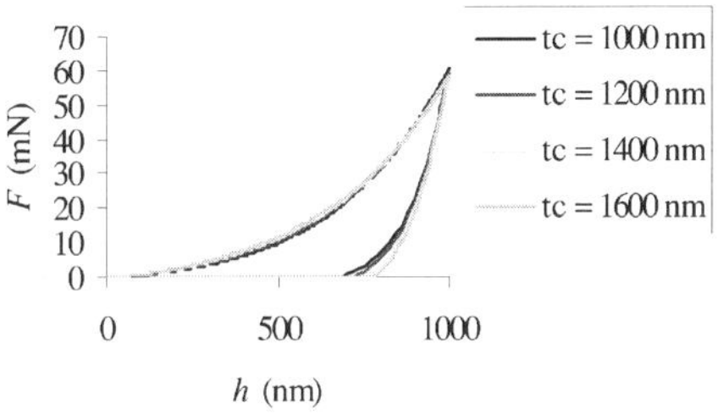

Figure 8. Load vs. displacement for stacks containing solid silica (t_c = 1000 nm).

Figure 9. Load vs. displacement for stacks containing solid silica (t_s = 500 nm).

The magnitudes of the hardness (H) and reduced modulus (E_r) of the thin film stacks are calculated from the corresponding loading-unloading curves (see Figures 6-9) by using the method of [2]. The dependence of H and E_r on t_c, t_s and the porosity of silica is illustrated in Figure 10. It is noted from Figure 10 that the varying trend of H is similar to that of the peak indentation load shown in Figures 6-9 since the hardness and yield strength of a metal film may be related with a linear relation [12]. From Figure 10(a) one can also observe that as t_s increases, E_r decreases for stacks with solid or porous silica layers, while the stiffness weakening is less for stacks containing solid silica layers than for stacks having porous silica layers with the same amount of increase in t_s. This may be attributed to the fact that the differences between the Young's modulus of the porous silica and those of other layers in the same stack are larger than the differences between the Young's modulus of the solid silica and those of the remaining layers. In addition, an inspection of Figure 10(b) indicates that as t_s rises, E_r increases for stacks with porous silica layers but has negligibly small variations for stacks having solid silica layers. This may be explained by the differences in the stiffness values of the film layers. Copper has a substantially higher Young's modulus than the porous silica. As a result, increasing the thickness of the copper layer and thus the volume fraction of copper in the film stack tends to strengthen the entire stack.

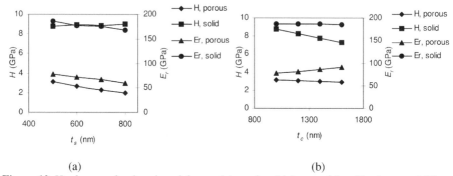

(a)   (b)

Figure 10. Hardness and reduced modulus vs. (a) varying thickness of the silica layer and (b) varying thickness of the copper layer.

CONCLUSION

A parametric FE study is conducted on the mechanical properties of multilayer thin film stacks containing solid or porous silica layers. Controlling parameters include the thicknesses of the dielectric (silica) layer and the interconnect (copper) layer. The numerical results indicate that stacks with porous silica layers have significantly lower hardness and reduced modulus than stacks having solid silica layers. Also, increasing the thickness of the silica layer lowers the hardness and reduced modulus of stacks containing porous silica layers, but has insignificant effect on the hardness and reduced modulus of stacks including solid silica layers. Additionally, as the copper layer is becoming thicker, stacks containing porous silica layers are strengthened in terms of reduced modulus but is slightly affected in terms of hardness, while stacks with solid silica layers have diminished hardness but their magnitudes of reduced modulus are insignificantly influenced.

REFERENCES

1. K. Maex, M. R. Baklanov, D. Shamiryan, F. Iacopi, S. H. Brongersma, Z. S. Yanovitskaya, J. Appl. Phys. **93**, 8793 (2003).
2. W. C. Oliver, G. M. Pharr, MRS Bull. **17**, 28-33 (1992).
3. F. Iskandar, M. Abdullah, H. Yoden, K. Okuyama, J. Appl. Phys. **93**, 9237 (2003).
4. J. A. Knapp, D. M. Follstaedt, S. M. Myers, J. C. Barbour, T. A. Friedmann, J. Appl. Phys. **85**, 1460 (1999).
5. F. Iacopi, S. H. Brongersma, B. Vandevelde, M. O'Toole, D. Degryse, Y. Travaly, K. Maex, Microelectron. Eng. **75**, 54 (2004).
6. K. Li, X.-L. Gao, G. Subhash, Int. J. Solids Struct. **42**, 1777 (2005).
7. P. Vena, D. Gastaldi, Compos.: Part B 36, **115** (2005).
8. H. Pelletier, J. Krier, A. Cornet, P. Mille, Thin Solid Films **379**, 147 (2000).
9. National Materials Advisory Board, *Coatings for high-temperature structural materials: trends and opportunities* (The National Academy of Sciences, 1996), p. 18.
10. H. E. Boyer, *Atlas of Stress-Strain Curves* (ASM, Metals Park, OH, 1987).
11. A. K. Bhattacharya, W. D. Nix, Int. J. Solids Struct. **24**, 1287 (1988).
12. A. A. Volinsky, W. W. Gerberich, Microelectron. Eng. **69**, 519 (2003).

Depth-Dependent Hardness Characterization by Nanoindentation using a Berkovich Indenter with a Rounded Tip

Ju-Young Kim[1], David T. Read[2], and Dongil Kwon[1]

[1]School of Materials Science and Engineering, Seoul National University, Seoul 151-744, South Korea

[2]Materials Reliability Division, National Institute of Standards and Technology, Boulder, CO 80305, USA

ABSTRACT

The height difference Δh_b between the ideally sharp Berkovich indenter tip and a rounded tip was measured by direct observation using atomic force microscopy (AFM). The accuracy of the indirect area function method for measuring Δh_b was confirmed. The indentation size effects (ISE) in (100) single crystal copper, (100) single crystal tungsten, and fused quartz were characterized by applying the ISE model considering the rounded tip effect. The model fits the data these materials well, even though fused quartz does not deform by dislocations. However, a very small value of the ISE characteristic length h' was obtained for fused quartz. The present h' value for (100) copper is 32% larger than a previously-measured value for polycrystalline copper. This may indicate that grain boundaries suppress the dislocation activity envisioned in the ISE model.

INTRODUCTION

The nanoindentation technique has been established as a powerful means to characterize the mechanical properties of small volume-material [1,2]. However, it has always been difficult to precisely determine the mechanical properties of micro/nano-materials by nanoindentation because of the well-known indentation size effect (ISE), that is, the hardness increase with decreasing indentation depth in the sub-micrometer range [3,4]. Nix and Gao clearly showed that the ISE in crystalline materials can be explained on the basis of the concept of geometrically necessary dislocations (GNDs) [5]. As more and more experimental results are reported, researchers find that the Nix-Gao model cannot accurately fit all the experimental results, especially for indentation depth less than about 100 nm [6]. Many researchers have extended the Nix-Gao model to address this problem; in particular, J.-Y. Kim *et al.* extended the ISE model by considering the rounded tip effect on ISE [7]. In the present study, we report an improved experimental method to accurately measure the bluntness of the rounded tip. This value is used in the extended ISE model to analyze experimental nanoindentation data on (100) single crystal copper, (100) single crystal tungsten, and fused quartz.

MODEL REVIEW AND EXPERIMENTAL DETAILS

The hardness is written as

$$\frac{H}{H_0} = \sqrt{1 + J\frac{h'}{h_c}},\tag{1}$$

where H is the hardness, h_c is the contact depth, H_0 is the macroscopic hardness, h' is the characteristic length for ISE, and J is the scaling factor for the ISE considering the rounded tip effect [7]. J is expressed as

$$J = \chi\left(1 + \frac{\Delta h_b}{h_c}\right)^{-2},\tag{2}$$

where Δh_b is the height difference between the ideally sharp tip and the rounded tip, and χ is the ratio of plastic to total contact depth. The χ term involves the elastic recovery effect of the tip bluntness, while the $(1 + (\Delta h_b / h_c))^{-2}$ term involves a geometric effect of the tip bluntness. χ is a function of contact depth, because the rounded indenter geometry is not self-similar at shallow contact depths. The $(1 + (\Delta h_b / h_c))^{-2}$ term reflects the correction of Δh_b for the distribution depth of GNDs. The bluntness constant Δh_b is needed for the indenter tip used in the measurement; H and χ values measured as functions of contact depth are needed to apply the model.

The main factor in the extended model is Δh_b, which was measured by atomic force microscopy (AFM) as shown in figure 1. For a Berkovich indenter, three height profiles, each tracing an edge, traversing the peak, and continuing down the opposite face were obtained by contact mode AFM scan. The profile locations were determined using a preliminary AFM scan over the whole indenter tip. Indenter tip shape parameters, Δh_b and radius curvature of the tip R were calculated from the AFM measurement results.

Three samples, (100) single crystal copper, (100) single crystal tungsten, and fused quartz, were prepared; the copper and tungsten specimens were mechanically polished and electro-polished. The average surface roughness, R_a, was measured by AFM. The Young's modulus was measured by an ultrasonic pulse-echo technique for copper, determined with reference to the previous research for tungsten [8], and measured by the Oliver-Pharr method [1]. The measured properties are shown in Table I. The analysis procedures for the nanoindentation results, such as contact area calculation, were the same as previously reported [7].

Table I. Surface roughness and Young's modulus of the samples.

Material	Surface roughness, R_a (nm)	Young's modulus, E (GPa)
(100) single crystal copper	7.05	113.1
(100) single crystal tungsten	5.85	414.8
Fused quartz	2.11	74.1

RESULTS AND DISCUSSION

Indenter tip shape

Δh_b was measured by a comparison between the measured height profile and that of an ideally sharp Berkovich indenter; the measured value is 18 nm as shown in figure 2. Nanoindentation experiments using the same indenter were conducted on a fused quartz standard specimen at various contact depths (3-400 nm). Δh_b was measured from the relation of the contact area and the contact depth analyzed by the Oliver-Pharr method; the measured value was 19 nm [7]. Thus, the indirect method to determine Δh_b using the Oliver-Pharr method compared well with the directly measured AFM results. The measured R was 319 nm, while a tip radius of 290 nm was calculated from the relation that tip radius is 16.13 times of Δh_b for an ideal sphero-conical indenter. This result indicates the indenter tip was worn slightly more flat than ideal spherical-conical geometry.

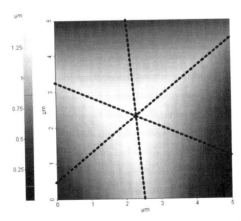

Figure 1. AFM scan on a Berkovich indenter, and three lines to obtain height profiles.

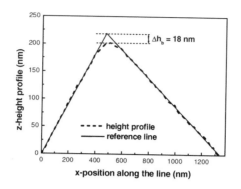

Figure 2. Height profile used for the calculation of Δh_b and R.

ISE characterization

Figure 3 shows the hardness variation with contact depth and curves fitted by the present model for (100) copper, (100) tungsten, and fused quartz. The physical relevance of the present modeling approach to fused quartz is considered below. The curves fit the hardness results over the full range of contact depth. The measured H_0 and h' values are 0.84 GPa and 612 nm for (100) copper, 3.48 GPa and 295 nm for (100) tungsten, and 6.89 GPa and 2 nm for fused quartz. The value of χ_0, the criteria for the onset of tip bluntness, was measured as about 90 nm for these three materials. To test what might happen when indentation depth is limited to small values, the model was applied to the hardness results only for contact depth less than 90 nm. The measured H_0 and h' values are 0.84 GPa and 683 nm for (100) copper, 3.63 GPa and 253 nm for (100) tungsten, and 6.87 GPa and 13 nm for fused quartz. These macroscopic hardness values and characteristic lengths for the ISE agree reasonably with those obtained using results from the full range of contact depth. h' values were calculated using the relation, $h'=81b\alpha^2 \tan^2 \theta \mu^2 / 2H_0^2$ [7], where μ is the shear modulus of 41.9 GPa for (100) copper, 160.0 GPa for (100) tungsten, and 31.7 GPa for fused quartz [9]. Measured and calculated characteristic values for ISE are presented in Table II. Calculated values of h' are in reasonably good agreement with the measured values, considering that only a rough estimate of the value of the geometrical factor α is available.

Comparing with previous results of H_0 of 0.846 GPa and h' of 463 nm for annealed polycrystalline copper, the h' value for (100) single crystal copper is 32% greater while the H_0 value for (100) single crystal copper is similar. The measured grain size of the annealed polycrystalline copper was 2913 nm, which is a few times greater than contact radius, a. It was

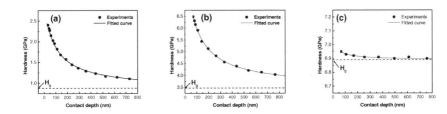

Figure 3. Hardness results with contact depth and curves fitted by the model for (a) (100) single crystal copper, (b) (100) single crystal tungsten, and (c) fused quartz.

assumed that the GNDs generated by indentation were distributed in the hemispherical volume with contact radius in the developed model. The existence of grain boundaries plays the role to suppress the ISE, even though the contact radius is sufficiently greater than grain size. One time pop-in occurred for (100) single crystal tungsten, at an indentation depth of about 70 nm. The pile-up effect was considered in calculating hardness for (100) single crystal tungsten, the pile-up phenomenon is not uniform just after pop-in, thus, the deviation of measured hardness values from the fitted curve is large for contact depths less than about 90 nm. The measured hardness values of fused quartz were quite uniform for the full range of contact depth. The ISE was negligible for the fused quartz. This was expected, since fused quartz does not deform by the generation and propagation of dislocations, and therefore its inhomogeneous plastic deformation at the surface caused by indentation is not explained by GND mechanism.

Table II. Measured and calculated characteristic values for ISE.

Material	Contact depths used for fitting	H_0 (GPa)	Measured h' (nm)	Calculated h' (nm)
(100) single crystal copper	Whole	0.84	612	826.3
	< 90 nm	0.84	683	826.3
(100) single crystal tungsten	Whole	3.48	295	771.0
	< 90 nm	3.63	253	708.6
Fused quartz	Whole	6.89	2	4.9
	< 90 nm	6.87	13	5.0

CONCLUSIONS

The bluntness constant Δh_b of a sharp Berkovich indenter tip was measured by direct AFM observation; the measured value was 18 nm. Δh_b measured by an indirect area function method based on the Oliver-Pharr method was 19 nm. This result provides confirmation that this indirect area function method is useful for measuring Δh_b. The ISEs of (100) single crystal copper, (100) single crystal tungsten, and fused quartz were characterized by applying the ISE model considering the rounded tip effect. Even though fused quartz does not deform by dislocations, the macroscopic hardness value H_0 extracted from the model fit to the hardness curve, restricted to shallow penetration depths, agreed well with the hardness at full penetration. From the comparison with previous results for polycrystalline copper, the existence of grain boundaries suppresses the ISE even though the contact radius is sufficiently greater than grain size.

ACKNOWLEDGMENTS

The authors would like to thank Dr. J.-H. Kim in KIMM for the AFM measurement. Support for this research was provided by the center for electrical component reliability design technology of the ministry of science and technology, South Korea under grant M1040300001304001000610 and is gratefully acknowledged.

REFERENCES

1. W.C. Oliver and G.M. Pharr, *J. Mater. Res.* **7**, 1564 (1992).
2. Y.-H. Lee and D. Kwon, *J. Mater. Res.* **17**, 901 (2002).
3. Q. Ma and D.R. Clarke, *J. Mater. Res.* **10**, 853 (1995).
4. K.W. McElhaney, J.J. Vlassak, and W.D. Nix, *J. Mater. Res.* **13**, 1300 (1998).
5. W.D. Nix and H. Gao, *J. Mech. Phys. Sol.* **46**, 411 (1998).
6. Y. Wei, X. Wang, and M. Zhao, *J. Mater. Res.* **19**, 208 (2004).
7. J.-Y. Kim, B.-W. Lee, D.T. Read, and D. Kwon, *Scripta Mater.* **52**, 353 (2005).
8. Y.-H. Lee and D. Kwon, *J. Kor. Inst. Met. & Mater.* **41**, 104 (2003).
9. W.D. Callister, Jr., *Materials Science and Engineering: An Introduction*, 3rd ed. (John Wiley & Sons, Inc., New York, 1994), p. 111.

Mechanical Behavior of
Nanostructured Films

Atomic-Scale Analysis of Strain Relaxation Mechanisms in Ultra-Thin Metallic Films

M. Rauf Gungor and Dimitrios Maroudas
Department of Chemical Engineering, University of Massachusetts, Amherst, MA 01003, U.S.A.

ABSTRACT

A comprehensive computational analysis is presented of the atomistic mechanisms of strain relaxation over a wide range of applied biaxial tensile strain in free-standing Cu thin films. The analysis is based on large-scale isothermal-isostrain MD simulations using slab supercells with cylindrical voids normal to the film plane and extending throughout the film thickness. Our analysis has revealed various regimes in the film's mechanical response as the applied strain level increases. Following an elastic response at low strain (< 2%), plastic deformation occurs accompanied by emission of screw dislocations from the void surface and threading dislocations from the film surfaces, in parallel with generation of vacancies due to slip of jogged dislocations. At the lower strain range following the elastic-to-plastic deformation transition (≤ 6%), void growth is the major strain relaxation mechanism, while at higher levels of applied strain (> 8%), a subsequent transition leads to a new plastic deformation regime where void growth plays a negligible role in the film strain relaxation.

INTRODUCTION

Nanometer-scale-thick films of ductile materials, such as metals, are used increasingly in modern technologies ranging from microelectronics to various areas of nanofabrication. Processing of such ultra-thin metallic films generates voids and other structural defects, which have detrimental effects on the function of components or devices that consist of or utilize these films; device interconnections in microelectronics is a typical example [1-4]. Fundamental understanding of strain relaxation mechanisms in ductile thin films through void growth is crucial for solving materials reliability problems in modern technological applications.

The dynamical growth of voids is an extremely important failure mechanism in ductile materials [5]. Extensive theoretical work on this problem has established a framework of analysis at the continuum scale [5,6]. However, more accurate and systematic modeling of materials mechanical behavior requires better mechanistic understanding of plasticity at the atomic and microstructural scales [7]. Toward this goal, molecular-dynamics (MD) simulations provide powerful means for direct monitoring of dynamical phenomena and detailed atomic-scale analysis [8,9]. In this paper, we explore the fundamental atomic-scale defect mechanisms that govern ductile void growth in metallic thin films based on large-scale MD simulations of the film's structural response to applied biaxial tensile straining.

COMPUTATIONAL METHODS

In our MD simulations, the classical equations of motion were integrated using a velocity Verlet algorithm[10] with a time step size of 8.12×10^{-16} s within an embedded-atom-method (EAM) parameterization for Cu [11]. For the results reported here, the simulation cell contained 1,443,036 atoms and had edge sizes of 616, 640, and 44 Å in the x, y, and z directions, respectively, at its unstrained state; the Cartesian x, y, and z axes are along the $[\bar{1}10]$, $[\bar{1}\bar{1}2]$, and $[111]$ crystallographic directions. A slab supercell was used with periodic boundary conditions applied in the x and y directions, while the boundaries in the z direction were traction-free surfaces. The choice of the film surface parallel to the (111) crystallographic plane was motivated by the common use of textured metallic thin-film interconnects with preferred <111> grain orientation [3].

The initial configuration was a metallic film with a perfectly cylindrical hole at the center of the cell with the axis of the cylinder along [111]; the void extended throughout the film thickness, which also is typical in metallic thin-film interconnects [12]. This choice of void configuration is extremely helpful in the analysis of the defect dynamics that govern plastic flow, void growth, and surface pattern formation. The initial void diameter was 92 Å; the supercell and initial void size were chosen so that a realistic void morphology could be simulated and characterized in detail during void size evolution, while interactions of a void with its periodic images had no considerable effects on the observed strain relaxation mechanisms. The simulations were performed at a constant temperature of 100 K by properly rescaling the atomic velocities at each time step. Initially, the supercell was unstrained, i.e., the cell edge sizes corresponded to the equilibrium density at the simulation temperature. The metallic film was strained biaxially in x and y up to the chosen strain level with a rate of 3×10^{13} s^{-1}. The effects of strain rate were examined over a one order-of-magnitude range; within this range, the atomistic mechanisms of ductile void growth were not affected by the strain rate.

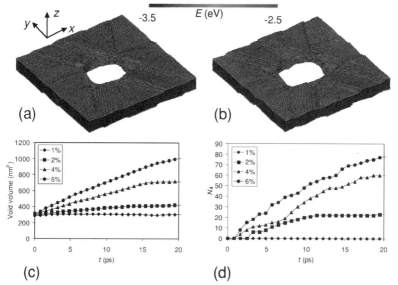

Figure 1. Configurations of strained Cu thin films with central cylindrical voids under applied biaxial strain $\varepsilon = 4\%$ (a) and 6% (b) at $t = 12.18$ ps; t is time in the MD simulation of biaxial straining at constant strain. In (a) and (b), atoms are colored according to their potential energy and the shown film surface area is 40×40 nm^2. Evolution of void size (c) and emitted screw dislocation population (d) for $\varepsilon = 1, 2, 4,$ and 6%.

THIN FILM MECHANICAL RESPONSE UNDER BIAXIAL STRAIN

At low strain ($\varepsilon = 1\%$), the film is purely elastically deformed, the void shape remains perfectly cylindrical, and the film surface is smooth. An elastic-to-plastic deformation transition occurs at a strain level between 1% and 2%. At strains higher than this critical strain, the mechanism of plastic deformation creates a distinct six-fold symmetric pattern on the film surface accompanied by void growth; this pattern formation is characteristic of the six-fold symmetry on {111} crystallographic planes, the slip-plane family of the FCC lattice structure. Representative atomic configurations of deformed Cu thin films under $\varepsilon = 4\%$ and 6% are shown in Figs. 1(a) and (b), respectively.

Plastic deformation is mediated by generation of perfect screw dislocation pairs with Burgers vectors a <110>/2, where a is the lattice parameter, pointing toward opposite surfaces of the film; the dislocation pairs are emitted from the void surface at six symmetrical locations and glide along <110> directions of the film plane on {111} slip planes. This FCC slip mechanism can be characterized primarily by the Thompson tetrahedron [13]. Through this plastic deformation mechanism, the initial cylindrical void grows and its morphology evolves into a hexagonal prismatic kind of shape. Furthermore, the film surface becomes rough from the formation of surface steps due to gliding screw dislocations as shown in Figs. 1(a) and (b). Increasing the applied strain level enhances void growth and the generation of screw dislocations emitted from the void surface; this is demonstrated in Figs. 1(c) and (d), respectively. This trend also is evident by comparing the surface step patterns of Figs. 1(a) and (b). Void growth and dislocation population reach a steady state eventually with initial transient periods that become longer as ε increases.

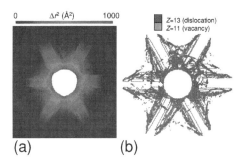

(a)   (b)

Figure 2. (a) Distribution of the atomic squared displacement, Δr^2, on the middle xy-plane of a strained Cu thin film. (b) Map of atomic coordination number, Z, shown as a top view for the entire thin film. Atoms with $Z = 12$ and $Z < 10$ are not shown in this map; only the atoms on the dislocation cores with $Z = 13$ and the atoms near vacancies with $Z = 11$ are indicated. Both snapshots are at $t = 12.18$ ps for $\varepsilon = 6\%$.

ATOMIC MOBILITY DURING PLASTIC DEFORMATION

Detailed analysis of the MD simulation results revealed formation of vacancies in addition to the primary screw dislocations; such point defects are responsible for the enhanced diffusional activity observed at the low simulation temperature. The distribution of atomic mobility around the void, as measured by the atomic squared displacement, $\Delta r^2(\mathbf{r}) \equiv [\mathbf{r}(t) - \mathbf{r}(0)]^2$ with $\mathbf{r}(t)$ denoting the atomic position vector at time t, is shown in Fig. 2(a) for $\varepsilon = 6\%$. This map demonstrates a perfect correlation of atomic mobility with plastic deformation; specifically, the mobility map exhibits a six-fold symmetric pattern that matches that of the dislocation slip zones shown in Fig. 1. Consistently with the trend demonstrated in Fig. 1, increasing ε enhances atomic mobility. Therefore, comparison of the results of Figs. 1 and 2(a) can be used to attribute atomic migration in the vicinity of the void to dislocation slip during ductile void growth.

Figure 2(b) shows the top view of the defect structure in the thin film through the distribution of the atomic coordination number, Z, after removing surface atoms and atoms with $Z = 12$, the coordination number for a perfect FCC lattice. Atoms in the dislocation cores and neighbors of a vacancy have $Z = 13$ and $Z = 11$, respectively; this Z distribution is used to clarify the correlation between dislocation slip and vacancy generation. Figure 2(b) shows clearly that vacancies follow the trace of the screw dislocations.

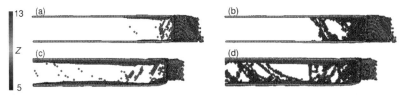

Figure 3. Cross-sectional views of the atomic coordination distribution in a strained Cu thin film under $\varepsilon = 6\%$ at $t = 4.06$ ps (a) and (b) and 11.37 ps (c) and (d). The region of the film shown corresponds to that within the yellow rectangular frame of Fig. 4. Atoms with $Z = 12$ are not shown and atoms with $Z = 13$ are not shown in (a) and (c).

VACANCY FORMATION AND MIGRATION DURING DUCTILE VOID GROWTH

We have analyzed the mechanisms of vacancy generation and migration systematically. Figure 3 shows the structural evolution of a thin film with a void under $\varepsilon = 6\%$; the atoms are colored according to their coordination numbers. The cross-sectional views are extracted from the region that is marked in the thin film in Fig. 4(a). This location was chosen for convenience, so that the dislocation slip direction coincides with the x-axis. Again, the atoms with $Z = 12$ are discarded from the coordination maps; the dislocations (visualized using atoms with $Z = 13$) are eliminated in the snapshots shown in Figs. 3(a) and (c). Comparing the corresponding snapshots in the two columns clarifies further the correlation of the vacancy distribution with that of the dislocation population.

Our analysis reveals that vacancies are generated during the slip of jogged screw dislocations. Subsequently, the vacancies migrate along dislocation cores of other gliding screw dislocations that approach the originally jogged one. In this low-temperature study, this pipe diffusion mechanism is the only one responsible for the atomic mobility that has been detected in Fig. 2. Under our simulation conditions, pipe diffusion is observed over the time scale captured by MD because the corresponding activation barrier is significantly lower than that for bulk diffusion, the vacancy migration velocity is increased significantly due to the external driving force for vacancy transport provided by the high applied tensile strain, and the vacancy migration length is indeed very short due to the small thickness of the film and the generation of the vacancies close to the film's free surfaces; the film surface acts as a vacancy sink and the vacancies are annihilated when they reach the surface.

NUCLEATION AND PROPAGATION OF THREADING DISLOCATIONS

The increase of surface roughness with higher ε increases the density of locations with higher local surface energy and stress concentration shown as "hot" (red) spots in Fig. 1(b). Such locations are favorable surface sites for nucleation of threading dislocations. In our MD simulations, threading dislocation nucleation is observed at the film surface for $\varepsilon > 4\%$. A representative case of threading dislocation dynamics is seen in Fig. 3(d) at a late stage of threading dislocation propagation; the propagating dislocation is visible as a long curved line at the right end of the group of dislocations shown on the left side of this cross-sectional view.

The top view of the surface of a biaxially strained Cu thin film is shown in Fig. 4(a). In addition to the surface steps that were shown in Figs. 1(a) and (b) as a result of gliding of screw dislocations emitted from the void surface, triangular surface features that sometimes appear as a triangular network of dislocation traces are clearly evident in the middle of the region framed by the yellow rectangle, as well as in Fig 1(b). The triangular patterns are related to the threading dislocations. The cross-sectional view of the thin-film region marked with the yellow rectangle is shown in Fig. 4(b); the atoms are colored according to their coordination number after removing the atoms with perfect coordination, $Z = 12$, to reveal the dislocation structure; two curved threading dislocations can be seen clearly in the middle of the region shown.

Each threading dislocation starts as a half loop with three main segments: the two segments that intersect the film surface, the threading arms, and a middle segment that lies parallel to the film surface and joins the threading arms; the two threading arms are $60°$ dislocations, while the middle segment has an edge dislocation character. All of the threading dislocation segments lie on the same {111} plane. The Burgers vector of the dislocation loop is $\mathbf{b} = a <112>/6$. When the half loop extends across the film thickness so that its middle segment "touches" the opposite film surface, it breaks up into two Shockley partial dislocations with $\mathbf{b} = a <112>/6$ through annihilation of its middle segment at the film surface. The leading Shockley partial (the one of the two farther from the void surface) can move parallel to the original perfect screw dislocations. However, the trailing Shockley partial cannot move in the same direction. Instead, the trailing partial dislocation slips to a neighboring {111} plane after undergoing the dissociation reaction $a[112]/6 \rightarrow a[211]/6 + a[\overline{1}01]/6$, generating a supersessile stair-rod partial [13] and another Shockley partial. The Shockley partial produced by this dissociation is free to move on the adjacent {111} plane, while the sessile stair-rod partial dislocation remains locked at the dissociation site. This explains how the propagation of threading dislocations in the film gives rise to the observed "secondary" pattern formation on the film surface consisting of triangular surface features, where the dislocation traces intersect at $60°$ angles.

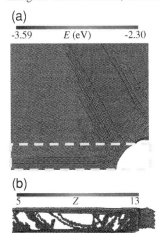

(a)

-3.59 E (eV)   -2.30

(b)

5   Z   13

Figure 4. (a) Top view of the upper left quadrant on the xy-plane of a strained Cu thin film surface with area equal to that captured in the configurations of Fig. 1. The atoms are colored according to their potential energy at $t = 10.88$ ps for $\varepsilon = 6\%$. (b) Cross-sectional view of the region within the yellow rectangular frame, shown as an atomic coordination map after removing the atoms with $Z = 12$. The dislocation cores ($Z = 13$) are shown in red.

CONCLUSIONS

The atomic-scale mechanistic information obtained by our MD simulations and analysis is not accessible to experiment and has important implications for the mechanical behavior of metallic thin films. In such films, ductile void growth is governed by a sequence of defect dynamical phenomena. At high strains, this sequence includes: screw dislocation emission from the void surface, dislocation propagation and film surface pattern formation, vacancy generation due to jogged dislocation slip, pipe diffusion of vacancies to the film surface, threading dislocation loop nucleation due to increased film surface roughness, and threading dislocation propagation and break-up leading to secondary film surface pattern formation. Furthermore, vacancy clustering ahead of the original void, such as that observed in Fig. 3(c), may lead to nanovoid nucleation, which can cause thin-film rupture under certain deformation conditions; such failure mechanisms have been known to be operative in bulk ductile materials [5,6].

Finally, we mention that we have carried out large-scale MD simulations of thin-film structural response to biaxial tensile straining at higher strain levels, $\varepsilon > 6\%$. Our simulations have revealed that for $\varepsilon > 8\%$, a new plastic deformation regime is operative, governed by a different mechanism of strain relaxation. This new relaxation mechanism is mediated by uniformly distributed dislocations in the thin film and it is independent of the void presence in the film. Specifically, expansion of the plastic zone around the void leading to ductile void growth is suppressed in this regime due to the interaction of the emitted dislocations from the void surface with the surrounding uniformly distributed defects that nucleate at the film surface. This leads to pinning of the dislocations emitted from the void surface and renders the role of the void in film strain relaxation negligible. Analysis of the thin-film structural response in this deformation regime will be presented in a forthcoming publication.

ACKNOWLEDGEMENTS

This work was supported by the National Science Foundation through Award Nos. CMS-0201319, CMS-0302226, and CTS-0205584.

REFERENCES

1. W. D. Nix, *Metal. Trans. A* **20A**, 2217 (1989).
2. R. J. Gleixner, B. M. Clemens, and W. D. Nix, *J. Mater. Res.* **12**, 2081 (1997).
3. C. V. Thompson and J. R. Lloyd, *MRS Bulletin* **18** (12), 19 (1993).
4. C.-K. Hu and J. M. E. Harper, *Mater. Chem. Phys.* **52**, 5 (1998).
5. L. B. Freund, *Dynamic Fracture Mechanics* (Cambridge University Press, Cambridge, 1998).
6. F. A. McClintock, *J. Appl. Mech.* **35**, 363 (1968); J. R. Rice and D. M. Tracey, *J. Mech. Phys. Solids* **17**, 201 (1969); A. Needleman, *J. Appl. Mech.* **39**, 964 (1972); J. S. Langer and A. E. Lobkovsky, *Phys. Rev. E* **60**, 6978 (1999).
7. See, e.g., H. Gao, Y. Huang, and W. D. Nix, *Naturwissenschaften* **86**, 507 (1999).
8. F. F. Abraham, *Adv. Phys.* **52**, 727 (2003); V. Bulatov, F. F. Abraham, L. Kubin, B. Devincre, and S. Yip, *Nature* **391**, 669 (1998); S. J. Zhou, D. L. Preston, P. S. Lomdahl, and D. M. Beazley, *Science* **279**, 1525 (1998); E. T. Seppälä, J. Belak, and R. E. Rudd, *Phys. Rev. B* **69**, 134101 (2004); J. Li, K. J. Van Vliet, T. Zhu, S. Yip, and S. Suresh, *Nature* **418**, 307 (2002); E. T. Lilleodden, J. A. Zimmerman, S. M. Foiles, and W. D. Nix, *J. Mech. Phys. Solids* **51**, 901 (2003).
9. M. R. Gungor, D. Maroudas, and S. J. Zhou, *Appl. Phys. Lett.* **77**, 343 (2000); M. R. Gungor and D. Maroudas, *Int. J. Fracture* **109**, 47 (2001); D. Maroudas and M. R. Gungor, *Comp. Mater. Sci.* **23**, 242 (2002).
10. M. P. Allen and D. J. Tildesley, *Computer Simulation of Liquids* (Oxford University Press, Oxford, 1990).
11. S. M. Foiles, M. I. Baskes, and M. S. Daw, *Phys. Rev. B* **33**, 7983 (1986).
12. See, e.g., O. Kraft and E. Arzt, *Acta Mater.* **45**, 1599-1611 (1997).
13. J. P. Hirth and J. Lothe, *Theory of Dislocations* (Wiley, New York, 1982).

Mater. Res. Soc. Symp. Proc. Vol. 875 © 2005 Materials Research Society

Investigation of the Deformation Behavior in Nanoindented Metal/nitride Multilayers by Coupling FIB-TEM and AFM observations

G. Abadias, C. Tromas, Y.Y. Tse, A. Michel
Laboratoire de Métallurgie Physique, UMR 6630, Université de Poitiers, SP2MI, Téléport 2, 86962 Chasseneuil-Futuroscope, FRANCE

ABSTRACT

Epitaxial TiN/Cu bilayers and multilayers with periods Λ between 5 and 50 nm have been grown by ultrahigh vacuum ion beam sputtering on Si and MgO(001) substrates at room temperature. The deformation modes induced by a Berkovich nanoindent have been imaged using Focused Ion Beam – Transmission Electron Microscopy (FIB-TEM) and Atomic Force Microscopy (AFM). The observations suggest that the mechanical response of the multilayers is essentially governed by an extensive plastic flow inside the Cu layers, which is confined by a bending of the more rigid TiN layers. This specific deformation behavior, with no contribution of the interfaces as a barrier for dislocation motion could explain the absence of significant hardness enhancement in this system.

INTRODUCTION

Transition-metal nitride thin films (such as TiN, ZrN, or VN) are a technologically important class of materials due to their mechanical properties as well as chemical stability. Combining different nitrides or coupling nitrides with metals in the form of nanocomposites or superlattices offers new opportunities to design superhard (hardness > 40 GPa) coatings [1-4]. Enhanced hardness with respect to the monolithic constituent materials has been reported in many multilayered structures, such as Mo/NbN [5,6], W/ZrN [4,5] or TiN/NbN [3,7,8], when the bilayer period (Λ) decreases down to the nanoscale. These experimental findings have been mainly explained by invoking Koehler's original model [9]. This theory, revisited later by Chu and Barnett [10], is based on dislocation hindrance at sharp interfaces between nanometer thin epitaxial layers of materials with a large difference in elastic shear moduli. Other analyses refer to deformation processes based on confined plasticity, where dislocations bow within individual layers, but this behavior concerns essentially the case of metal/metal multilayers [11-13]. However, the mechanical deformation modes in nitride nanolayers during indentation remain poorly understood, especially regarding the role of dislocation motion. Only few studies [14-17] have reported direct experimental observations of dislocation activity in these films to validate theoretical mechanisms. This is mainly due to the difficulty in preparing samples with appropriate imaging conditions in the transmission electron microscope (TEM), due to small grain sizes and high defect densities in the as-deposited films. Recently, it has been shown that the Focused Ion Beam (FIB) is a well suited technique for the preparation of TEM cross-section specimens from samples that have been locally deformed by nanoindentation [18]. Additionally, hardening is not always observed in nanoscale multilayers [15,19,20], suggesting that the deformation across the interfaces is not prevented. Therefore, new insights should be gained to understand the exact deformation mechanisms operating at small scales under nanoindentation tests. In particular, the case of nanolayers combining elastically soft metals (Cu or Ag) and hard nitrides remains to be investigated. Previous results have shown no significant hardness enhancement in TiN/Cu nanolaminates [20]. The aim of the present study is to provide a direct observation of the deformation modes induced in these layers by a Berkovich indenter, by combining for the first time FIB-TEM and Atomic Force Microscopy (AFM) observations.

EXPERIMENTAL PROCEDURES

~200-nm-thick TiN/Cu films were deposited at room temperature (RT) by ultrahigh vacuum dual ion beam sputtering (DIBS). Deposition occurred in a NORDIKO 3000 sputtering device (base pressure 10^{-8} Torr) by sputtering alternatively pure Cu (99.9%) and Ti (99.99%) targets using a primary Ar ion beam accelerated under 1.2 kV and a current of 80 mA. A secondary ion beam, extracted from a Ar + N_2 plasma, accelerated under a voltage of 25 V and current of 40 mA, was used during Ti deposition to form the crystalline titanium nitride phase in the cubic B1 structure. Two samples sets are investigated: i) TiN/Cu and Cu/TiN bilayers with equal thickness ($h_{Cu}=h_{TiN}=100$ nm) grown on (001) MgO substrates and ii) a multilayer with bilayer period $\Lambda=46.3$ nm and $h_{Cu}/h_{TiN}\sim1.5$, grown on two types of substrate: a Si wafer covered with a thin (~2 nm) native oxide and (001) MgO single crystals. The microstructure, preferred orientation and interface morphology have been characterized by X-ray Diffraction (XRD) in various configurations: low angle and high angle symmetric θ–2θ scans, ω-scans (rocking curve measurements) and ϕ–scans. The stress state has been determined by XRD using the $sin^2 \psi$ method, where ψ is the angle between the sample surface and a given (*hkl*) pole or reflection, and ϕ the corresponding in-plane azimutal angle. Details on the diffractometer set-up may be found elsewhere [21]. Nanoindentation tests were performed using a NHT nanoindenter (CSM Instrument) with a Berkovich tip. For hardness measurements, arrays of indents have been performed with loads ranging from 0.3 to 20 mN. Hardness values have been averaged over 10 tests with equal loads. The true penetration depth h_c has been determined using the Oliver and Pharr method [22]. Specific arrays with alternate loads of 4 and 8 mN were made for FIB-TEM sample preparation. Very high load indents have been performed for optical localization of the area of interest. Samples were prepared in a Philips FIB 200 TEM system operating at 30 keV with Ga ions, and using the lift-out technique. A 20×1.5 μm^2 area was chosen such as to include 3-4 indents. To limit the damage to the sample caused by the ion beam during thinning, ~100 nm C and 1 μm Pt protective overlayers were deposited on the sample surface, prior to FIB milling. Staircase shaped cuts were then milled out on either side of the region of interest, by decreasing gradually the tilt angle as well as the beam current, so that the sample eventually obtained is electron transparent (<80 nm thickness), while the other dimensions are 20 μm in length and 3.5 μm in height. The final specimen is then placed on a carbon membrane TEM grid. TEM observations were done using a JEOL 3010 microscope, using the centred bright field imaging mode to enhance the contrast between the layers.

RESULTS AND DISCUSSION

Growth and Microstructure of TiN/Cu bilayers and multilayers

Figure 1.a. shows θ–2θ scans recorded on the bilayer samples grown on (001)MgO. For the MgO/TiN/Cu bilayer, only (002) Bragg reflections are observed for Cu and TiN. Although the lattice mismatch between fcc Cu and Na-Cl-type TiN materials is huge (15.9%), a cube-on-cube epitaxial growth of Cu on TiN is clearly evidenced, as shown in Figure 1.b., displaying ϕ-scan XRD measurements. The same in-plane orientational relationship with a 4-fold symmetry is found for both (111) Cu and (111) TiN Bragg reflections, as four sharp peaks are observed separated from each other by 90° in φ. The corresponding epitaxial relationship is thus [110] $(001)_{Cu}$ // [110] $(001)_{TiN}$ // [110] (001) $_{MgO}$. For the MgO/Cu/TiN bilayer sample, the growth of polycrystalline layers, with (111) and (002) preferred orientations for Cu and TiN, respectively, is observed (see fig.1.a). In this case, the film/substrate interface is incoherent.

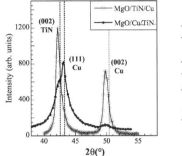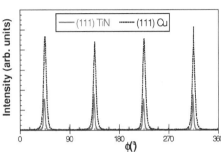

Figure 1 : a) θ–2θ scans of the TiN/Cu and Cu/TiN bilayers grown on (001) MgO. Vertical dashed lines correspond to bulk reflections. Sample surface has been tilted of 1° to avoid the substrate reflection located at 2θ=42.92°, b) φ-scans XRD spectra recorded on the MgO/TiN/Cu bilayer for the (111) TiN and (111) Cu Bragg reflections at ψ=54.74° tilt angle.

The growth and structural characterization of TiN/Cu multilayers has been previously reported as a function of Λ and h_{Cu} [20, 23]. We recall here only the main features for the multilayer with Λ=46.3 nm. A (001) fiber-texture growth is obtained, with a mosaic spread of ~6° and ~9° for TiN and Cu layers, respectively. This (001)-preferred orientation of the Cu layers is achieved due to TiN texture inheritance, as shown above for the MgO/TiN/Cu bilayer sample. The same preferred orientation was also observed by Ljungcrantz for TiN/Cu superlattices grown by magnetron sputtering on (001) MgO substrates [24]. TEM observations confirm the cube on cube epitaxial relationship but reveal the presence of several interfacial and growth defects in Cu layers, which have been introduced to relieve the misfit strain: misfit dislocations and micro-twins, small angle grain boundary dislocations, surface steps [20,23].
The state of stress, as determined from XRD measurements using the sin^2 ψ method, is found to be largely compressive in TiN sublayers (–5.8 GPa), while the Cu sublayers exhibit low tensile stress (+ 0.1 GPa). The magnitude of compressive stress in TiN is essentially accounted for by growth-induced point defects due to high energetic particles (sputtered Ti atoms or backscattered Ar neutrals) involved in the DIBS process [21]. The tensile stress found in Cu layers suggests the presence of an additional source of stress, namely coherency stress, due to observed epitaxy.

Nanoindentation results

The measured hardness H of the multilayers with Λ ranging from 5 to 46.3 nm has been reported elsewhere [20]. The values, extracted from the observed "plateau" of the H vs. h_c/T curves (T is the total film thickness), are representative of the film hardness only. In all samples, the H values fall in a range between the hardness of TiN (30 GPa) and Cu (4 GPa) monolithic films, which attests that no hardness enhancement occurs for this system. For the MgO/TiN/Cu bilayer, a hardness of ~ 5.7 GPa is found for low h_c values (close to pure Cu value), and then increases with h_c due to progressive influence of TiN and MgO. For MgO/Cu/TiN, a progressive decrease of the hardness is observed with increasing h_c, but in that case the determination of the true h_c value is altered by the occurrence of buckling in the TiN surface layer during loading (due to large compressive stress exhibited by this layer). This point will be discussed further below.

FIB-TEM observations of induced plastic zones under nanoindentation imprints

Fig. 2 shows a TEM image around the indent (8 mN load) of the Λ=46.3 nm [TiN/Cu]$_4$ multilayer deposited on Si. Here, the TiN and Cu layers appear as bright and dark, respectively, due to their electronic density contrast. Strong contrasts are visible in the Si substrate, attesting the presence of a stress field, although the corresponding indentation depth was approximately half the film thickness. In all regions examined by FIB-TEM, the multilayered film remains adherent to the Si substrate. This was confirmed from nanoindentation tests, for which no delamination of the film/substrate interface could be detected from loading/unloading curves recorded at high loads.

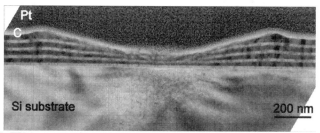

Figure 2: FIB-TEM cross-section image showing a general view of the indented region of the TiN/Cu multilayer with Λ=46.3 nm deposited on a Si substrate. The applied load was 8 mN.

A strong plastic flow is clearly visible inside the Cu layers, starting from the indenter tip apex toward its edges, as attested by the significant variations in Cu thickness: whereas the thickness measured far from the indents is 26.0 ±1.5 nm for the bottom Cu layers, it decreases down to ~15 nm close to the indent apex, which represents ~40% decrease. For the topmost Cu layer, the maximum thickness at the hillock reaches 42± 6 nm depending on the observed indent. On the contrary, no thickness variation is observed for the TiN layers within the measurement error bars. Rather, the TiN layers are forced to accommodate the extensive plastic deformation of the Cu layers by bending. This is more clearly seen in Fig. 3.a. taken below the hillock region. Whether this bending is elastic or plastic is difficult to answer, due to strong defects already present in the as-grown layers. Note that the vertical fringes observed in the TiN layers are related to a strong compressive stress-field (mentioned above) and were also observed before indentation. A higher resolution TEM micrograph taken from a region below the hillock is shown in fig. 3.b. In this region, the Cu layers remain crystalline, as seen from Moiré fringes between surrounding grains. A columnar alignment of Cu grains with (111) preferred orientation (dark contrast) is observed below the hillock, suggesting that possible grain rotations have taken place during nanoindentation, as also observed by Laughlin et al. in Cu single-crystals [25].

Therefore, it may be concluded that plastic deformation leads to pronounced accumulation of material inside the Cu layers that leads to the characteristic pile-up pattern observed at the surface, while the more rigid TiN layers accommodate deformation by bending.

Figure 3 : a) FIB- TEM cross-section image showing TiN bending and Cu plastic flow below the hillock region, b) High resolution TEM micrograph of TiN/Cu interface [insert is shown in a)].

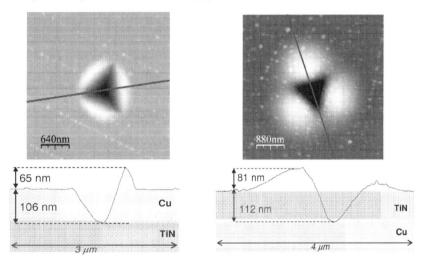

Figure 4 : AFM images of (a) MgO/TiN/Cu and (b) MgO/Cu/TiN bilayers for similar value of h_c (~ half the film thickness). Below are shown the corresponding sections along the blue line.

AFM observations of surface deformation around nanoindentation imprints

To correlate the information obtained from FIB-TEM observations, AFM images were taken around the indents for the two bilayer samples. The results for the MgO/TiN/Cu bilayer are shown in fig.4.a. The hillock profile is relatively abrupt and corresponds to plastic flow of Cu along the indenter edges. Since the volume of the hillock is the same as that of the indenter, it may be concluded that no plastic deformation occurred in TiN. For higher loads, plastic deformation is transmitted to MgO via the TiN layer, as single or multiple pop-in events are observed on the loading curve. Similar trends are observed for the Λ=46.3 nm TiN/Cu multilayer sample (not shown here), for which the surface layer is Cu. The other situation, i.e., when the surface layer is TiN, has also been examined. Results are reported in fig.4.b. In this case, the hillock profile extends much further laterally, due to confinement of Cu plastic flow by the more rigid TiN layer. The hillock volume (0.10 μm^3) is three times higher than that of the indenter (~0.03 μm^3), indicating that buckling of the TiN layer has occurred during loading. For higher loads, TiN buckling increases and once a local curvature threshold is reached, cracking takes place. These results suggest that deformation behavior in TiN/Cu multilayers is governed by plastic flow inside the Cu layers, which is confined by the more rigid TiN layers that tend to accommodate deformation by bending or buckling. This could explain the lack of hardness enhancement observed in this system.

ACKNOWLEDGMENTS

The authors want to thank A. Charaï and W. Saikaly (CP2M, Marseille) for the use of FIB facility.

REFERENCES

1. S. Veprek, M.G.J. Veprek-Heijman, P. Karvankova, J. Prochazka, *Thin Solid Films* **476**, 1 (2005)
2. J. Patscheider, *MRS Bulletin* **28**, n°3, 180 (2003)
3. W.D. Sproul, *Science* **273**, 889 (1996)
4. B.M. Clemens, H. Kung, S.A. Barnett, *MRS Bulletin* **24**, n°2, 20 (1999)
5. S.A. Barnett and A. Madan, *Scripta Mater.* **50**, 739 (2004)
6. A. Madan, Y Wang, S.A Barnett, C. Engstrom, H. Ljungcrantz, L. Hultman, M. Grimsditch, *J. Appl. Phys.* **84**, 776 (1998)
7. X. Chu, M.S. Wong, W.D. Sproul, S. L. Rohde, S.A. Barnett, *J. Vac. Sci. Technol. A* **10**, 1604 (1992)
8. M. Shinn, L. Hultman, S.A. Barnett, *J. Mater. Res.* **7**, 901 (1992)
9. J.C. Koehler, *Phys. Rev. B* **2**, 547 (1970)
10. X. Chu and S.A Barnett, *J. Appl. Phys.* **77**, 4403 (1995)
11. D.E. Kramer, T. Foecke, *Phil. Mag. A* **17/18**, 3375 (2002)
12. J.D. Embury, J.P. Hirth, *Acta. metall. mater.* **42**, 2051 (1994)
13. A. Misra, J.P. Hirth, H. Kung, *Phil. Mag. A* **82**, 2935 (2002)
14. L. Hultman, C. Engström, M. Oden, *Surf. Coat. Technol.* **133-134**, 227 (2000)
15. J.M. Molina-Aldareguia, S.J. Lloyd, M. Oden, T. Joelsson, L. Hultman, W.J. Clegg, *Phil. Mag. A* **82**, 1983 (2002)
16. A. Minor, E.A. Stach, J.W. Morris, I. Petrov, *J. Electron. Mater.* **32**, 1023 (2003)
17. D.E. Kramer, M.F. Savage, A. Lin, T. Foecke, *Scripta Mater.* **50**, 745 (2004)
18. L.W. Ma, J.M. Cairney, M. Hoffman, P.R. Munroe, *Surf. Coat. Technol.* **192**, 11 (2005)
19. H. Ljungcrantz, C. Engstrom, L. Hultman, M. Olsson, X. Chu, M.S. Wong, W.D. Sproul, *J. Vac. Sci. Technol. A* **16**, 3104 (1998)
20. Y.Y. Tse, G. Abadias, A. Michel, C. Tromas, M. Jaouen, *Mat. Res. Soc. Symp. Proc.* **778**, U6.8., (2003)
21. G. Abadias and Y.Y. Tse, *J. Appl. Phys.* **95**, 2414 (2004)
22. W.C. Oliver and G.M. Pharr, *J. Mater. Res.* **7**, 1564 (1992)
23. G. Abadias, Y.Y. Tse, A. Michel, C. Jaouen, M. Jaouen, *Thin Solid Films* **433**, 166 (2003)
24. H. Ljungcrantz, PhD Thesis, Linköping University, Sweden, (1995)
25. K.K. McLaughlin, N.A. Stelmashenko, S.J. Lloyd, J. Vandeperre, W.J. Clegg, *Mater. Res. Soc. Symp. Proc.* **841**, R1.3.1, (2005)

Mater. Res. Soc. Symp. Proc. Vol. 875 © 2005 Materials Research Society

Thermal Plasma Chemical Vapor Deposition of Superhard Nanostructured Si-C-N Coatings

Nicole J. Wagner[1], Megan J. Cordill[2], Lenka Zajickova[1], William W. Gerberich[2] and Joachim V. R. Heberlein[1]
[1] Mechanical Engineering, University of Minnesota
Minneapolis, MN 55455, USA
[2] Chemical Engineering and Materials Science, University of Minnesota
Minneapolis, MN 55455, USA

ABSTRACT

A triple torch plasma reactor was used to synthesize Si–C–N composite films via the thermal plasma chemical vapor deposition process. The argon-nitrogen plasma provided atomic nitrogen to carbon- and silicon-based reactants, which were injected through a central injection probe and ring configuration. Films were deposited with variations of the total nitrogen flow through the torches (1.5-4.5slm), reactant mixture (silicon tetrachloride and acetylene or hexamethyldisilazane) and substrate material (silicon and molybdenum). Micro X-ray diffraction was used to determine that both α-Si$_3$N$_4$ and β-Si$_3$N$_4$ were dominant in most of the depositions. Composites of silicon nitride and silicon carbide were synthesized on molybdenum. The bonding of amorphous phases was investigated using Fourier transform infrared spectroscopy, which indicated the presence of N–H, CH$_x$ and C≡N in various films. Indentation tests on the polished film cross-sections determined that large variations in hardness and elastic modulus existed for minor changes in film composition. Correlations between indentation results and scanning electron and optical microscope images showed that the mechanical properties greatly depend on the film morphology; the denser, smoother, and more crystalline films tended to display enhanced mechanical properties.

INTRODUCTION

Materials that protect against erosion, wear, and other harmful degradation are of great interest for various industries to coat, for example, automotive engine parts and machining tools. Nanostructured composite coatings offer enhanced mechanical properties that would reduce and possibly eliminate the necessity of costly and hazardous coolants for such applications.

Nanostructured composite films consisting of silicon nitride and silicon carbide have attracted great interest for their enhanced mechanical properties [1]. It is believed that the composite displays enhanced mechanical properties over its individual components [2]. The ideal design of this composite material is composed of nano-crystals dispersed in an amorphous nitride matrix [3].

Various methods are utilized for the preparation of the silicon nitride-silicon carbide nanocomposite, including polymer pyrolysis [4], powder mixing [5], and high-pressure nitridation and sintering [6]. In this study, a thermal plasma chemical vapor deposition (TPCVD) process was used to grow the Si-C-N composite on a silicon or molybdenum substrate. This process allows the possibility of using versatile multiphase reactants and producing a controllable film structure at high growth rates.

EXPERIMENTAL

Film Deposition

Nanostructured Si-C-N films were deposited in a triple torch thermal plasma reactor (figure 1) through a chemical vapor deposition process. This system consisted of three dc plasma torches that are angled such that the emerging jets converged to create a chemically reactive region at the substrate with a well-entrained reactant mixture. The argon-nitrogen plasma provided dissociated nitrogen, while the silicon tetrachloride ($SiCl_4$) and hexamethyldisilazane ($C_6H_{19}NSi_2$) were vaporized through a bubbler system and injected through the central injection probe. The carbon-based reactant acetylene (C_2H_2) was injected simultaneously through the injection ring for those experiments using $SiCl_4$. Due to the corrosive nature of $SiCl_4$, the reactant was substituted with $C_6H_{19}NSi_2$ to compare the choice of reactant(s) to film composition and quality. In addition, the $C_6H_{19}NSi_2$ is believed to assist the transport of carbon to the substrate surface where it would be incorporated into the growing film.

The silicon substrates had been manufactured with a mirror polish, while the molybdenum substrates were roughened prior to deposition with 120grit SiC paper to improve film adhesion. The chamber pressure was maintained at 130Torr throughout the experiment with a liquid-ring pump. A Raytek Marathon series pyrometer was used to record the substrate temperature. The deposition rate was approximately 60µm/hr.

Structural Characterization

The chemical composition of the films was analyzed by Fourier transform infrared spectroscopy (FTIR) and micro X-ray diffraction (XRD), while the film microstructure was observed through optical and scanning electron microscopy (SEM). Due to the roughness of the deposited films, diffuse reflectance IR spectra were taken on a Nicollet Magna-IR 750 measured from 650 to 4000nm^{-1}. The X-ray diffraction patterns were taken with a Bruker-AXS Microdiffractometer with a 2.2 kW sealed copper X-ray source. The spot size was collimated to 0.8mm to obtain data from a representative region, while studying the radial variations across the film. To enhance the resolution of the peaks corresponding to silicon nitride and silicon carbide, the Siemens area-detector was placed at a distance of 29.8cm from the sample. With this setting, four frames of 15 degrees were taken from 2θ = 20-80 degrees to construct the diffraction pattern. Scanning electron micrographs were taken to observe the morphology of the deposition surface. High-resolution SEM images were examined to verify that the films were nanostructured. Finally, optical microscope images were taken of the film cross-sections to determine film thickness and obtain a qualitative understanding of film roughness.

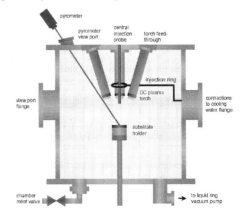

Figure 1: Triple torch plasma reactor.

Mechanical Properties

The hardness and elastic modulus were evaluated through depth-sensing indentation techniques using a Hysitron, Inc. Triboindenter. The films were indented with a Berkovich pyramid diamond indenter with an applied load of 10mN. The load-displacement curves were then analyzed according to the Oliver and Pharr method [7]. To maintain consistent conditions for the mechanical properties investigations, cross-sections of the samples were prepared for indentation. For this, the samples were quartered, mounted in epoxy resin, ground and polished with SiC paper, diamond suspension and a final 20nm colloidal silica suspension, and cleaned with ethanol.

RESULTS AND DISCUSSION

Films were deposited on silicon or molybdenum with $SiCl_4$ and C_2H_2 or $C_6H_{19}NSi_2$ with variations in the total nitrogen flow rate through the torches. The films were then analyzed for their chemical composition, morphology and mechanical properties. Correlations were drawn between the deposition conditions and these characterization results.

Chemical Composition

The infrared analysis encompasses many absorption peaks throughout the entire spectrum from 650-4000cm^{-1}. Figure 2 displays the spectra of films deposited on silicon substrates with $C_6H_{19}NSi_2$ at various total flow of nitrogen through the torches. The absorption regions around 1200, 2200, 2900 and 3350cm^{-1} correspond to the presence of Si-NH-Si [8], C≡N [9], CH$_x$ [9] and N-H [9], respectively. The relative intensities of the peaks around 1200, 2200 and 3350cm^{-1} tend to increase, while those of the peaks around 2900cm^{-1} decreases with increasing nitrogen flow through the torches. A summary of these peak assignments is given in table I.

Structural Analysis

X-ray diffraction patterns taken of the depositions (figure 3) confirm the presence of crystalline α- and β-Si_3N_4 [10, 11]. Very little to no amorphous background is present. The

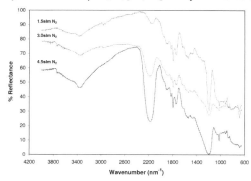

Figure 2: FTIR spectra of films on silicon deposited with $C_6H_{19}NSi_2$.

Table I: FTIR peak assignments.

Type of bond	Wavenumber [cm^{-1}]
N−H	3300-3500 [9]
C−H	2700-3300 [9]
C≡N	2220-2260 [9]
Si−NH−Si	1110-1250 [8]
C−N	1020-1230 [9]

background tends to be stronger for depositions on molybdenum. Silicon carbide diffraction peaks are detected only in films deposited with $C_6H_{19}NSi_2$ on molybdenum substrates. The polymorph of SiC is not specified, since most phases of SiC produce the diffraction peaks indicated in figure 3. It is believed that the C_2H_2 injected through the ring during experiments run with $SiCl_4$ was not properly entrained in the jet flow to reach the substrate. It is possible that an amorphous carbon-based compound is present, and FTIR spectra that indicate some carbon bonding in these depositions support this possibility.

Film Morphology

Scanning electron microscope images of depositions on silicon and molybdenum synthesized from $SiCl_4$ and C_2H_2 or $C_6H_{19}NSi_2$ are shown in figure 4. From figures 4a and c, it is seen that films deposited on silicon are more faceted than those on molybdenum (figures 4b and d). In addition, the films deposited with $SiCl_4$ and C_2H_2 have finer faceted structures (figures 4a and b), while the films grown from $C_6H_{19}NSi_2$ (figures 4c and d) tend to have larger features. The high resolution image (figure 4e) indicates that the larger features of the film deposited with $C_6H_{19}NSi_2$ on molybdenum is composed of nano-sized structures.

Optical cross-sectional images for films grown on silicon or molybdenum are shown in figure 5. These images represent the columnar growth that occurs through perturbations in the substrate surface. It is observed that the roughening of the molybdenum substrates created more columns, while depositions on silicon substrates had a relatively smooth initial growth layer with less columnar growth.

Mechanical Properties

The dependence of nitrogen flow on film hardness and elastic modulus is depicted in figure 6. The depositions on silicon substrates using $C_6H_{19}NSi_2$ tend to have an increase in hardness and elastic modulus with a decrease in nitrogen flow, while the highest hardness was achieved with $SiCl_4$ and C_2H_2 on silicon at a total of 4.5slm nitrogen through the torches. While the films deposited on silicon showed an increase in hardness and elastic modulus with decreasing nitrogen flow, those on molybdenum display an increased hardness and elastic modulus at both extremes of nitrogen flow.

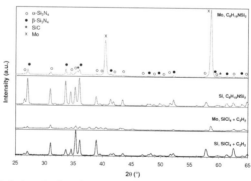

Figure 3: Indexed micro X-ray diffraction patterns of films on silicon or molybdenum deposited with $SiCl_4$ and C_2H_2 or $C_6H_{19}NSi_2$.

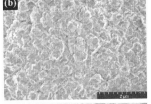

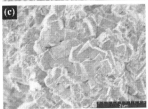

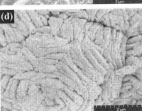

Characterization Summary

The FTIR spectra indicated a relative increase in the band corresponding to C≡N around 2200cm^{-1} with increased nitrogen flow, which corresponded to a decrease in the mechanical properties of the films. This result is consistent with studies [12] showing that the presence of the C≡N in films is detrimental to the hardness. Rather the C-N bond, which was not observed to vary with deposition conditions, enhances film mechanical properties. In addition, studies [13] have shown that an increase in the formation of CH_x has an unfavorable effect on hardness and elastic modulus. However, our indentation results have shown that films deposited with increased nitrogen flow displayed increased CH_x content and provided higher hardness values.

The X-ray diffraction patterns indicate that an increased amorphous background tends to be related to lower hardness values for either reactant(s). Scanning electron microscope images confirm this observation correlating the faceted growths with higher hardness. In addition, the presence of silicon carbide in the films deposited with $C_6H_{19}NSi_2$ on molybdenum did not provide an increase in mechanical properties. However, since silicon carbide was only detected in films deposited on molybdenum, other factors such as porosity and amorphous content are believed to play a large role in depressing the hardness and elastic modulus. Representative optical images showed that films on silicon undergo smoother growth, while those on molybdenum have less dense and columnar structures.

CONCLUSIONS

The chemical composition, morphology and mechanical properties of Si-C-N films deposited with varying reactants, nitrogen flow rates and substrate material were investigated. The highest hardness and elastic modulus values were provided from films synthesized from $SiCl_4$ and C_2H_2 on silicon. Films

Figure 4: SEM images of films deposited with $SiCl_4$ and C_2H_2 or $C_6H_{19}NSi_2$ on (a,c) silicon and (b,d) molybdenum, respectively, and (e) with $C_6H_{19}NSi_2$ on molybdenum at high resolution.

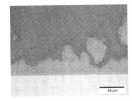

Figure 5: Optical microscope images of polished cross-sections of films deposited on silicon (left) and molybdenum (right).

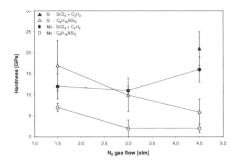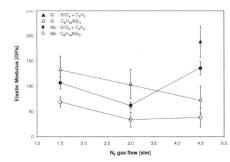

Figure 6: Hardness (left) and elastic modulus (right) as a function of total nitrogen flow through the torches.

grown from $C_6H_{19}NSi_2$ displayed increasing mechanical properties with decreasing nitrogen flow rates through the torches. Finally, although Si_3N_4-SiC composites are expected to possess enhanced mechanical properties compared to the individual components, this study has shown that the combination of Si_3N_4 and SiC does not alone provide an increase in film hardness. Rather, other factors including chemical composition and film morphology play a dominating role in determining film mechanical properties.

ACKNOWLEDGEMENTS

Funding was provided by the Department of Energy Grant No. DE-FG02-85ER13433 Ao15, the National Science Foundation Integrated Graduate Education and Research Traineeship Grant No. DGE-0114372 and the National Science Foundation North Atlantic Treaty Organization Postdoctoral Fellowship Grant No. DGE-0312210.

REFERENCES

[1] Lange, F. F., *J. Am. Ceram. Soc.* **56** 445-450 (1973).
[2] Badzian, A., *J. Am. Ceram. Soc.* **85** 16-20 (2002).
[3] Veprek, S., *J. Vac. Sci. Technol. A* **17** (5) 2401-2420 (1999).
[4] Riedel, R., Seker, M. and Becker, G., *J. Eur. Ceram. Soc.* **5** 113-122 (1989).
[5] Pezzotti, G. and Sakai, M., *J. Am. Ceram. Soc.* **77** (11) 3039-3041 (1994).
[6] Poorteman, M., Descamps, P., Cambier, F., Plisnier, M., Canonne, V. and Descamps, J. C., *J. Eur. Ceram. Soc.* **23** 2361-2366 (2003).
[7] Oliver, W. C. and Pharr, G. M., *J. Mater. Res.* **7** (6) 1564-1583 (1992).
[8] Benitez, F. and Esteve, J., *Soc. Vac. Coat.* **505** 280-285 (2002).
[9] Bruice, B. Y., "Organic Chemistry," Prentice Hall, NJ (1995) p.541.
[10] X-ray powder diffraction file #01-071-0570.
[11] X-ray powder diffraction file #01-082-0697.
[12] Walters, J. K., Kühn, M., Spaeth, C., Dooryhee, E. and Newport, R. J., *J. Appl. Phys.* **83** 3529-3534 (1998).
[13] Jedrzejowski, P., Cizek, J., Amassian, A, Klemberg-Sapieha, J. E., Vlcek, J. and Martinu, L., *Soc. Vac. Coat.* **505** 530-534 (2003).

Mater. Res. Soc. Symp. Proc. Vol. 875 © 2005 Materials Research Society   O3.11/BB2.11

Depth Profiling of Mechanical Properties on the Nanoscale of Single-Layer and Stepwise Graded DLC Films by Nanoindentation and AFM

C. Ziebert, S. Ulrich, M. Stüber
Forschungszentrum Karlsruhe, Institut für Materialforschung I, Hermann-von-Helmholtz-Platz 1, 76344 Eggenstein-Leopoldshafen, Germany

ABSTRACT

The strong ion bombardment, applied during sputter deposition of diamond-like carbon films (DLC), which is needed to promote the growth of the sp^3-bonded hard phase, inevitably is accompanied by compressive stress generation, thus limiting their maximum thickness. Gradient coatings with gradients in composition, constitution or properties are a well-known concept to manage such stress problems. The stepwise graded layer concept adjusts a graded constitution of the growing carbon film by a stepwise increase of the ion energy, i.e. the substrate bias voltage, during magnetron sputtering. To study the influence of the layer thickness on the expansion of the interface regions between the layers deposited with different bias voltage, samples with increasing deposition time of the top layer and thus thickness ratio were investigated by using the small angle cross-section nanoindentation method (SACS). It was revealed that the thickness of the interface regions is linearly dependent on the thickness ratio of the graded layers, which might be an evidence for stress-induced diffusion and relaxation processes in the carbon network. By using microindentation with a Berkovich indenter and ex-situ AFM-imaging it was found that all graded films exhibited higher Berkovich thresholds for crack development and thus better crack resistance than the hardest single-layer film and kept a high hardness value of about 4000 HV0.005.

INTRODUCTION

The properties of hydrogen-free diamond-like carbon (DLC) coatings can be adjusted in a wide range in dependence of the deposition method and the process parameters [1]. If a high hardness is desired, usually a strong ion bombardment has to be applied to promote the growth of the sp^3-bonded hard phase, which inevitably is accompanied by compressive stress generation. This usually limits the thickness of hard DLC single-layer films. To solve this problem, multilayer designs with alternating DLC and metallic layers [2] or alternating hard and soft layers [3], functional grading [4], subsequent deposition and annealing steps [5], and metal doping [6,7] have been used. In this work the stepwise bias-graded layer design reported by Stüber et al. [8] was applied to adjust a graded constitution of the growing film by a stepwise increase of the ion energy, i.e. the substrate bias voltage, during magnetron sputtering. This concept allowed to deposit adherent DLC films with thickness up to 10 μm and hardness values up to 5300 HV0.05 [8]. Further optimization requires a depth profiling of the mechanical properties particularly at the interface regions between the layers. Recently we have shown that the small angle cross-section nanoindentation method (SACS) [9] allows to record depth profiles of the hardness with nanometer resolution and to determine expansions of interface regions [10]. This study aimed at investigating the dependence of the expansions of interface regions on the layer thickness ratio. Another interesting aspect was to find out, whether the film constitution, adjusted by the stepwise bias-graded layer concept, could improve the crack resistance. Therefore the crack development in single and stepwise graded DLC films was investigated by microindentation and AFM-imaging to correlate it to ion bombardment intensity and to the expansions of the interface regions.

sample	U_B (V)	$D_1{:}D_2{:}D_3$	D_{tot} (μm)	HV0.005	σ_{tot} (GPa)
S1	0	-	4.8	1650	-0.6
S2	-100	-	4.7	2400	-1.5
S3	-300	-	1.6	4100	-
G1	0/-150/-300	1:1:0.5	3.3	3700	-1.5
G2	0/-150/-300	1:1:1	4.0	3800	-1.8
G3	0/-150/-300	1:1:1.5	4.6	4000	-2.0

Table I. Sample parameters and properties.

EXPERIMENTAL DETAILS

Single-layer and stepwise bias-graded DLC films were deposited onto (100) silicon wafer and polished commercial M15 WC-Co hard metal cutting inserts by non-reactive d.c. magnetron sputtering of a pure C target in argon (power density 11.32 W/cm^2, argon pressure 0.6 Pa, target-substrate distance 5 cm, substrate bias voltage U_B = 0 to -300 V) in a Leybold Z550 sputter facility. In case of the graded films the deposition started with a substrate bias voltage of 0 V to initiate a high adhesion of the growing film. Then the substrate bias voltage was increased to -150 V and finally to -300 V to produce a hard film surface.

Both the layer thickness D_i (i = 1,2,3) and the total thickness D_{tot} were measured by a calotester. To determine the residual stress values σ_i of every layer the deposition process was stopped after every layer and the residual stress was measured by the wafer bending method before the deposition of the next layer with increased bias voltage started. Table I summarizes parameters and properties of the three single-layer (S1-S3) and the three graded (G1-G3) DLC films used in this study.

To record the hardness depth profiles and to determine expansions of interface regions the small angle cross-section nanoindentation method (SACS) [9], was applied on a CSIRO UMIS2000 system with a Berkovich tip at 2 mN maximum load. At first this method drastically enlarges the area to be investigated by nanoindentation on the differently graded layers by preparing a cross-section under a very small angle α of about 0.05 to 0.15° using the nanogrinding method [10]. At second the distance passed by the nanoindenter, while it performs a nanoindentation linescan over the different layers, can be transformed into the depth information using a simple geometric formula [11]. Due to the very small angle of the cross section, a sufficient number of indentations can be made in each single layer and even in the interface regions of the DLC films, which considerably increases resolution. At a distance s_i of the interface between two graded layers on a small angle cross section, the effective layer thickness d_i, which is the maximum available depth for the indentation in the layer under investigation without reaching the underlying layer, is given by

$$d_j = s_j \sin \alpha \qquad (1)$$

Due to the small value of α the effective layer thickness decreases strongly close to the interface and the indenter finally penetrates into the softer layer which is situated under the layer under investigation even at low loads of 2 mN. Consecutively, the calculated hardness value becomes too small, which leads to a broadening of the measured hardness profile. This broadening can be theoretically approximated by a weighting of the effective area fractions of the indenter in both layers and their averaged hardness values and it can be separated from the measured hardness profile as described in [11].

To investigate the cracking behavior, the surface topography of the samples after indentation at higher loads of 100 mN to 1 N was recorded using a Dimension 3100 AFM (Digital Instruments) with a Si tip (Nanosensors LFM; tip radius < 10 nm) in contact mode.

RESULTS AND DISCUSSION

Influence of layer thickness ratio on expansions of interface regions

Figure 1 shows the measured depth profile of hardness (◊) at the two interface regions between the layers of a graded DLC film (G2), together with theoretical broadening (bold lines). Starting from the right side the stepwise decrease of the hardness with increasing depth h for the three graded layers deposited with bias voltages of -300 V (3. layer, H = 29.8 GPa), -150 V (2. layer, H = 25.4 GPa) and 0 V (1. layer, H = 16.5 GPa) can be clearly recognized. The calculated profile indicated by the bold lines, which is broadened due to the exceeding of the effective layer thickness close to the interface region, as described above in the experimental details section, gives values for the theoretical broadening. This values have to be subtracted from the measured expansions of the interface regions, indicated by the two shaded regions, d_{12} = 60 nm (interface region between 1. and 2. layer) and d_{23} = 60 nm (interface region between 2. and 3. layer), giving corrected expansions of the interface regions of d_{12} = 28 nm and d_{23} = 39 nm for sample G2. In figure 2, the corresponding corrected measured interface expansions (filled symbols) for all three graded DLC films are displayed with increasing layer thickness ratio D_3/D_2, which indicate a linear increase of d_{23} with the layer thickness ratio and no significant dependence of d_{12} on this parameter.

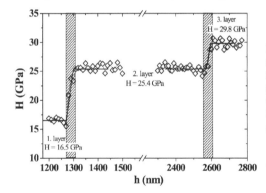

Figure 1. Measured depth profile of hardness (◊) at the interface regions between the layers of a graded DLC film (G2), together with theoretical broadening (bold lines) and measured interface expansions (shaded regions).

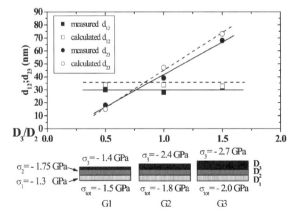

Figure 2. Comparison of measured (filled symbols) and calculated (open symbols) expansions of interface regions of three graded DLC-films with increasing ratio D_3/D_2 together with least squares fits (lines).

To explain these results the following simple model was used. If one assumes that the stress values of the different layers can be summed to the total stress σ_{tot} according to their thickness ratio D_i/D_{tot}:

$$\sigma_{tot} = \sum_{i=1}^{3} \sigma_i \frac{D_i}{D_{tot}}, \qquad (2)$$

and that the expansions of the interface regions d_{ij} ($i = 1,2$; $j = 2,3$) are proportional to the stress difference between the layers and the layer thickness values D_i and D_j:

$$d_{ij} = \alpha \frac{|\sigma_j - \sigma_i| D_i D_j}{D_{tot}}, \qquad (3)$$

the theoretical expansions of the interface regions can be calculated by inserting the measured stress and thickness values, as shown in figure 2 (open symbols). The dependence on the layer thickness values comes from the idea that the thicker the layer at the same residual stress level, the more the system gets bended, which leads to an increased interface expansion. The fact that a relation based just on the residual stress does not give similar agreement to the data, gives a first evidence for that assumption. In this way a good qualitative as well as quantitative agreement between theoretical and calculated interface expansions was found for a proportional constant $\alpha = 0.1$ for DLC. This suggests that one of the processes leading to the formation of the interface regions in stepwise graded amorphous DLC films might be a stress-induced diffusion of carbon atoms across the interface during the deposition. This would lead to stepwise graded distances between the neighbored carbon atoms in the network and thus to the stress managing effect. However it has to be checked in future by theoretical studies whether the assumptions can be applied to amorphous multilayer systems and whether a stress-induced carbon self-diffusion process would be fast enough [12] according to the related temperature and time scale in the sputter deposition of graded layers.

Crack investigation by AFM and microindentation

After the stress managing effect studied above, which allowed to overcome the thickness limitation, it was investigated, whether the constitution, adjusted by the stepwise bias-graded layer concept, could improve the crack resistance of DLC films. Figure 3 compares the evolution of load vs. indentation depth curves with increasing load of a 1.6 μm thick hard single-

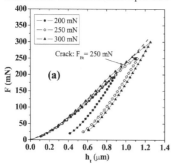
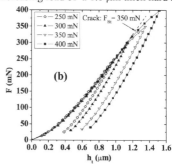

Figure 3. Crack development investigated by load vs. indentation depth curves: (a) in a single-layer DLC-film (S3) (b) in a graded DLC-film (G2).

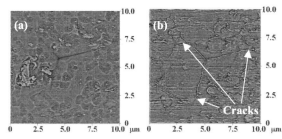

Figure 4. High-pass filtered AFM topography images of a single-layer DLC-film (S3) after indentation at 200 mN (a) and 250 mN load (b).

layer DLC film (a) and a graded DLC film with similar thickness of the hard top layer (b). A marked decrease in the slope appears for the single layer film (S3) at an indentation load of 250 mN, which gives evidence for the development of cracks as discussed in [13]. To check that cracks have been really produced, the topography of the samples was recorded by using AFM in contact mode after the indentations with increasing load have been performed. As an example figure 4 shows AFM topography images of the single-layer DLC-film S3 after indentation at 200 mN (a) and 250 mN load (b). The images were high-pass filtered to enhance edges and improve the visibility of the crack features. While no cracks can be seen at an indentation load of 200 mN, at 250 mN radial cracks, starting at the corners of the Berkovich indenter, have been formed in deed, as indicated by the arrows in figure 4(b). Similar radial cracks were found by Li et al. in amorphous carbon films, which were deposited by arc evaporation onto Si [14]. As explained in their study, the advances of the indenter during the radial cracking are not big enough to form steps in the loading curve, but generate small discontinuities, which lead to the observed slope change in the loading curve with increasing indentation loads. The corner cracks are not straight lines as in brittle materials, but stepped lines, suggesting that their generation happens slowly and gradually and not suddenly and rapidly [15]. However, to get more information about the geometry of the cracks below the surface, it is thought of using a focused ion beam in future to prepare cross sections, which can be investigated by electron microscopy or AFM.

By applying the same type of measurements for all samples, the comparison of the Berkovich threshold for crack development F_{Bt} in single-layer DLC-films, deposited with different bias voltage (S1-S3), and graded three-layer DLC-films, with different layer thickness ratio (G1-G3), shown in Figure 5, was created.

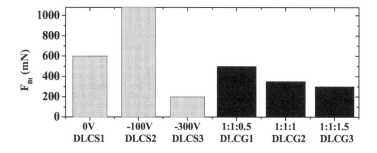

Figure 5. Comparison of Berkovich threshold for crack development F_{Bt} in single-layer DLC-films deposited with different bias voltage (S1-S3) and graded three-layer DLC-films with different layer thickness ratio (G1-G3).

This allows to correlate F_{Bt} to the intensity of the ion bombardment during sputter deposition of the single-layer films and to the expansion of the interface regions between the graded layers. The single-layer film S2, which was deposited at a bias voltage of -100 V exhibits the highest value of F_{Bt} of more than 1000 mN (maximum load of UMIS system). Increasing the bias voltage to -300 V leads to an hardness increase to 4100 HV0.005 whereas the Berkovich threshold goes down to only 200 mN. This suggests that for crack resistance there exists an optimum configuration of the carbon atoms in the single-layer network generated by the densification by subplantation [16] of the impinging particles, the repulsive part of the interatomic potential [17] and the elastic behavior of the network [18]. The comparison with the graded films reveals that at first all graded films exhibit higher Berkovich thresholds and thus better crack resistance than the hardest single-layer film S3. At second all graded films keep the high hardness of 3700-4000 HV0.005. Looking back to figure 2 it can be deduced that, the smaller the expansion of the interface between the 2. and the 3. layer d_{32}, the higher F_{Bt}, which reaches its maximum value among the graded films of 500 mN for sample G1.

CONCLUSIONS

The investigations of stepwise bias-graded DLC films by the small angle cross section nanoindentation method have shown, that the expansions of their interface regions are linearly dependent on the thickness ratio of the graded layers and the difference of the residual stresses. This first evidence of stress-induced diffusion processes in the carbon network will have to be checked in future by further theoretical treatments using e.g. the subplantation model and molecular dynamics simulations.

Concerning the possible advantages of such graded films for applications it was found, that the smaller the expansion of the interface between the 2. and the 3. layer d_{32}, the higher the Berkovich threshold for crack development. In addition it was revealed that all graded films exhibit higher Berkovich thresholds F_{Bt} and thus better crack resistance than the hardest single-layer DLC film and keep a high hardness value of 3700 to 4000 HV0.005. The next steps will be to study different indenter geometries such as cube corner and sphere to find out more about crack propagation, fracture toughness and stress development and to optimize the interfaces between the graded layers to further enhance the temperature stability, the maximum achievable thickness and the crack resistance of DLC films.

REFERENCES

[1] J. Robertson, *Mater. Sci. Eng.* R **37**, 129 (2002).
[2] H. Ziegele, H.J. Scheibe, B. Schultrich, *Surf. Coat. Technol.* **97**, 385 (1997).
[3] M. Gioti, S. Logothetidis, C. Charitidis, *Appl. Phys. Lett.* **73**, 184 (1998).
[4] H. Holleck, and M. Stüber, "Method of manufacturing a composite material structure", US Patent No. 6, 110, 329 (2000); EU Patent No. EP 0912774B1 (2002).
[5] T.A. Friedmann, J.P. Sullivan, J.A. Knapp, D.R. Tallant, D.M. Follstaedt, D.L. Medlin, P.B. Mirkarimi, *Appl. Phys. Lett.* 71, 3820 (1997).
[6] H. Dimigen, H. Hübsch, R. Memming, *Appl. Phys. Lett.* **50**, 1056 (1987).
[7] C. Bauer, H. Leiste, M. Stüber, S. Ulrich, and H. Holleck, *Diamond Rel. Mater.* **11**, 1139 (2002).
[8] M. Stüber, S. Ulrich, H. Leiste, A. Kratzsch, and H. Holleck, *Surf. Coat. Technol.* 116-119, 591 (1999).
[9] S. Ulrich, C. Ziebert, M. Stüber, E. Nold, H. Holleck, M. Göken, E. Schweitzer, and P. Schloßmacher, *Surf. Coat. Technol.* **188-189**, 331 (2004).

[10] H.H. Gatzen, J.C. Maetzig, *Prec. Eng.* **21**, 134 (1997).

[11] C. Ziebert, C. Bauer, M. Stüber, S. Ulrich, and H. Holleck, *Thin Solid Films* **482**, 63-68 (2005).

[12] J.D. Hong, R.F. Davis, *J. Am. Ceram. Soc.* **63**, 546 (1980).

[13] S.V. Hainsworth, M.R. McGurk, T.F. Page, *Surf. Coat Technol.* **102**, 97(1998).

[14] X. Li, D. Diao, B. Bhushan, *Acta mater.* **45**, 4453 (1997).

[15] A. Karimi, Y. Wang, T. Cselle, M. Morstein, *Thin Solid Films* **420-421**, 275 (2002).

[16] D. Marton, K. Boyd, J. Rabalais, Y. Lifshitz, *J. Vac. Sci. Technol.* A **16**, 455 (1998).

[17] D. Brenner, *Phys. Rev.* B. **42**, 9458 (1990); **46**, 1948 (1992).

[18] P.C. Kelires, *Phys. Rev. Lett.* **73**, 2460 (1994).

Mechanical Properties of
Thin Films—Testing and Analysis

Mater. Res. Soc. Symp. Proc. Vol. 875 © 2005 Materials Research Society

Experiments on the Elastic Size Dependence of LPCVD Silicon Nitride

Yuxing Ren and David C. C. Lam
Department of Mechanical Engineering, The Hong Kong University of Science and Technology,
Clear Water Bay, Kowloon, Hong Kong,
People's Republic of China

ABSTRACT

Recent experimental observations showed significant elastic size effects in small scales. While surface stress theories are used to describe nanometer scale size effects, strain gradient theories describe size effect observed in micron sized epoxy. Size effects in single crystalline silicon and epoxy make it unclear whether there is size dependence of the elastic behaviors of widely used LPCVD silicon nitride thin films in submicron scale. In this paper, submicron thick LPCVD silicon nitride beams were fabricated and bending tests were conducted on the beams. Results showed fluctuating normalized bending rigidities in the beams with different thickness. XPS and XRD analyses were used to analyze the material consistency of the beams. The fluctuations maybe related to varying crystalline phase fractions in the thin films. The beams were annealed and bending tests were conducted to investigate possible correlation between the fluctuations and crystalline phase fractions. Results showed similar level of fluctuations in normalized bending rigidities before and after annealing while XRD results of the annealed films showed increase in crystalline phase fractions for all thicknesses. While LPCVD silicon nitride may have size dependence in the nanometer scale, size dependence of normalized bending rigidity of LPCVD silicon nitride appears to be insignificant in submicron scale.

INTRODUCTION

In conventional plane strain cantilever elastic bending, the bending rigidity, which contains contributions from geometry and material properties, can be normalized and the normalized bending rigidity depends only on material properties. Recent experiments and simulations showed size effects on elastic properties in a variety of materials in small scales. These effects are delineated from analyses of elastic properties without contributions from geometry. There are theories to describe size effects in small scales and surface stress theories are representative in nanometer scale. Streitz et al [1]and Wolf [2]showed size effect in atomic scale by computational simulations. In experimental observations, Li and Ono et al found that the elastic modulus of silicon cantilevers decreased from 170GPa to 53GPa as the thickness decreased from 300nm to 12nm [3]. The findings are described using surface stress theories.

In the researches on carbon nanotubes, analytical models were built and computational simulations were conducted to describe the elastic properties of carbon nanotubes and size dependence of Young's modulus were predicted [4-7]. The effects induced by surface stress and observed in nanotubes vanish when the structural dimensions are in micron or larger scales. However, experimental observations in elastic bending tests of epoxy micro-cantilever beams revealed that the normalized bending rigidities of the beams are size dependent with a 2.4 times increment as the thickness is reduced from 115μm to 20μm [8]. The size effect is described using strain gradient theories. Although size effects observed in single crystalline silicon and carbon nanotubes are in nanometer scale, the size effect of micron sized epoxy makes it unclear that whether conventional elasticity is sufficient to describe the elastic deformation behaviors of other amorphous thin films like submicron thick LPCVD silicon nitride. In MEMS applications, silicon nitride thin films are used as elastic deformable structures like AFM sensing probes[9], their elastic behaviors are essential for design of the devices. LPCVD silicon nitride is a polymorphous crystalline compound containing α and β crystal modifications. In as-deposited silicon nitride thin films, amorphous contents are major structures and crystalline contents can be modified with annealing [10]. Elastic properties of silicon nitride films are affected by crystalline contents in the films and are dependent on the ratio of α/β contents [11]. Crystallinity of the films is essential to be investigated prior to the elastic analysis. In this paper, to investigate whether size effect exists in submicron thick silicon nitride, high resolution bending tests, and precise geometry measurements are introduced to delineate normalized bending rigidities of silicon nitride beams. Elemental compositions and crystallinity of the beams are discussed to analyze the effect of material consistency on the elastic behaviors.

EXPERIMENTAL

Fabrication and dimension measurement

Low stress silicon nitride thin films were fabricated using standard MEMS device fabrication processes [12]. Residual stress usually exists in the deposited thin films and makes the released structures curved, which introduces difficulties in testing and analyses. By adjusting the deposition conditions, residual stress in LPCVD silicon nitride changes between tensile and compressive state [13-15]. Since a tensile-compressive crossover exists, silicon nitride thin films with low residual stress can be fabricated. Silicon nitride thin films were deposited using identical gas ratio, DCS: NH_3: N_2 = 1000: 16: 100 (sccm), at 840°C and 200mTorr of pressure. The deposited silicon nitride thin films were patterned by photolithography and dry etched. The TMAH solution etched the bulk silicon and released the silicon nitride cantilevers. The final configurations of the beams are shown in Figure 1a. The beams remained straight after released from the substrates. Undercuts and fillets induced by the processes were observed in the support areas of the beams (Figure 1b). Accurate determination of the beam's geometry is critical for the accurate determination of the beam's bending behaviors. Thickness of silicon nitride thin films were measured with ellipsometry [13, 16]. The values and errors of the thicknesses are listed in Table 1. The widths, lengths and undercuts of the selected series of the beams were determined from SEM micrographs. The values and measuring errors are listed in Table 1. The radii of fillets in the anchor areas of the beams were around 2 to 3μm.

Elemental and micro-structural analyses of the as-fabricated silicon nitride thin films

Materials with different elemental compositions or crystalline phase content have different elastic behaviors. Elemental and crystallographic characterizations were conducted to determine the consistencies of the beams' elemental compositions and crystallinity. As-fabricated silicon nitride thin film samples with different thicknesses were examined using X-ray photoelectron spectroscopy (XPS) elemental analysis. Results showed that there is little

(a)   (b)

Figure 1. (a) Silicon nitride cantilever beams with low-level residual stress (b) Undercuts and fillets in the support areas of the beams

Table 1. Dimensions and loading distances of as-fabricated beams

Thickness (nm)	302.7±4.1	458.6±6.0	678.9±13.7
Width (μm)	45.02±0.16	44.94±0.13	45.49±0.19
Length (μm)	100	100	100
Undercut (μm)	3.60	3.34±0.08	3.63±0.19
Loading distance (μm)	14±0.5	20±0.5	29.2±0.5

difference in elemental contents of thin films with different thicknesses.   Crystalline phase fractions can be determined from the intensities of silicon nitride characteristic peaks using X-ray diffraction analysis (XRD).   The intensities of the peaks were non-uniform with higher intensities in thicker films (Figure 2a).   This implied that the crystalline phase fractions in thick films were larger.   The differences in crystalline phase fractions may potentially affect the elastic properties of the as-fabricated silicon nitride thin films.

Bending tests and data analyses

Bending tests on the as-fabricated beams were performed in a Triboindenter with displacement resolution of 0.1nm and load resolution of 0.1μN.   To minimize the elastic support effect on the bending rigidities, the loading distances from the edges of the supports were set at distances larger than 40 times the thicknesses of the beams and listed in Table 1.   In order to enhance the position accuracy, a Berkovich tip was used.   By using Berkovich tip, there should be small indentations on the surface of the cantilevers.   The indentations may have affected the bending rigidities.   To investigate the effect, an indentation test was conducted on a silicon nitride thin film on the silicon substrate.   For a load smaller than 20μN, which is close to the maximum load used in the bending tests, the displacement was around 1nm indicating the compliance of bending was hundreds of times that of indentation.   The effect of indentation in the bending rigidities is negligible.

To investigate size dependence of elastic behaviors, normalized bending rigidities are delineated from bending data.   In this study, the contributions from geometries cannot be normalized with simple analytical models because of the non-ideal undercuts and the fillets in the support areas and the point load mechanism.   Finite element method (FEM) analyses were used to analyze the mechanical behaviors of structures [17-19]using the commercial software ANSYS.   FEM models were built with the measured geometries and the applied point load mechanism.   Normalized bending rigidities of the beams can be obtained by taking the ratios of the bending rigidities from experiments and FEM models.   The normalized bending rigidities of the as-fabricated beams are plotted in Figure 3.

For each thickness, the standard deviations of normalized bending rigidities of the three tested beams are under 20% with less in thicker beams. There is a 25% difference between the maximum and the minimum median values of normalized bending rigidities for all thicknesses. While fluctuations in normalized bending rigidities for beams with different thicknesses are typical, possibility remains that the fluctuations are not statistical, but is correlated with the varying crystalline phase fractions in the thin films.

ANNEALING OF THE BEAMS AND EXPERIMENTS ON THE ANNEALED BEAMS

Experimental results indicated that the normalized bending rigidities, within data scatter, are essentially unchanged from thick to thin beam. If indeed, the crystalline phase fraction has a significant effect on the normalized bending rigidities, the normalized bending rigidities of thick beams should be higher than that of the thin beams. The fact that the normalized bending rigidities remained approximately the same may because the effect of the crystalline phase content is small such that rigidities are not affected. If crystalline phase content were non-negligible, then thin beams can exhibit the same level of normalized bending rigidities only if there are additional mechanisms increasing the bending rigidities of the thin beams, i.e., stiffening size effect. To investigate if there is a significant correlation between crystalline phase content and normalized bending rigidities, the crystalline phase fraction in the beams can be increased by annealing. Crystallization of CVD silicon nitride thin films occurs at above $1100^{\circ}C$ [10], and annealing the as-fabricated beams at $1100^{\circ}C$ can increase the crystalline phase fractions. In this study, three samples of different thickness were annealed in a tube furnace at $1100^{\circ}C$ for two hours with ramp rates of $5^{\circ}C/min$. Nitrogen forming gas with 5% hydrogen was used in the sealed tube furnace to limit oxidation. After annealing, the dimensions of the annealed silicon nitride thin films were measured and are listed in Table 2.

Table 2. Dimensions and loading distances of the annealed beams

Thickness (nm)	314.0±5.4	473.8±8.4	653.7±13.3
Width (μm)	45.02±0.16	44.94±0.13	45.49±0.19
Length (μm)	100	100	100
Undercut (μm)	3.60	3.34±0.08	3.63±0.19
Loading distance (μm)	14±0.5	19.6±0.5	28.4±0.5

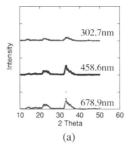 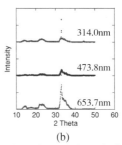

(a)   (b)

Figure 2. XRD spectra of (a) as-fabricated and (b) annealed silicon nitride thin films

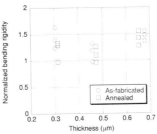

Figure 3. Normalized bending rigidities of the as-fabricated and annealed beams

XPS results indicated that the elemental contents were similar before and after annealing. The XRD results indicated that the intensities of the silicon nitride peaks of all annealed beams have increased with the thickest film having the highest intensity peak (Figure 2b). Bending tests on the annealed silicon nitride beams of all thicknesses were performed with the same setting and conditions as those performed on the as-fabricated beams. The loading distances were slightly adjusted due to the thickness changes of the beams. FEM analyses were used to analyze the data. The normalized bending rigidities are plotted in Figure 3 together with the as-fabricated set. Results showed that data scatter in the normalized bending rigidities of the annealed set did not increase, but remained similar to that of the as-fabricated set despite the increment of the crystal phase fraction. The standard deviations of normalized bending rigidities in all thicknesses are less than 20% with less in thicker beams. There is a 29% difference between the maximum and the minimum median values of normalized bending rigidities for all thicknesses. Since there is no significant contribution from difference in crystalline phase content, it can be concluded that there is no size stiffening, i.e., no size effect in silicon nitride beams in the submicron thickness range.

CONCLUSIONS

High resolution load-displacement testing and measurement were conducted to enable the precise analyses of the elastic behaviors of the silicon nitride beams. Geometric measurement errors of the beams are less than 2% in thickness and 1% in the width. The accuracy of the loading positions was limited by the resolution of the Triboindenter, which induced the errors in the bending rigidities for different beams. Analyses of the data using FEM revealed that the normalized bending rigidities of the as-fabricated beams with different thicknesses can be grouped within a bracketed range of values. While elemental analyses indicated that the beams had essentially identical elemental compositions, X-ray diffraction analyses revealed that the beams have a weak trend of increasing crystal content with thickness of the amorphous silicon nitride beams. However, the analyses on annealed beams revealed that correlation between crystalline phase content and normalized bending rigidities can be concluded to be insignificant. The changes of crystalline contents in the annealed beams are insignificant and the beams remained amorphous. Unlike epoxy beams, size dependence can be concluded to be insignificant for amorphous silicon nitride at submicron scale. The scales in which size effects exist are not consistent for different groups of materials and the mechanism of the size effects might be different. Surface stress and strain gradient effects may control the size dependence jointly in small scales. More or less significance of different mechanisms on size effect might depend on the nature of materials.

ACKNOWLEDGMENTS

This research work was supported by the Research Grant Council of the Hong Kong government (Project No.:HKUST6190/03E; HKUST6080/00E).

REFERENCES

1. F. H. Streitz, R. C. Cammarata, and K. Sieradzki, *Phys. Rev. B-Condensed Matter* **49**, 10699-10706, (1994).
2. D. Wolf, *Appl. Phys. Lett.* **58**, 2081-2083, (1991).
3. X. Li, T. Ono, Y. Wang, and M. Esashi, *Technical Digest. MEMS 2002 IEEE International Conference. Fifteenth*, 427-430, (2002).
4. V. N. Popov, V. E. Van Doren, and M. Balkanski, *Sol. St. Comm.* **114**, 395-399, (2000).
5. C. Li and T.-W. Chou, *Int. J. Solids & Structures* **40**, 2487-2499, (2003).
6. C. Goze, L. Vaccarini, L. Henrard, P. Bernier, E. Hernandez, and A. Rubio, *Elsevier. Synth. Met.* **103**, 2500-2501, (1999).
7. T. Chang and H. Gao, *Journal of the Mechanics and Physics of Solids* **51**, 1059-1074, (2003).
8. D. C. C. Lam, F. Yang, A. C. M. Chong, J. Wang, and P. Tong, *Journal of the Mechanics and Physics of Solids* **51**, 1477-1508, (2003).
9. R. J. Grow and S. C. Minne, *J. Microelectromech. Syst.* **11**, 317-321, (2002).
10. V. I. Belyi and L. L. Vasilyeva, *Silicon nitride in electronics*, (Amsterdam; New York: Elsevier, 1988).
11. H. Kawaoka, T. Adachi, T. Sekino, Y-H. Choa, L. Gao, K. Niihara, *J. Mat. Res.* **16**, 2264-2270, (2001)
12. M. J. Madou, *Fundamentals of microfabrication*. (Boca Raton, Fla: CRC Press, 1997).
13. Y. Toivola, J. Thurn, R. F. Cook, G. Cibuzar, and K. Roberts, *J. Appl. Phys.* **94**, 6915, (2003).
14. B. Rousset, L. Furgal, P. Fadel, A. Fulop, D. Pujos, and P. Temple-Boyer, *EDP Sciences. Journal de Physique IV* **11**, Pr3-937-944, (2001).
15. J. M. Olson, *Mater. Sci. Semicond. Process.* **5**, 51-60, (2002).
16. I. G. Rosen, T. Parent, B. Fidan, C. Wang, and A. Madhukar, *IEEE Trans. on Control Systems Technology* **10**, 64-75, (2002)
17. T.-Y. Zhang, M.-H. Zhao, and C.-F. Qian, *J. Mat. Res.* **15**, 1868-1871, (2000).
18. N. Lobontiu and E. Garcia, *J. Microelectromech. Syst.* **13**, 41-50, (2004).
19. J. A. Knapp and d. B. M. P., *J. Microelectromech. Syst.* **11**, 754-764, (2002).

Mater. Res. Soc. Symp. Proc. Vol. 875 © 2005 Materials Research Society   O4.5

THERMOMECHANICAL BEHAVIOR AND PROPERTIES OF PASSIVATED PVD AND ECD Cu THIN FILMS

M. Gregoire [1], S. Kordic [2], P. Gergaud [3], O. Thomas [3], and M. Ignat [4]

[1] STMicroelectronics, Crolles2 Alliance, 850 rue Jean Monnet, 38926 Crolles, France
[2] Philips Semiconductors, Crolles2 Alliance, 860 rue Jean Monnet, 38920 Crolles, France
[3] TECSEN, CNRS, Université of Aix-Marseille III, Faculté St Jérôme, 13397 Marseille, France
[4] LTPCM-INPG, CNRS, Domaine Universitaire, BP 75, 38402 St Martin d'Hères, France

ABSTRACT

The thermomechanical behavior is investigated of SiCN-encapsulated blanket Physical Vapor Deposited (PVD) and Electrochemically Deposited (ECD) Cu films. At lower ECD Cu film thicknesses an anomalous shape and a tail of the stress-temperature curve are observed, which are not caused by impurities at the interfaces, but are correlated to highly textured microstructure. Repeated thermal cycling of up to 400 °C does not markedly change the texture of the films, but a significant texture change takes place with increasing ECD Cu thickness. Thermal cycling induces grain growth for thicker films only. Impurity content and distribution in the PVD films do not change due to cycling.

INTRODUCTION

Copper thin films fabricated using Physical Vapor Deposition (PVD) and Electrochemical Deposition (ECD) are widely used in modern integrated circuits (IC's), yet the majority of studies up until now deal with sputtered Cu films. In contrast to sputtered films, Cu films fabricated with ECD techniques have a high content of contaminants, which influence film properties.

It is well known that high internal stresses in films used in IC's can lead to important failure mechanisms such as dielectric cracking, interfacial delamination, and stress voiding [1, 2]. In addition, IC's (Figure 1) are subjected to thermal cycles during manufacturing, and during the

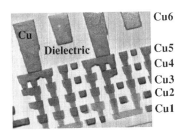

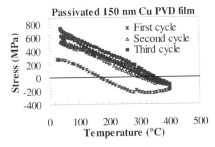

Figure 1. A scanning electron microscope cross section of a 65 nm design rule, 6 ECD Cu metal level interconnect, with a low relative permittivity dielectric (ε_r =3).

Figure 2. Stress as a function of temperature determined using substrate curvature technique for the encapsulated PVD Cu stack. Sample was subjected to three thermal cycles.

normal IC use. Prediction and improvement of thin film reliability require detailed knowledge of their thermomechanical behavior.

To determine the stress in films during thermal cycling, the substrate curvature technique is used [3]. The impact of passivation [1, 4-5], texture [6-8], and thickness [8-9] on the stress-temperature behavior have been studied on sputtered Cu films, and marked changes in the stress-temperature curve have been observed if Cu was intentionally contaminated with oxygen [5, 10].

In this paper, the thermomechanical behavior of encapsulated PVD and ECD Cu thin films is investigated, which were deposited on a PVD Ta/TaN diffusion barrier layer. Different thicknesses of ECD Cu films were analyzed. The Cu microstructure and the grain growth were characterized depending on the thermal treatment and the film thickness. The impurity content and the distribution within the films were measured as well.

EXPERIMENTAL

The studied PVD film stack consisted of three blanket layers: Cu/Ta/TaN with respective thicknesses of 150 nm, 15 nm, and 10 nm. These films were deposited on a Si (100) substrate. The films were encapsulated with a 40 nm-thick SiCN layer deposited using Plasma Enhanced Chemical Vapor Deposition (PECVD).

The ECD stack starts in the same way as the PVD stack described above. The PVD Cu deposition is followed by ECD Cu, with thicknesses ranging from 0.4 μm to 1.8 μm. The ECD stack was also encapsulated with SiCN. Both PVD and ECD stacks were fabricated using the same processing conditions, which, along with the thicknesses, are usual in IC manufacturing. All films were deposited at room temperature, and were not annealed.

Substrate curvature technique was used on the above films to determine the evolution of biaxial stress as a function of the temperature. The samples were subjected to three thermal cycles, from ambient temperature to 400 °C, with a constant heating and cooling rate of 10 °C/min. The stress evolution was determined using Stoney's formalism [11], and verified by calculations which consider the whole stack [12]. After removing the SiCN layer in an SF_6 plasma the ECD Cu microstructure was characterized by Electron Back-Scattered Diffraction (EBSD), and the PVD Cu texture was analyzed using θ-2θ X-Ray Diffraction (XRD). The orientation of the ECD Cu grains was indexed with a precision of better than 15°. Mean Cu grain size was determined from EBSD mappings. Secondary Ion Mass Spectroscopy (SIMS) was used in the investigation of the impurity content. The Cu thickness was measured by X-Ray Fluorescence (XRF).

RESULTS

Stress-temperature behavior

Figure 2 shows the stress-temperature $\sigma(T)$ evolution over three thermal cycles of the 150 nm-thick PVD Cu stack. At ambient temperature films exhibit a tensile stress of ~300 MPa. With temperature increase the tensile stress in the films decreases and goes into compression due to the constraining effect of the substrate. A stress plateau is observed from 280°C to 400°C. This behavior can be attributed to a change in the microstructure such as grain growth [4]. Before cooling, during the isothermal hold at 400 °C during 15 min, stress relaxation was observed. During subsequent cooling to ambient temperature the stress reaches high tensile values of ~550 MPa. During the second and the third cycle a stabilized thermomechanical response is observed with a slight hysteresis.

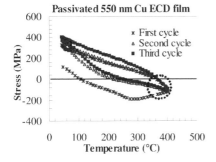

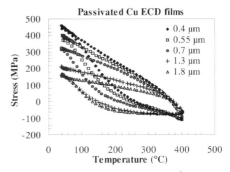

Figure 3. Stress as a function of temperature for the encapsulated ECD Cu stack. Sample was subjected to three thermal cycles.

Figure 4. Stress as a function of the 3rd cycle temperature for passivated ECD Cu stacks of different thicknesses.

In Figure 3 the $\sigma(T)$ curve is shown for the 550 nm-thick ECD Cu stack over three thermal cycles. The first thermal cycle is similar for both PVD and ECD Cu stacks, except that the zero-stress temperature is lower, and there is more pronounced plastic yielding at higher temperatures for ECD Cu.

During the second and the third heating cycle the hysteresis is more pronounced compared to the PVD stack, and the "compressive" or "negative" yielding begins in the tensile regime at ~175 MPa. Above ~350 °C stress evolution exhibits an additional increase in compression. This results in a "tail" of the $\sigma(T)$ loop encircled in Figure 3. Several authors attribute the negative yielding and the tail to the segregation of oxygen to the interfaces of encapsulated Cu layers [5, 10].

In the following only the third cycle for the different stacks will be discussed.

Figure 4 shows the thermomechanical behavior of passivated ECD Cu layers with thickness ranging from 0.4 µm to 1.8 µm. It can be seen that the "tail" and the "negative" yielding are observed only for layers with a thickness inferior to 0.7 µm. For thicker films, the hysteresis shape exhibits the normal behavior: the plastic yielding appears at compressive stress levels during the heating, and at higher temperatures the "anomalous" tail is not observed [9].

Microstructure

Figure 5 shows EBSD grain orientation maps of ECD Cu after the removal of the SiCN capping layer. These samples have not been subjected to thermal cycling. From the EBSD mappings the surface fractions of grain crystallographic orientations were quantified (Table 1a).

From Figure 5 and Table 1a it can be seen that a significant texture change takes place with the increasing Cu thickness: the surface fraction of <111> orientations and the associated <511> twin orientation decreases, while there is an increase in the surface fraction of the <001> orientation and the associated <221> twin orientation. As expected, the PVD Cu layer exhibits a strong <111> texture (Table 1b).

In the same way the surface fractions of different crystallographic grain orientations for thermally cycled samples were determined. A difference in the textures was not detectable between the non- cycled and the cycled samples. The size of Cu grains was examined before and after the

Figure 5. EBSD orientation mappings of the grains in the ECD Cu films before thermal cycling. The thickness of ECD Cu film is **(a)** 0.4 µm, **(b)** 0.55 µm, **(c)** 0.7 µm, **(d)** 1.3 µm, and **(e)** 1.8 µm. The contrast in the images follows the unit triangle **(f)**.

cycling, and it can be seen from Figure 6 that grain growth takes place only in thicker ECD films: for a 1.8 µm-thick ECD Cu layer, the grain size increases by over 40 %.

Impurities content

Oxygen and carbon depth profiles have been measured for PVD (Figure 7) and ECD (Figure 8) stacks before and after cycling. It can be seen from Figure 7 that impurity content and distribution in the PVD stack does not change significantly due to cycling. Two ECD stacks with Cu thicknesses of 0.4 µm and 1.8 µm were analysed. For the non-cycled stacks the maximum in O concentration is located at the interface of PVD Cu and Ta films, while the maximum in C concentration is found at the interface between PVD and ECD Cu films.

In each case, the thermal cycling induces a redistribution of O and C content in the Cu film: the carbon depth profile for cycled stacks shows an increase in the concentration at the Cu/Ta interface. There is a global increase of the oxygen concentration in the film, but the interface O concentration does not changed.

Since in all ECD stacks oxygen and carbon are present in Cu and at the Cu/Ta interface, and the oxygen concentration at the interface does not change with cycling, the presence of these impurities cannot be the cause of the tail and the negative yielding in the $\sigma(T)$ curves. The tail and the negative yielding are present in thin ECD Cu films, but not in the thicker ones, which is in contrast to the observations made on sputtered films [5, 10]. Our observations show that the tail and the negative yielding in the $\sigma(T)$ curves are correlated to the thickness and the texture of the ECD film: 22% <111> for the 1.8 µm, and 89% for the 0.4 µm film (Table 1a).

Table 1. (a) EBSD surface fractions of different grain crystallographic orientations for ECD Cu films. **(b)** Volume fractions of crystallographic orientations in a 150 nm-thick PVD Cu film measured using XRD.

Thickness of Cu ECD layers (µm)	Surface fraction (%)								150nm-thick Cu PVD layer	
	<111> and <511>	<001> and <221>	<101>	<211>	<310>	<311>	<331>			Volume fraction (%)
									<111>	99.3
0.4	89	5	0	5	1	0	0			
0.55	71	11	1	8	3	2	5		<200>	0.5
0.7	52	19	2	9	6	3	2		<220>	0.04
1.3	22	26	8	8	8	3	14			
1.8	22	27	9	9	7	3	14		<311>	0.11

(a)   **(b)**

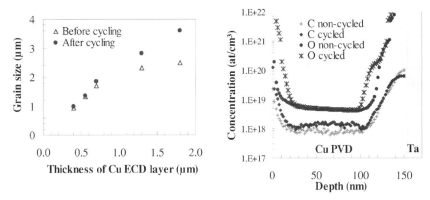

Figure 6. Copper grain sizes determined by EBSD as a function of the ECD Cu film thickness before and after three thermal cycles.
Figure 7. Oxygen and carbon depth profiles for a 150 nm-thick PVD Cu film before and after cycling.

DISCUSSION AND CONCLUSIONS

Two different thermomechanical behavior modes were observed in ECD Cu stacks. When the ECD Cu thickness was inferior to 0.7 µm, an "anomalous" shape of the $\sigma(T)$ curve was observed.

Figure 8. Oxygen (**a**) and carbon (**b**) depth profiles for non-cycled and cycled ECD Cu films.

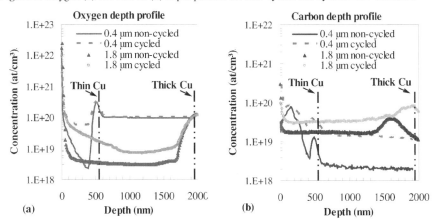

In this case the texture and the grain size did not evolve during thermal cycling. When ECD Cu thickness was over 1.3 µm the hysteresis loop did not exhibit the unusual shape, and Cu grain size increased during cycling. It is reported in the literature that the driving force for the anomalous behavior is the misfit dislocation energy recovery. The segregation of impurities during thermal cycling may modify the film adhesion and its plastic yielding. This is the reason that the segregation of oxygen at the interfaces during the cycling has been associated to these observations [10, 13]. We have, however, observed that the anomalous shape and the tail of the $\sigma(T)$ curve depend on the ECD Cu film thickness, and are correlated to the Cu texture.

The electrochemically deposited films have larger impurity content compared to the sputtered films. The electrolyte is composed of $CuSO_4/H_2O$ and H_2SO_4 solution and various other additives are included. The influence of impurities such as sulphur and hydrogen during cycling should be investigated in future work.

REFERENCES

1. W.D. Nix, *Metall. Trans. A* **20A**, 2217 (1989).
2. M. Gregoire, S. Kordic, M. Ignat, X. Federspiel, P. Vannier, and S. Courtas, *Int. Interconnect Tech. Conf.* (2005).
3. P.A. Flinn, *MRS Symp. Proc.* **130**, 41 (1988).
4. R.M. Keller, W. Sigle, S.P. Baker, O. Kraft, and E. Arzt, *Mat. Res. Soc. Symp. Proc.* **436**, 221 (1997).
5. S.P. Baker, R.M. Keller, A. Kretschmann, and E. Arzt, *Mat. Res. Soc. Symp. Proc.* **516**, 287 (1998).
6. E.M. Zielinski, R.P. Vinci, and J.C. Bravman, *Appl. Phys. Lett.* **67** (8), 1080 (1995).
7. W.M. Kuschke, A. Kretschmann, R.-M. Keller, R.P. Vinci, C. Kaufmann, and E. Arzt, *J. Mater. Res.* **13**, 2962 (1998).
8. S.P. Baker, A. Kretschmann, and E. Arzt, *Acta. Mater.* **49**, 2145 (2001).
9. R.M. Keller, S.P. Baker, and E. Arzt, *J. Mater. Res.* **13**, 1307 (1998).
10. J.B. Shu, S. B. Clyburn, T.E. Mates, and S.P. Baker, *J. Mater. Res.* **18** (9), 2122 (2003).
11. G.G. Stoney, *Proc. Roy. Soc. Lond.* **A82**, 172 (1909).
12. P.H. Towsend, D.M. Barnett, and T.A. Brunner, *J. Appl. Phys.* **62** (11), 4438 (1987).
13. S.P. Baker, R.M. Keller-Flaig, and J.B. Shu, *Acta. Mater.* **51**, 3019 (2003).

A Model for Curvature in Film-Substrate System

G. Vanamu[1], T. A. Khraishi[2] and A. K. Datye[1]
[1]Department of Chemical and Nuclear Engineering, University of New Mexico, Albuquerque, NM 87131, U.S.A.
[2]Mechanical Engineering Department, University of New Mexico, Albuquerque, NM 87131, U.S.A.

ABSTRACT

Growth of lattice mismatched films creates bending in the whole structure. There has been great interest in the study of these curvatures in epitaxially-grown materials. An analytical solution for the radius of curvature produced by stresses developed in growing lattice mismatched materials has been obtained. The analyses were based on beam bending theory and strain partitioning theory introduced by our group earlier. The expressions for radius of curvature were obtained for a two-layer heterostructure. The variation of the radius of curvature with the relative thicknesses, relative lattice constants, and relative elastic constants of the layers was determined. The model was verified by applying it to a symmetric tri-laminate structure. The above model can also be extended to determine the curvature for multi-layered heterostructures.

INTRODUCTION

Growth of lattice mismatched films; combining very dissimilar materials have generated much interest because a variety of novel properties are realized. Heteroepitaxial structures (such as Si-Ge alloys) are gaining prominence for advanced electronic [1-2] and optoelectronic devices [3-5]. In order to produce relaxed Si_xGe_{1-x} material on a Si substrate, the most promising scheme is the linearly graded buffer layer (Si_xGe_{1-x}) [6-10]. Growth of lattice mismatched structures causes wafer bending or curvature. Curvature in an epitaxially grown heterostructure is a very critical quantity. First, stress and strain fields can be determined using curvature measurements of the multilayered system. Second, there is a need to decrease the curvature in a heterostructure as it could pose a problem for device fabrication [11]. In recent years there has been great interest in the study of stresses and strains in epitaxially grown multilayers. These elastic fields can be determined using curvature measurements of the multilayer system. It is necessary to find the variables that affect the curvature. The calculations for the curvature were done in two ways. In first case authors [12-15] used force and moment balance equations and in second case authors [16-17] used minimization of the total elastic energy of the system. This paper calculates the curvature using the strain formulations obtained from the model developed in our previous work [18]. In this work we have developed a model to calculate curvature for a two-layer structure. We have verified our model by applying it to a sandwich beam, which consists of two equally-thick layers of the same material on either side of a middle core.

THEORETICAL MODEL

Consider a planar heterostructure made up of two layers with cubic unit cells as shown in Figure 1. This figure describes the heterostructure after growth, where the

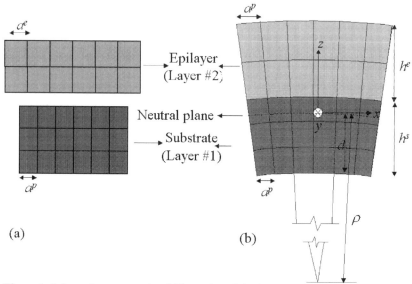

Figure 1. Schematic cross-sectional illustration of the heterostructure (a) before growth, (b) after growth.

curvature in the structure is obvious. The strained layers are assumed to only have normal in-plane stresses (perpendicular to the growth direction). The two orthogonal normal stresses, σ_{xx} and σ_{yy}, are equal in magnitude. The strains in the x and y directions (ε_x and ε_y) are also equal in magnitude but the out-of-plane strain (ε_z) is different due to Poisson effect. The bending strain, ε_x^b in any layer is given by [19],

$$\varepsilon_x^b = \frac{z}{\rho} \tag{1}$$

where ρ is the radius of curvature, z is the out-plane direction. For the case of isotropic materials, the stress-strain equations (Hooke's law) for the two layers are given as [20],

$$\sigma_x^b = E\varepsilon_x^b = \frac{Ez}{\rho} \tag{2}$$

where σ_x^b is the bending stress x-direction and E is the Young's modulus. The position of the neutral plane (distance "d" from the bottom) was found using the condition that the resultant axial force acting on the cross section is zero; therefore,

$$\int_s \sigma_x^b dA + \int_e \sigma_x^b dA = 0 \tag{3}$$

For a unit depth in y direction equation 3 reduces to,

$$\int_{-d}^{h^s-d} \frac{E^s z}{\rho} dz + \int_{h^s-d}^{h^s+h^e-d} \frac{E^e z}{\rho} dz = 0 \tag{4}$$

where h^s and h^e are the thicknesses, E^s and E^e are Young's moduli of the substrate and epilayer respectively. Solving equation 4 for d we obtain,

$$\frac{d}{h^s} = \frac{1 + 2E_r h_r + E_r h_r^2}{2(1 + E_r h_r)} \tag{5}$$

Where $E_r = \dfrac{E^e}{E^s}$, $h_r = \dfrac{h^e}{h^s}$. The above equation sets limits for "d/h^s", $(1/2) < (d/h^s) <$ $(1 + h_r/2)$. The dependence of d/h^s on h_r is plotted in Figure 2. It can be seen that d/h^s approaches ½ as h_r approaches 0. The relationships between the bending moment, the stresses and the radius of curvature can be derived as below [19]. The total bending moment,

$$M = \int_s \sigma_x^s z dA + \int_e \sigma_x^e z dA \tag{6}$$

Substituting Equation 2 into above Equation we get total bending moment per unit depth

$$M = \frac{E^s}{\rho} \int_s z^2 dA + \frac{E^e}{\rho} \int_e z^2 dA \tag{7}$$

Therefore rearranging the above equation we get $\kappa = \dfrac{1}{\rho} = \dfrac{M}{E^s I^s + E^e I^e}$ (8)

where κ is the curvature, ρ is the radius of curvature, and I is the moment of inertia per unit depth given using parallel-axes theorem as,

$$I^s = \int_s z^2 dA = \frac{1}{12}(h^s)^3 + h^s\left(d - \frac{h^s}{2}\right)^2, \quad I^e = \int_e z^2 dA = \frac{1}{12}(h^e)^3 + h^e\left(\frac{h^e}{2} + h^s - d\right)^2 \tag{9}$$

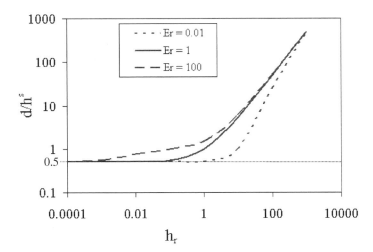

Figure 2. Normalized distance of the neutral plane from the bottom (d/h^s) plotted against the normalized height (h_r)

The bending moments of epilayer and substrate around the neutral plane were calculated,

$$M^e = \sigma^e h^e (\frac{h^e}{2} + h^s - d), \; M^s = \sigma^s h^s (d - \frac{h^s}{2}) \tag{10}$$

where σ^s and σ^e are the residual or the growth stresses. The total bending moment

$$M = M^s - M^e \tag{11}$$

The stresses and strains in each layer are used from the work of our group [18],

$$\sigma^s = B^s \varepsilon^s, \; \sigma^e = B^e \varepsilon^e \text{ and } \sigma^s h^s = -\sigma^e h^e \tag{12}$$

where $B^s = E^s/(1 - v^s)$, $B^e = E^e/(1 - v^e)$, v is Poisson's ratio,

$$\varepsilon^e = \frac{a_r^{-1} - 1}{1 + E_r \bar{v}_r^{-1} h_r a_r^{-1}}, \; \varepsilon^s = \frac{a_r - 1}{1 + E_r^{-1} \bar{v}_r h_r^{-1} a_r}, \; a_r = \frac{a^e}{a^s} \text{ and } \bar{v}_r = \frac{1 - v^e}{1 - v^s} = \frac{\bar{v}^e}{\bar{v}^s}$$

Here a^e and a^s represent the bulk lattice constants of the epilayer and substrate materials, respectively. Solving equations 10, 11 and 12 for total bending moment

$$M = \frac{B^e}{2} \left(\frac{a_r^{-1} - 1}{1 + E_r \bar{v}_r^{-1} h_r a_r^{-1}} \right) (h^e)^2 (1 + h_r^{-1}) \tag{13}$$

Substituting equations 9 and 13 into equation 8 we get the normalized radius,

$$\frac{\rho}{h^s} = \frac{\bar{v}^e \left\{ E_r^{-1} \left(\frac{1}{6} + 2 \left(\frac{d}{h^s} - \frac{1}{2} \right)^2 \right) + \left[\frac{1}{6} h_r^3 + 2 h_r \left(\frac{h_r}{2} - \frac{d}{h^s} + 1 \right)^2 \right] \right\}}{h_r (1 + h_r) \left(\frac{a_r^{-1} - 1}{1 + E_r \bar{v}_r^{-1} h_r a_r^{-1}} \right)} \tag{14}$$

where d/h^s is given by Equation 5. The same analysis from before applies to the y-direction as well. This model for two layers can be easily extended to three or four layers. For homoepitaxy ($a_r = 1$), it can be seen that ρ/h^s is infinite which means the radius of curvature is infinite or the curvature is zero. Figure 3 plots ρ/h^s against h_r as a function of a_r at constant $E_r = 1$, $\bar{v}_r = 1$, and $\bar{v}^e = 0.75$. When the ratio $a_r > 1$, the radius becomes negative. Since the absolute value of ρ/h^s is what is important, the figure plots just that. We notice in the figure that there is a global minimum value for ρ/h^s for a given a_r curve. This value increases as a_r approaches unity. Figure 4 plots ρ/h^s against h_r as a function of E_r at constant $a_r = 0.9$, $\bar{v}_r = 1$ and $\bar{v}^e = 0.75$. It can be seen that the radius increases as h_r increases or decreases. We notice that the position of the global minimum shifts to the left with increasing E_r.

In order to verify the above approach it is applied to a sandwich beam. Equation 5 is now extended to a three-layer model and the expression for "d" can be obtained as,

$$d = \frac{E^1 (h^1)^2 + E^2 (h^2)^2 + E^3 (h^3)^2 + 2E^2 h^1 h^2 + 2E^3 h^2 h^3 + 2E^3 h^3 h^1}{2E^1 h^1 + 2E^2 h^2 + 2E^3 h^3} \tag{15}$$

where h^1, h^2, h^3 and E^1, E^2, E^3 are the thickness and Young's moduli of first, second and third layers respectively. In the case that $E^1 = E^2$, the last equation reduces to Equation 5 for a two-layer structure. Substituting $E^1 = E^3$ and $h^1 = h^3$ (sandwich beam) into Equation 5:

$$d = h^1 + \frac{h^2}{2} \tag{16}$$

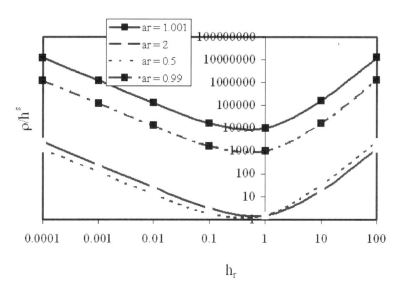

Figure 3. Normalized radius of curvature (ρ/h^s) plotted against the normalized height (h_r) as a function of a_r at $E_r = 1$, $\overline{V}_r = 1$, $\overline{V}^e = 0.75$.

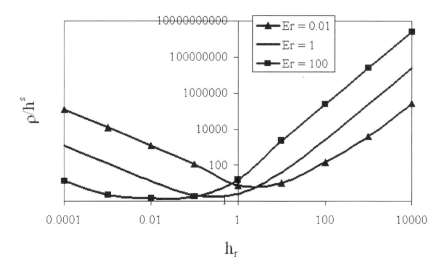

Figure 4. Normalized radius of curvature (ρ/h^s) plotted against the normalized height (h_r) as a function of E_r at $a_r = 0.9$, $\overline{V}_r = 1$, $\overline{V}^e = 0.75$.

Equation 16 means that the neutral axis is in the center of the sandwich, which is what to be expected. In addition, for a symmetric sandwich, using $\sigma^1 = \sigma^3$ and $h^1 = h^3$, one can show that the moment around the Neutral axis of the beam is equal to zero. This is expected due to symmetry of the layers.

CONCLUSIONS

In summary, we have developed a model to calculate the curvature of a two-layer heterostructure using beam bending theory and strain partitioning theory. We have actually determined the effect of E_r and a_r on the radius of curvature. So in principle such a model can be applied to find an optimal combination of h_r, E_r, a_r and $\bar{v}_r$ that would result in the lowest possible curvature. The model was verified by applying it to a symmetric tri-laminate structure.

REFERENCES

1. P. M. Mooney, J. L. Jordansweet, K. Ismail, J. O. Chu, R. M. Feenstra, F. K. Legoues, Appl. Phys. Lett. **67**, 2373 (1995).
2. K. Ismail, M. Arafa, K. L. Saenger, J. O. Chu, B. S. Meyerson, Appl. Phys. Lett. **66**, 1077 (1995).
3. E. A. Fitzgerald, Y. H. Xie, D. Monroe, P. J. Silverman, J. M. Kuo, A. R. Kortan, F. A. Thiel, B. E. Weir, J. Vac. Sci. Technol. **B 10**, 1807 (1992).
4. M. T. Currie, S. B. Samavedam, T. A. Langdo, C. W. Leitz, E. A. Fitzgerald, Appl. Phys. Lett. **72**, 1718 (1998)
5. F. Scarinci, M. Fiordelisi, R. Calarco, S. Lagomarsino, L. Colace, G. Masini, G. Barucca, S. Coffa, S. Spinella, J. Vac. Sci. Technol. **B 16**, 1754 (1998).
6. F. Schaffler, Mater. Res. Soc. Symp. Proc. **220**, 433 (1991).
7. K. Ismail, B. S. Meyerson, P. J. Wang, Appl. Phys. Lett. **58**, 2117 (1991).
8. Y. J. Mii, Y. H. Xie, E. A. Fitzgerald, D. Monroe, F. A. Thiel, B. E. Weir, L. C. Feldman, Appl. Phys. Lett. **59**, 1611 (1991).
9. S. C. Jain, W. Hayes, Semicond. Sci. Technol. **6**, 547 (1991).
10. A. R. Powell, R. A. Kubiak, T. E. Whall, E. H. C. Parker, D. K. Bowen, Mater. Res. Soc. Symp. Proc. **220**, 277 (1991).
11. N. A. Elmasry, S. A. Hussien, A. A. Fahmy, N. H. Karam, S. M. Bedair, Materials letters, **14**, 58 (1992).
12. R. H. Saul, J. Appl. Phys. **40**, 3273 (1969).
13. F. K. Reinhart and R. A. Logan, J. Appl. Phys. **44**, 3171 (1973).
14. G. H. Olsen and M. Ettenberg, J. Appl. Phys. **48**, 2543 (1977).
15. J. Vilms and D. Kerps, J. Appl. Phys. **53**, 1536 (1982).
16. E. du Tre´molet de Lacheisserie and J. C. Peuzin, J. Magn. Magn. Mater. **136**, 189 (1994).
17. P. M. Marcus, J. Appl. Phys. **79**, 8364 (1996).
18. D. Zubia, S. D. Hersee, T. A. Khraishi, Appl. Phys. Lett. **80**, 740 (2002).
19. J. M. Gere and S. P. Timoshenko, *Mechanics of Materials*, (PWS-KENT, Boston, 1990), 3rd ed., p. 254.
20. J. P. Hirth and A.G. Evans, J. Appl. Phys. **60**, 2372 (1986).
21. G. Vanamu, J. Robbins, T. A. Khraishi, A. K. Datye and S. H. Zaidi, (Journal of Electronic Materials, in press).

Mater. Res. Soc. Symp. Proc. Vol. 875 © 2005 Materials Research Society   O4.7

Elastic Behavior of Fibre-textured Gold Films by Combining Synchrotron X-ray Diffraction and In-situ Tensile Testing

D. Faurie, P.-O. Renault, E. Le Bourhis, P. Goudeau
Laboratoire de Métallurgie Physique, UMR 6630 CNRS - Université de Poitiers, SP2MI, Bd Marie et Pierre Curie, BP 30179, 86962 Futuroscope Chasseneuil Cedex, France.

ABSTRACT

The elastic behavior of gold thin films deposited onto Kapton substrate has been studied using in-situ tensile tester in a four-circle goniometer at a synchrotron beam line (LURE facility, France). Knowing the stress tensor in the film, the strong $\{111\}$ fibre texture was taken into account using the Crystallite Group Method (CGM). CGM strain analysis allows predicting a non linear relationship between strain and $\sin^2 \Psi$ obtained for the thin films due to the strong anisotropy of gold. In contrast, the average of strains in longitudinal and transversal directions varies linearly with $\sin^2 \Psi$. The evolution of the slope of these curves as a function of the applied stresses in the film allowed determining the single-crystal elastic constant s_{44} of thin gold films.

INTRODUCTION

Synchrotron X-ray diffraction is well-known as a trustworthy and powerful tool to determine the stress state of small-sized crystalline materials such as thin films [1]. It is a phase selective and a non destructive technique which allows determining both the mechanical behavior and the microstructure of diffracting phases. The elastic change in interplanar distance is used to determine the residual or applied stresses, owing that the elastic constants are known [2,3]. Indeed, thin film elastic constants may differ from the bulk references as a result of the particular microstructure (nanometer grain size, high defects density, constraints caused by the substrate) [4-6]. Combining X-ray diffraction and in-situ uniaxial tensile tester, it is possible to determine X-ray elastic constants of diffracting phases in a thin film deposited on substrate. The first experiments were performed using metallic substrates [7,8]. More recently, the elastic-plastic behaviors of Cu and Au thin films have been characterized using compliant polyimide substrates (Kapton®) [1,9,10]. Also, Badawi *et al.* [11] have determined elastic constants of non textured tungsten thin films studying the linear d-$\sin^2 \Psi$ curves as a function of applied load, assuming an isotropic elastic behavior of both the compliant substrate and the thin film. For a fibre-texture thin film that is made of elastically anisotropic crystals, the Crystallite Group Method (CGM) [3], based on Reuss assumptions, can be used to determine the stresses from X-ray diffraction, provided the dispersion of texture is not elevated (not higher than 10°) [12]. This method consists in idealizing the texture as a set of crystallites with the same orientation [3,13]. When we consider the case of a uniaxial tensile load applied on a film/substrate system, transversal stress is induced by the difference between the Poisson's ratio of the thin film and that of the substrate. Therefore, the film is submitted to a non equi-biaxial stress field ($\sigma_{11} \neq \sigma_{22}$). It has been shown theoretically (using the CGM) that for fibre-textured polycrystals, a non equi-biaxial stress state combined with an elastic anisotropy results in a non-linear ε-$\sin^2 \Psi$ curves [14]. In the present paper, we report synchrotron X-ray diffraction measurements on gold thin films deposited onto Kapton® substrate loaded in-situ under tensile stresses. We show how the CGM can be used to describe the elastic behavior of the deposited thin films (which exhibit a strong fibre texture), that are under non-equibiaxial stress state.

EXPERIMENTAL DETAILS

Specimen preparation: 700 nm thick Gold thin films have been deposited by physical vapor deposition. The substrate was a 127.5 μm thick polyimide (Kapton®) dogbone substrate; the in-plane sample dimensions were 14*6 mm². The substrate was cleaned with ethanol before deposition. The base pressure of the growth chamber was $7x10^{-5}$ Pa while the working pressure during film growth was approximately 10^{-2} Pa. Gold deposition was carried out at room temperature with an Ar+-ion-gun sputtering beam at 1.2 keV.

Pole figures are shown on Fig. 1 with Ψ defined as the angle of inclination of the specimen surface normal with respect to the diffraction vector and Φ the rotation angle of the specimen around the specimen surface normal. For the {111} family (Fig. 1a), the plot of the intensity as a function of angle Ψ shows only two maxima at approximately 0° and 70.5° which are characteristic of a {111} texture. For the {400} family (Fig. 1b), one peak is observed at 54.7°. For the {331} family (Fig. 1c), three peaks are observed at 22°, 48.5° and 82.4°. All these peaks are defined by the angle between the (111) planes along the normal to the growth direction and the planes of the considered families. These peaks show also the pole directions of each family used for strain measurements. Besides, no other peak is observed. Thus only one component of the texture is observed and we conclude that the thin films exhibit a {111} texture. Moreover, for a given angle Ψ, the X-ray intensity is independent on the angle Φ because of the rotational symmetry around the growth direction (fibre texture). The degree of scatter of this texture can be given by the FWHM of the peaks of a Ψ scan for the {111} plane family: 8°.

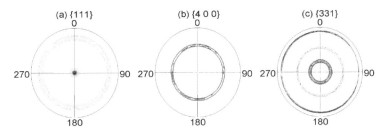

Figure 1: X-rays pole figures of a 700 nm thick Au film deposited on a 125 μm polyimide substrate on the plane families **a)**{111}, **b)**{400}, **c)**{331}.

Tensile testing and X-ray diffraction: The external load was applied to composite (film + substrate) dogbone specimens by means of a 300 N *Deben*^{TM} tensile module. The tensile tester is equipped with a 75 N load cell enabling force measurements with a precision better than 0.3 N. X-ray diffraction measurements were performed using a four-circle goniometer on the DW22 beam line at the French synchrotron radiation facility LURE (Orsay). A wavelength ($\lambda = 0.161$ nm) was chosen to analyze five independent families of hkl planes (Table 1). Under each applied load, the measurements have been performed in the pole directions for the longitudinal ($\Phi = 0°$)

and the transversal ($\Phi = 90°$) directions. It should be noted that the position $\Phi = 0°$ is taken as to refer to the σ_{11}^f direction.

DIFFRACTING PLANE	(222)	(331)	(311)	(420)	(331)	(400)	(311)	(222)	(420)
Angle Ψ (deg)	0	22	29.5	39.23	48.53	54.74	58.52	70.53	75.04
Sin^2 Ψ	0	0.14	0.24	0.4	0.56	0.67	0.73	0.89	0.93

Table 1: Angle Ψ for different hkl families used in this study.

Under our experimental conditions, an increment in force $\Delta F = 1$ N corresponds to increments in longitudinal and transversal stresses (in the films) of $\Delta\sigma_{11}^f = 20.5$ MPa and $\Delta\sigma_{22}^f = 5.6$ MPa respectively. In order to calculate these values, a mechanical approach has been used to describe the thin film-substrate composite elastic behavior. For more details on the computational procedure, the reader could refer to [15]. Six forces were applied ranging from 0 N to 6.6 N. From here onwards, we shall refer to the unloading state as T0 while T1 to T5 are related to loading states under increasing loads.

For each experimental data point, the strain is calculated using the unloading state as a reference state:

$$\varepsilon_{\Phi\Psi} = \ln\left(\frac{\sin\theta_{\Phi\Psi}^{T0}}{\sin\theta_{\Phi\Psi}^{TX}}\right) \quad (1)$$

where $\theta_{\Phi\Psi}^{T0}$ is the angular position of the considered diffraction peak for the unloading state T0 and $\theta_{\Phi\Psi}^{TX}$ the corresponding angles for the loading state TX. It should be noted that the unloading state T0 corresponds to the as deposited state and hence, the film is submitted to compressive residual stresses (about 200 MPa) but not to applied stresses. For each applied load, the measurements have been performed in the respective pole directions. In both cases, fitting functions (Pearson VII) have been used to extract peak positions.

RESULTS AND DISCUSSION

The measured data can be represented as a function of sin^2 Ψ. The experimental ε-sin^2 Ψ curves in the case of the loading state T5 are shown in Fig. 2a (full lines and closed symbols) for the longitudinal ($\Phi = 0°$) and transversal ($\Phi = 90°$) measurement directions that are $\varepsilon_{0,\psi}$ and $\varepsilon_{90,\psi}$ respectively. The anisotropic elastic behavior of the material is observed on the ε-sin^2 Ψ plots and characterized by a non linearity of the curves as predicted by CGM. Indeed, in the case of a state of biaxial stresses, the relationship between strain and stresses for a {111}fibre texture is given by [15] :

$$\varepsilon_{\Phi\Psi} = \left(\sigma_{11}^{f} - \sigma_{22}^{f}\right) \cdot \left[\frac{s_{11} - s_{12} + s_{44}}{6} \cdot \cos 2\Phi \cdot \sin^{2}\Psi + \left(-s_{11} + s_{12} + \frac{s_{44}}{2}\right)\frac{\sin(3\beta + \Phi)\sin 2\Psi}{3\sqrt{2}}\right]$$

$$+ \left(\sigma_{11}^{f} + \sigma_{22}^{f}\right) \cdot \left[\frac{2s_{11} + 4s_{12} - s_{44}}{6} + \frac{s_{44}}{4}\sin^{2}\Psi\right] \quad (2)$$

where s_{11}, s_{12}, s_{44} are the single-crystal elastic constants of the material. For $\sigma_{11}^{f} \neq \sigma_{22}^{f}$, the ε-$\sin^2 \Psi$ relation is not linear because of its dependence on $\sin 2\Psi$. This non-linearity is characteristic of an elastically anisotropic material, since for an isotropic material the $\sin 2\Psi$ prefactor reduces to zero ($s_0 = s_{11} - s_{12} - \frac{s_{44}}{2} = 0$) yielding a linear ε-$\sin^2 \Psi$ relationship. Noticeably, the average of the strains in the longitudinal ($\Phi = 0°$) and transversal ($\Phi = 90°$) directions reduces to a simple linear expression as a function of $\sin^2 \Psi$:

$$\frac{\left(\varepsilon_{0,\Psi} + \varepsilon_{90,\Psi}\right)}{2} = \left(\sigma_{11}^{f} + \sigma_{22}^{f}\right) \cdot \left[\frac{2s_{11} + 4s_{12} - s_{44}}{6} + \frac{s_{44}}{4}\sin^{2}\Psi\right] \quad (3)$$

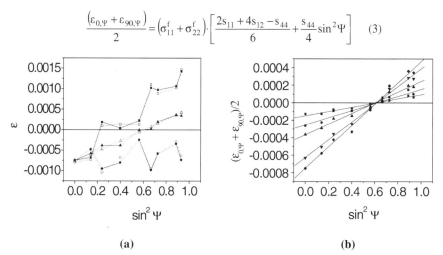

(a)   (b)

Figure 2: (a) Experimental (full symbols) and theoretical (empty symbols) strains for the longitudinal (square symbols) and transversal (circle symbols) measurement directions as a function of $\sin^2 \Psi$ (loading state T5). The triangle symbols are the average of the two measured strains. **(b)** Experimental average strain $\left(\varepsilon_{0,\Psi_0} + \varepsilon_{90,\Psi_0}\right)/2$ as a function of $\sin^2 \Psi$ for all loading states. The linear fits (continuous lines) of the experimental data are obtained from Eq. (3). We observe a common meeting point of all the fitted lines in good agreement with expected behavior.

As predicted by Eq. (3), the average of $\varepsilon_{0,\Psi}$ and $\varepsilon_{90,\Psi}$ strains is determined to increase linearly with $\sin^2 \Psi$. The very slight dispersion of the average strain curve shows firstly, the good accuracy of the experimental results using synchrotron radiation and secondly, emphasizes the non-linearity of the ε- $\sin^2 \Psi$ curves. Fig. 2b shows the average of $\varepsilon_{0,\Psi}$ and $\varepsilon_{90,\Psi}$ strains as a function of $\sin^2 \Psi$ for all loading states. As discussed in the case of the T5 loading state, the experimental values obtained under each loading states are all well fitted by straight lines. Furthermore, an intersecting point Ψ_0 is predicted to be located at $\left(\varepsilon_{0,\Psi_0} + \varepsilon_{90,\Psi_0} \right) \Big/ 2 = 0$ while linear fits of the experimental data yield very close intersecting points at $\sin^2 \Psi_0 = 0.61 \pm 0.02$ and $\left(\varepsilon_{0,\Psi_0} + \varepsilon_{90,\Psi_0} \right) \Big/ 2 = 0.00000 \pm 0.00002$. Using Eq. (3), this intersection point is defined by :

$$\sin^2 \Psi_0 = \frac{-4s_{11} - 8s_{12} + 2s_{44}}{3s_{44}} \qquad (4)$$

and allows obtaining a linear combination of the single-crystal elastic constants of the material. Moreover, if we define P as the slopes of the $\left(\varepsilon_{0,\Psi_0} + \varepsilon_{90,\Psi_0} \right) \Big/ 2$ -$\sin^2 \Psi$ curves, the plot of P as a function of $\left(\sigma_{11}^f + \sigma_{22}^f \right)$ is expected to be linear as confirmed by Fig. 3. The slope of this curve is obtained as $P^* = s_{44}/4$ ($P = \left(\sigma_{11}^f + \sigma_{22}^f \right) \cdot s_{44}/4$, by deriving Eq. (3), and allows determining the s_{44} single-crystal compliance to be about 28.6 ± 2.1 TPa^{-1} that is higher than its bulk counterpart (23.81 TPa^{-1} [16]). In order to determine the two other single crystal elastic constants, we must adjust all the experimental data using Eq. (2) and hence take into account the non linearity of $\varepsilon_{0,\Psi}$ and $\varepsilon_{90,\Psi}$ strains as a function of $\sin^2 \Psi$ (Fig. 2a). This work will be the subject of another publication [17].

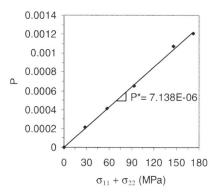

Figure 3: Plot of P (closed symbols) as a function of $\left(\sigma_{11}^f + \sigma_{22}^f \right)$. The linear fit (dotted lines) allows to obtain $P^* = s_{44}/4$ by deriving Eq. (3). The resulting single crystal compliance s_{44} is determined to be about 28.6 TPa^{-1}.

It should be noted that CGM that is used in the present work is based on the two main following assumptions. Firstly, we considered that the film texture could be represented by an ideal fibre texture and this would require further refinements in the future. However, the texture of the gold films under study is relatively sharp (dispersion being less than 8°). Thereafter, assuming an ideal fibre texture should not affect the analysis severely. Secondly, using CGM we assumed a Reuss-type grain interaction between the Au crystallites. This hypothesis means that the stress tensor components are the same in all crystallites. This second assumption is more severe than the first one [18] and work is in progress to consider a more realistic representation of grain interaction using homogenization methods.

CONCLUSIONS

We have studied the elastic behavior of thin gold films deposited onto Kapton® substrate using in-situ tensile tester in a four-circle goniometer at a synchrotron beam line (LURE facility, France). Knowing the stress tensor in the film, the strong {111} fibre texture was taken into account using the Crystallite Group Method (CGM). CGM strain analysis forecasts a non linear relationship between strain and $\sin^2 \Psi$ for anisotropic single crystals and allows representing the behavior of thin films. Moreover, it has been shown theoretically and experimentally that the average of strains in longitudinal and transversal directions varies linearly with $\sin^2 \Psi$. The evolution of the slope of these curves as a function of the applied stresses in the film allowed determining the single-crystal elastic constant s_{44} of thin gold films with good accuracy (five loading states, nine diffracting planes and two measurement directions).

REFERENCES

1. J. Böhm, P. Gruber, R. Spolenak, A. Stierle, A. Wanner, E. Arzt, Review of Scientific Instruments, **75**, 1110 (2004).
2. I.C. Noyan and J.B. Cohen, Residual Stress Measurement by Diffraction and Interpretation , Springer, New York, 1987.
3. V. Hauk, Structural and residual stress analysis by non destructive methods: evaluation, application, assessment, Elsevier, Amsterdam, 1997.
4. H. Huang and F. Spaepen, Acta. Mater. **48**, 3251 (2000).
5. A. J. Kalkmann, A. H. Verbruggen, and G. L. A. M. Jaussen, Appl. Phys. Lett. **78**, 2673 (2001).
6. J. Schiotz, T. Vegge, F. D. Di Tolle, and K. W. Jacobsen, Phys. Rev. B **60**, 11971 (1999)
7. I. C. Noyan, G. Sheikh, J. Mater. Res. **8**, 764 (1992).
8. P.-O. Renault, K.F. Badawi, L. Bimbault, and Ph. Goudeau, E. Elkaïm and J.P. Lauriat, Appl. Phys. Lett. **73**, 1952 (1998).
9. M. Hommel, O. Kraft, Acta Mater. **49**, 3935 (2001).
10. O. Kraft, M. Hommel, E. Arzt, Mater. Sci. Eng. **A288**, 209 (2000).
11. K. F. Badawi, P. Villain, Ph. Goudeau, P.-O. Renault, Appl. Phys. Lett. **80**, 4705 (2002).
12. P. Gergaud, S. Labat, O. Thomas, Thin Solid Films **319**, 9 (1998).
13. S. Labat, P. Gergaud, O. Thomas, B. Gilles, A. Marty, J. Appl. Phys. **87**, 1172 (2000).
14. K. Tanaka, Y. Akiniwa, T. Ito, K. Inoue, JMSE Series A **42**, 224 (1999).
15. P. -O. Renault, E. Le Bourhis, P. Villain, Ph. Goudeau, K. F. Badawi, D. Faurie, Appl. Phys. Letters **83**, 473 (2003).
16. J. C. Smithells, Metals Reference Book, 5th edition (Butterworths, London, 1976).
17. D. Faurie, P. -O. Renault, E. Le Bourhis, P. Goudeau, in preparation.
18. U. Welzel, private communication (2004).

Mater. Res. Soc. Symp. Proc. Vol. 875 © 2005 Materials Research Society   O4.8

A Microtensile Set Up for Characterising the Mechanical Properties of Films

B. Cyziute[1], L. Augulis[1], J. Bonneville[2], P. Goudeau[2], B. Lamongie[2], S. Tamulevicius[1], C. Templier[2]

[1] Fizikos katedra, Kauno Technologijos Universitetas, Studentu 50, 3028, Kaunas, Lithuania
[2] Laboratoire de Métallurgie Physique, Université de Poitiers, UMR-CNRS 6630, BP 30179, 86962 Futuroscope, France

ABSTRACT

A computer control deformation set-up has been specifically developed for measuring the elastic and plastic properties of thin films. It combines a piezo-actuated microtensile-testing device, based on an original tripod design, with an optical image acquiring and analysis system for measuring specimen strains. The paper will be partly devoted to describe the experimental deformation set up and its performance through mechanical tests of polyimide and aluminum samples. The Young's moduli, which are deduced from the stress-strain curves, are in good agreement with reported bulk average values. The results confirmed the ability of the equipment for the measurements of very small load and displacement levels, which are a prerequisite for such type of investigations.

Keywords: Thin film, tensile test, mechanical properties, load relaxation.

I. INTRODUCTION

Mechanical properties of low dimensional materials are not yet well understood. A great amount of instrumental and theoretical works have been devoted since a lot of years for analyzing the mechanical behavior of nanostructured materials prepared through very different elaboration procedures. Evaluating the mechanical properties, dynamic behaviors of such components, is still a challenge in mechanical engineering.

We have developed a computer control deformation set up specifically designed for investigating the elastic and plastic properties of thin films. The equipment performances have been first evaluated by preliminary tensile tests on 30 μm thick aluminum sheets of commercial purity and on polymer foils with thickness 25 μm. The results confirmed the ability of the equipment for the measurements of very small load and displacement levels, which are a prerequisite for such type of investigations.

The paper will be partly focused on the experimental deformation set up. In particular, the tensile set-up will be presented together with the specimen holder. The deformation measurements, based on an optical analysis, will be briefly described. Finally, preliminary results concerning Young's modulus, Poisson's ratio and activation volume are reported and tentatively discussed in connection with published data available in the literature.

II. EXPERIMENTAL DETAILS

II.1. Tensile testing device: A prototype of the testing apparatus has already been presented elsewhere [1]. A fundamental modification concerns the deformation measurement, that was originally determined by electronic speckle pattern interferometry and which has now been replaced by an optical digital image correlation method (described below).

As shown in figure 1, a tripod configuration is used and the displacement of the moving plate (2) is obtained by a single piezoceramic actuator (4). This simple, but particular, configuration leads to a small misalignment when the specimen is elongated. Finite element calculations have established that the corresponding shear components were negligible [2]. Two piezoceramic actuator locations allow for a change of the beam ratio in the range 1/2 to 1/4, which permit a total displacement of the moving grip of 140 µm and 280 µm, respectively.

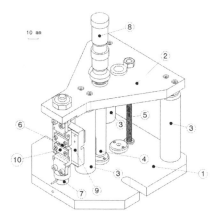

Figure 1. Schematic representation of the tensile equipment: 1-baseplate, 2-moving plate, 3-supporting columns, 4-piezoceramic actuator, 5-maitaining spring, 6-specimen, 7-load cell, 8- micrometer, 9- electromagnetic holder, 10-grips.

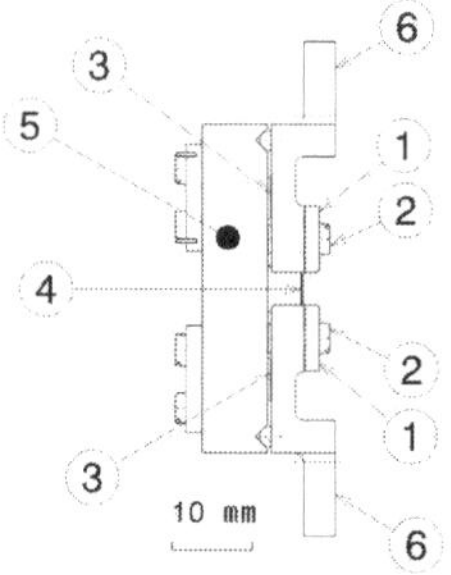

Figure 2. Schematic representation of the electromagnetic support: 1-grooved plates, 2-screws, 3-electromagnets, 4-specimen, 5-alignment guides, 6-head holders.

Other improvements consist in the location of the load cell having a sensitivity of $2.5 \ 10^{-4}$ N, which is now rigidly clamped on the base, and in the adding of a micrometer on the moving plate, which allows a fine adjustment of the zero load level applied to the specimens prior to deformation. The specimen holder is presented in figure 2. Each specimen head is screwed in with a locking grooved plate on a holder. During specimen fixing and mounting, the two head holders are rigidly kept together with an electromagnetic support, which ensures a perfect specimen alignment and avoids unwanted deformation.

II.2. Tensile specimens: Tensile specimens have a dog bone shape with a total length l_T of 4 mm between the heads. The shoulder radius is $R = 1$ mm, which yields a length $l_c = 2$ mm of constant cross-section. Then, the gauge length, which would correspond to the length of an equivalent specimen that have a constant cross-section in between the two sample heads, is $l_e = 3.51$ mm.

Aluminum specimens of commercial purity with thicknesses of 30 μm were deformed together with Kapton HN™ samples with a thickness of 25 μm. The imposed strain rates $\dot{\varepsilon}_a$ were of the order of $10^{-4}\,s^{-1}$ and were increased by one order magnitude when performing strain-rate jumps. Load relaxation experiments were also performed by stopping the moving grip, that is the output voltage applied to the piezoceramic actuator was kept constant.

II.3.Strain measurement: The displacement field is determined using a digital image processing method from the commercial software DEFTAC [3-5]. A main advantage of using this technique is that no patterning or marking of the specimen surface is needed as "natural dust" acts as marker. The optical set up is composed of a 1392*1030 CCD camera equipped with an objective and the sample surface is illuminated with a glancing focused light; a filter avoids heating of the sample surface. The acquisition image is computer controlled owing to a specific interface card and software [6].

Figure 3 gives an example which has been taken from measurements on polyimide films. Displacements are measured during the tensile test at four various spots or windows (see fig.3) of the film surface, which allows for the extraction of strains along the tensile axis and also in the perpendicular direction to this axis. The Poisson's ratio can be then deduced directly from the ratio of this two strain direction measurements while the Young's modulus extraction needs to plot the strain-stress curve.

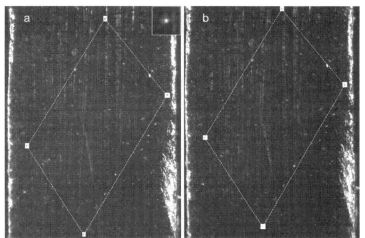

Figure 3: Optical image of the Kapton HN™ sample surface analyzed using the Deftac® software [6]. The four squares (or windows) on image (a) on image (a) enclose white 'dust' markers a few pixels in size. The contrast of these windows has been inversed to show where the markers are located and a magnification of the top marker is presented on the upper right part of the image. The fours white lines that are connecting the markers correspond to the four sides of the parallelogram which deformation is used to calculate the strain. Image b corresponds to 1.1% applied strain. The specimen width is 1 mm.

III. EXPERIMENTAL RESULTS AND DISCUSSION

III.1. Kapton HN ™ specimens: A typical stress-strain obtained for a 25 μm thick specimen is shown in figure 4. It must be noticed that the jerky aspect of the curve arises from the resolution of the optical imaging analysis system (better than 10^{-4}) and does not result from piezoceramic actuator instabilities. A linear fit (see dashed line in figure 4) of the early part of the curve yields a Young modulus value E around 4.8 GPa. The Poisson's ratio ν deduced over the same strain range is about 0.34. The yield stress σ_y, which is defined at the first departure from linearity, corresponds to a stress level of 30 ± 4 MPa. A direct comparison with reported values in the literature is not easy since published values are rather scarce and because it has been observed that for such a polymer both E and σ_y can be specimen size dependent. Therefore, while the obtained ν value is in good agreement with generally admitted values for Kapton, the present value of E is definitively larger than the recommended manufacturer value (between 2 and 3 GPa) and slightly lower than the one reported for 127 μm specimen, E ≈ 5.2 GPa [7,8]. Similarly, σ_y is found higher than for 50 μm thick specimens, σ_y ≈ 24 MPa [1], which is in agreement with the expected size dependence for σ_y. It must be also emphasized that this difference in E and σ_y may also result from the different specimen preparation routes, which for such small specimens may drastically influenced the deformation path history and, in a correlative manner, the initial recorded flow stress.

Load relaxation experiments were performed at various stress (strain) levels of the deformation curve to investigate the strain-rate sensitivity of the flow stress. During such an experiment, the load (stress) in recorded as a function of time at a fixed position of the moving grip. The time dependence of the stress is given by the following relation [9, 10]:

$$\sigma - \sigma_0 = -\frac{kT}{V} \ln \left(\frac{t}{c} + 1\right), \tag{1}$$

Where c is integration constant, k is Boltzman's constant, T is absolute temperature and V is activation volume. A load relaxation test performed at σ_0 = 64 MPa and ε_p ≈ 0.5 % is presented in figure 5, together with the corresponding calculated fit using equation 1. A very nice agreement is obtained, which strongly supports the proposed analysis. Activation volume values determined using this technique are reported as a function of stress in figure 6. The values are rather large, which indicates that the flow stress is poorly thermally activated, at least for room temperature deformation conditions. This has been verified by performing strain-rate jumps, where the stress variations associated with abrupt changes of one order of magnitude in applied strain-rate were not measurable.

It is also observed that the activation volume decreases with increasing stress. The stress dependence of the activation volume is often used for characterizing the transition between microplastic and macroplastic stages [11]. However, this would require a more complete study, not yet available, where the preceding activation volume values are known over a larger plastic strain domain and corrected from eventual artifacts that may arise from work hardening and plastic strain microstructural evolution [12]. Work is in progress to establish if such a transition stress can be determined for the Kapton HN^{TM}.

III.2. Aluminum specimens: A stress-strain curve obtained with a 30 μm thick specimen is shown in figure 7. The jerky aspect of the curve is now clearly visible and the scattering of the data is indicative of the limits of our strain deformation measurement technique, at least in its present development. The elastic slope extends over a strain which is smaller than 0.13 %. In order to avoid possible artifacts from specimen misalignment and microplastic deformation, the Young's modulus E is measured on the unloading portion of the stress-strain curve. A linear fit of the corresponding data in figure 7 yields a value of E ≈ 60 GPa. In addition, strain measurement perpendicular to the tensile axis leads to a Poisson's ratio of nearly 0.3. Both values are in good agreement with commonly accepted values for aluminum. We estimate that, for similar load condition, the signal to noise ratio of the strain data allows us to estimate E values that would be three times larger than the present measured value, which ensures us the possibility of investigating most of pure crystalline materials.

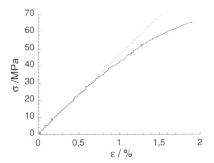

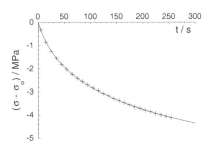

Figure 4. Stress-strain curve obtained with a Kapton HN^{TM} 25 μm thick specimen deformed at room temperature and a strain rate of 0.014% s^{-1}. The dashed line corresponds to a linear fit (see text).

Figure 5. Stress relaxation curve obtained on a Kapton HN^{TM} 25 μm thick specimen deformed at room temperature. Initial stress $\sigma_0 = 64$ MPa and $\varepsilon_p \approx 0.5\%$. The crosses are the experimental data and the line corresponds to a fit according to equation 1.

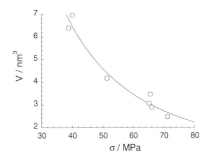

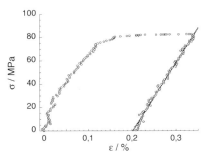

Figure 6. Stress dependence of the activation volume for the Kapton HN^{TM} 25 μm thick specimen.

Figure 7. Stress-strain curve obtained with Al 25 μm thick specimen deformed at room temperature and a nominal strain rate of 0.014% s^{-1}.The dashed line in the unloading part corresponds to a linear fit (see text).

Few load relaxation tests were also performed at various stress levels during constant strain-rate experiments. As a rule, the initial portion of the stress decrease nicely follows the logarithmic time dependence predicted by equation 1. However, the time duration over which a good correlation is observed never exceeds few tenths of seconds, which corresponds to stress drop smaller than 2 MPa. The activation volumes are rather large, of the order of few hundredths of atomic volumes. To date, our results are too scarce to draw decisive conclusions, in particular concerning the long time relaxation behavior. The preliminary results suggest that, akin to bulk specimens, athermal processes, such as for instance forest dislocation intersection mechanisms, control the low temperature deformation of polycrystalline aluminum thin foils. The observed short time (stress) correlation can be understood in the framework of a thermally activated process that includes a large athermal stress component.

IV. CONCLUDING REMARKS

Capabilities of an original new tensile testing apparatus devoted to free standing thin films have been tested with 25 μm kapton HN and 30 μm aluminum foils. Elastic properties as well as relaxation effects have been investigated. The good quality of the obtained results indicates that the tensile machine is now ready for working in a lower load range characteristic of very thin films.

The next step will consist in studying metallic coated polyimide foils before doing tensile testing directly on free standing films.

ACKNOWLEDGMENTS

The authors would like to thank 'Region Poitou-Charentes' and EGIDE (for the French-Lithuanian exchange program Gilibert) and the French embassy in Lithuania for financial supports. Thanks are also due to T. Kruml for valuable comments on the manuscript and to Y. Diot and F. Berneau for technical assistance. Authors are indebted to the DEFTAC program's developers for the efficiency of their hot-line.

REFERENCES

1. L. Augulis, S. Tamulevicius, R. Augulis, J. Bonneville, P. Goudeau, C. Templier, *Optics and Lasers in Engineering* **42**, 1 (2004)
2. J. Colin, private communication.
3. H. A. Bruck, S. R. McNeill, M.A. Sutton and W. H. Peters, *Experimental Mechanics* **29**, 261 (1989)
4. S. Choi and S.P. Shah, *Experimental Mechanics* **29**, 307 (1997)
5. P. Doumalin, PhD thesis, Polytechnic High School, Palaiseau, France (2000)
6. J.-C. Dupre, V.-C. Valle, F.-J. Bremand and F. Hesser, Deftac 2004 Version 4.0, Copyright©2004. All right reserved, http://www-lms.univ-poitiers.fr, France
7. P. Villain, PhD, 2002, Poitiers University, France
8. K.F. Badawi, P. Villain, P. Goudeau and P.-O. Renault, *Appl. Phys. Lett.* **80**, 4705 (2002)
9. J. Bonneville, P. Spätig, J.-L. Martin, *Material Research Soc. Symp. Proc.- High-Temperature Ordered Intermetallic Alloys VI* **364**, 369 (1995)
10. F. Guiu and P. L. Pratt, *Phys. Stat. Sol.* **6**, 111 (1964)
11. J-L. Farvacque, J. Crampon, J-C. Doukhan and B. Escaig, *Phys. Stat. Sol.* **14**, 623 (1972)
12. L. P. Kubin, *Phil. Mag.* **30**, 705 (1974)

Characterization of Stress Relaxation, Dislocations and Crystallographic Tilt Via X-ray Microdiffraction in GaN (0001) Layers Grown by Maskless Pendeo-Epitaxy

R.I Barabash[1]*, G.E. Ice[1], W. Liu[1], S. Einfeldt[2], D. Hommel[2], A. M. Roskowski[3], R. F. Davis[3]

[1]Metals and Ceramics Div., Oak Ridge National Laboratory, Oak Ridge, USA
[2]Institute of Solid State Physics, University of Bremen, Germany
[3]Materials Science and Engineering Department, North Carolina State University, Raleigh, USA

ABSTRACT

Intrinsic stresses due to lattice mismatch and high densities of threading dislocations and extrinsic stresses resulting from the mismatch in the coefficients of thermal expansion are present in almost all III-Nitride heterostructures. Stress relaxation in the GaN layers occurs in conventional and in pendeo-epitaxial films via the formation of additional misfit dislocations, domain boundaries, elastic strain and wing tilt. Polychromatic X-ray microdiffraction, high resolution monochromatic X-ray diffraction and finite element simulations have been used to determine the distribution of strain, dislocations, sub-boundaries and crystallographic wing tilt in uncoalesced and coalesced GaN layers grown by maskless pendeo-epitaxy. An important parameter was the width-to-height ratio of the etched columns of GaN from which the lateral growth of the wings occurred. The strain and tilt across the stripes increased with the width-to-height ratio. Tilt boundaries formed in the uncoalesced GaN layers at the column/wing interfaces for samples with a large ratio. Sharper tilt boundaries were observed at the interfaces formed by the coalescence of two laterally growing wings. The wings tilted upward during cooling to room temperature for both the uncoalesced and the coalesced GaN layers. It was determined that finite element simulations that account for extrinsic stress relaxation can explain the experimental results for uncoalesced GaN layers. Relaxation of both extrinsic and intrinsic stress components in the coalesced GaN layers contribute to the observed wing tilt and the formation of sub-boundaries.

INTRODUCTION

Stress is present in almost all III-Nitride heterostructures. Lattice mismatch and high defect density produce intrinsic stress. The mismatch in the thermal expansion coefficients between the substrate and the nitride material structures produces an additional extrinsic stress component. Stress relaxation in the GaN layers causes formation of dislocations, boundaries and strain. Dislocations and tilt boundaries are known to impair the performance of GaN-based light emitting devices [1, 2]. Lateral epitaxial overgrowth (LEO) techniques such as pendeoepitaxy [3 – 5] have recently been shown to greatly reduce the density of threading dislocations. In the pendeoepitaxy

---

* Contact author: Rosa Barabash <barabashr@ornl.gov>

process, GaN wings are grown from the sides of GaN columns (Fig.1). The top of the column may or may not be covered with a mask (masked or maskless pendeoepitaxy) to inhibit GaN growth. When the LEO process is continued until the overgrowth of a mask, threading dislocations are observed to form a grain boundary along the edge of the mask [6]. In the case of maskless pendeoepitaxy [7 – 9] for which the growth starts from the sidewalls and *simultaneously* from the surface of unmasked GaN columns, the wing tilt depends on the *c*-axis strain in the columns. Despite extensive characterization, the basic structure of LEO films including such critical information as the direction of the wing tilt remains controversial. The application of powerful X-ray white beam microdiffraction (WBMD) to this important problem together with finite element (FE) simulation and high resolution X-ray diffraction (HRXRD) has enabled us to provide detailed new information on the structure of these materials.

EXPERIMENTAL DETAILS

GaN stripes having a rectangular cross section and oriented along $\left[1\,\overline{1}\,00\right]$ were etched from 1 µm thick GaN single crystal layers grown by metalorganic vapor phase epitaxy on 0.1 µm thick AlN buffer layer deposited on 6H-SiC (0001) substrate. The subsequent overgrowth process was stopped before the wings coalesced. A schematic geometry of the stripe, with two wings on SiC substrate is shown in Fig. 1. Details of the growth process have been reported separately [8].

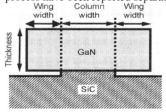

Figure 1 The schematic sample geometry.

Two samples with different GaN stripe geometries were analyzed. The GaN thickness, the column width and the wing width were 1.46 µm, 3.80 µm, and 0.98 µm for sample 1 and 3.53 µm, 3.46 µm and 1.74 µm for sample 2, as determined by cross-sectional scanning electron micsroscopy.

The x-ray synchrotron measurements with WBMD has been carried out at the beamline ID-34-E at the Advanced Photon Source, Argonne IL. WBMD uses a modified Laue diffraction method based on polychromatic radiation [10, 11]. This approach allows for true 3D mapping of the crystalline phase, orientation and plastic deformation with 1 µm spatial resolution. A region of interest is identified using an optical microscope and the polychromatic x-ray beam is then focused onto the sample. The orientation at each position of the crystal is precisely determined by an automated indexing program. Details on the experimental setup and data collection can be found elsewhere [10, 11]. Dimensions of the beam were 0.5 by 0.5 microns with a penetration depth of ~ 50 -100 microns (depending on the material of the film and substrate). The Synchrotron Laue diffraction was done with white beam; therefore this method is sensitive to the changes in local lattice orientation and allows to determine the lattice curvature at any probed location. In our experimental setup the microfocusing optics introduce a small ≤ 1*mrad* convergence to the incident beam. This convergence angle has a negligible effect on broadening of the Laue spot (an order of magnitude smaller then the broadening due to lattice curvature and wing tilt observed in the experiment). HRXRD measurements were performed using a Philips X'Pert MRD diffractometer equipped with a four bounce Ge(220) monochromator with the wavelength 1.540598 Å and a three bounce Ge (220) analyzer. FE simulations were conducted using the software FELT. Details of the calculations can be found in Ref. 9.

RESULTS AND DISCUSSION

The white microbeam was scanned perpendicular to the stripe axis, and Laue patterns were recorded at 1000 different points with a spacing of 0.25 µm. Several (usually 10-20) parallel line scans were recorded to obtain a 2D map of the sample. Due to the small size of the microbeam ($0.5 \mu m$) it was possible to obtain separate images from different locations on the column and the wings. Because the high-energy x-ray beam passes through the GaN and the AlN layers, the SiC substrate contributes to the observed Laue patterns. A typical line scan moves the sample under the beam in the X axis which exposes different parts of the column, wing and substrate structure. Two Laue reflections belonging to the SiC and one GaN reflection are shown at Figs. 2 b, c. The SiC reflections do not change position with sample translation. They were used as a reference to determine the position of the GaN $\left(01\,\bar{1}\,8\right)$ Laue reflection. When the probing spot is moved along the positive X axis and first hits the left wing, the GaN reflection appears at the right side of the SiC reflections, as shown in images A of Figs. 2b, c. As the probe spot is moved across the wing to the wing/column interface, the split GaN reflections can be observed for sample 1 (image B of Fig. 2c) – one from the wing and one from the column. When the probe is in the center of the column, the GaN reflections are aligned with the SiC reflections for both samples, as shown (C of Figs. 2 b and c). With further displacement towards the right wing the GaN reflections appear to the left of the SiC reflections (compare image D and E of Figs. 2b and c). The shift of the GaN reflections along with the geometry of the structure (Fig.2 a), clearly indicates that the wings are tilted upward.

An upward-tilt of the wings is also predicted by FE simulations. For the calculations it was assumed that the GaN is under biaxial tensile stress in the c-plane due to the mismatch in thermal expansion coefficients between the GaN layer and the SiC substrate [9]. FE simulations of displacements in the Z directions (Fig. 3 a) for the geometry corresponding to both samples also show that the wing tilt tends to increase with an increasing width-to-height ratio of the column. The WBMD data presented in Fig. 3 b demonstrate that the crystallographic tilt changes quite abruptly at the column/wing interface for sample 1. In sample 2, however, this change is more gradual, and extends over the whole column between the right and left wing. Sample 1 has a width-to-height ratio of the column of 2.60 and a wing tilt of 0.16°, whereas the corresponding numbers for sample 2 are 0.98 and 0.08°, respectively. This finding is also in good agreement with the results of HRXRD. The broadening of the GaN Laue reflections at different locations of the wings and column is analyzed at corresponding positions along the WBDM line scans, as shown in Figs. 2 d, e. The broadening is generally more pronounced for sample 1 with the larger width-to-height ratio of the GaN column. For sample 1 the linescans in Fig. 2 d show that the GaN reflection is particularly broadened at the column/wing interface where it splits into two distinct peaks.

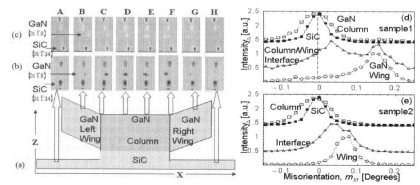

Figure 2: Scheme of the sample geometry (a) together with a series of Laue images showing the $(01\bar{1}8)$ GaN and $(01\bar{1}24)$ SiC spots. The Laue images for sample 2 (b) and 1 (c) correspond to different locations of the probing beam across the stripe: (A) center of the left wing; (B) left wing/column interface; (C) center of the column; (D) column/right wing interface; (E) center of the right wing. Line scans through the $(01\bar{1}8)$ GaN Laue spot as a function of misorientation, m_x, for the column, wing and column/wing interface for sample 1 (d) and sample 2 (e) The SiC line scan is also shown for comparison.

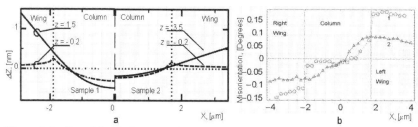

Figure 3: Profiles of the displacement along the Z-axis (a) as determined from FE simulations for sample 1 (left side) and 2 (right side), respectively and relative crystallographic orientation (b) of the GaN c-axis across the stripe with distance from the column center as determined from WBMD analysis for sample 1 and sample 2. The height Z=0 corresponds to the GaN/SiC interface.

It should be noted that it is this sample for which the wing tilt changes abruptly at this interface (Fig. 2 b). The splitting profile shows clearly resolved peaks, which indicates that wings and column have their own well-defined orientation, and that a small angle tilt boundary has formed at the column/wing interface in sample 1. Figure 2 e shows that the crystallographic orientation gradually changes across the stripe for sample 2.

The misorientation between the column and wings at their interface was quantified from the splitting of the GaN Laue spots Using the measured misorientation at the column/wing interface, the spacing between dislocations in the boundary can be derived using the Burgers relation for a small angle tilt boundary: $\vartheta = b / h$. Here ϑ is the measured tilt angle at the interface, b is the Burgers vector modulus and h is the spacing between dislocations in the boundary. The tilt angle at the column/wing interface was always smaller then the total tilt between wing and column (especially for the sample 2). This means that thermal stresses were only partially relieved by the interface tilt boundary and some elastic strain remained. For the measured tilt angle at the interface between column and wing Laue spots of $\vartheta = \vartheta_1 = 0.12^0$ for sample 1 and $\vartheta = \vartheta_2 = 0.04^0$ for sample 2 and for $b = \frac{1}{3}\langle 11\overline{2}0\rangle$ we obtained the distance between dislocations in the tilt interface boundary $h_1=98c$ for sample 1 and $h_2=294c$ for sample 2. Here c is the interatomic distance along the Z axis of the GaN. For a GaN layer thickness of 1.46 μm or 3.53 μm in either sample 1 or sample 2 this amounts to 29 dislocations in the interface tilt boundary for sample 1 versus 23 dislocations for sample 2. The dislocation density in the boundary for GaN layers grown with maskless pendeoepitaxy is much smaller than for layers grown using the mask [6].

In the coalesced films the FE simulation accounted only for extrinsic stress relaxation due to the mismatch in the thermal expansion coefficients between the GaN and the SiC. These simulations predicted almost zero wing tilt due to compactness of a coalesced pendeostructure (Fig.4a, top). However HRXRD measurements of reciprocal space maps[13] and WBMD results (Fig. 4 b) showed the same features as for uncoalesced films: i.e. both wing tilt and a difference in strain between stripes and the wings were observed. If small notches were assumed to form on the bottom side of the coalescence front, the FE simulations indicated a tendency for downward tilt, as schematically shown in the bottom of Fig.4a. However WBMD measurements similar to ones shown in Fig.2 indicated that the wing tilt in coalesced films is upward, as in the case of uncoalesced films. This means that further understanding of the processes of stress relaxation controlling the formation of the wing tilt is necessary.

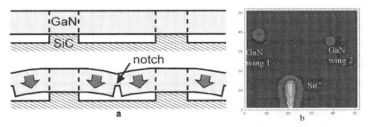

Figure 4. Schetch of the compact coalesced film (a) without wing tilt (top) and with the downward wing tilt, as illustrated by the arrows, due to small notches at the bottom side of the film (bottom); WBMD Laue image with central SiC Laue spot and two distinct Laue spots from two wings demonstrating wing tilt (b).

CONCLUSIONS

X-ray WBMD experiments and FE simulations have shown that the crystallographic tilt of the wings in GaN layers grown on SiC substrates by maskless pendeoepitaxy is upward. The strain, stress and tilt across the stripe increased with the width-to-height ratio. For small width-to-height ratios the tilt is small, and it changes gradually in the region of the column. For larger width-to-height ratios the tilt is higher, and it changes abruptly at the wing/column interface. The evaluation of the local broadening of the reflections and their intensity profiles revealed that the density of defects was reduced and/or the strain was homogenized in the wings as compared to the column. A low-angle tilt boundary is formed at the column/wing interface in samples with larger width-to-height ratio of the GaN column. The estimated dislocation spacing at the column/wing interface is approximately three times smaller in sample 1 than in sample 2. Finite element simulations that accounted for extrinsic stress relaxation provided information that allowed an explanation of the experimental results for uncoalesced GaN layers. Relaxation of both extrinsic and intrinsic stress components in the coalesced GaN layers contribute to the observed wing tilt and formation of sub-boundaries.

ACKNOWLEDGEMENT

The research was supported in part by the U. S. Department of Energy, Division of Materials Sciences and Engineering through a contract with the Oak Ridge National Laboratory. Oak Ridge National Laboratory (ORNL) is operated by UT-Battelle, LLC, for the U.S. Department of Energy under contract DE-AC05-00OR22725. The research at NCSU was sponsored by the Office of Naval Research via contract N00014-98-1-0654 (Harry Dietrich, monitor). Measurements were performed on the Unicat beamline 34-ID at the Advanced Photon Source.

REFERENCES

[1] S.D. Lester, F.A. Ponce, M.G. Craford, and D.A. Steigerwald., Appl. Phys. Lett. **66**, 1249 (1995).

[2] T. Metzger, R. Hopler, E. Born, Phil. Mag. A **77**, 1013 (1998).

[3] S. Einfeldt, Z.J. Reitmeier, and R.F. Davis, J. Cryst. Growth **253**, 129 (2003).

[4] O.-H. Nam, M.D. Bremser, T.S. Tsvetanka, and R.F. Davis, Appl. Phys. Lett. **71**, 2638 (1997).

[5] K. Linthicum, T. Gehrke, D. Thomson, E. Carlson, P. Rajagopal, T. Smith, D. Batchelor, and R. Davis, Appl. Phys. Lett. **75**, 196 (1999).

[6] A. Sakai, H. Sunakawa, and A. Usui, Appl. Phys. Lett. **73**, 481 (1998).

[7] T. Zheleva, S. Smith, D. Thomson, T. Gehrke, K. Linthicum, P. Rajagopal, E. Carlson, W. Ashmawi, and R. Davis, MRS Internet J. Nitride Semicond. Res. **4S1**, G3.38 (1999).

[8] A.M. Roskowski, P.Q. Miraglia, E.A. Preble, S. Einfeldt, T. Stiles, R.F. Davis, J. Schuck, R. Grober, and U. Schwarz, Phys. Stat. Sol. (a) **188**, 729 (2001).

[9] S. Einfeldt, A.M. Roskowski, E.A. Preble, and R.F. Davis, Appl. Phys. Lett. **80**, 953 (2002).

[10] R. Barabash, G. Ice, and F. Walker, J. Appl. Phys. **93**, 1457 (2003)

[11] R. Barabash, G. Ice, B. Larson, G. Pharr, K. Chung, and W. Yang, Appl. Phys. Lett. **79**, 749 (2001).

[12] R. Barabash, G. Ice, W. Liu, S. Einfeldt, A. Roskovsky and R.F. Davis, J. Appl. Phys., **97**, 013504-1 (2005)

[13] S. Einfeldt, D. Hommel, and R.F. Davis, in "Vacuum Science and Technology: Nitrides as seen by the technology", Eds. T. Paskova, B. Monemar, 147 (2002)

Mater. Res. Soc. Symp. Proc. Vol. 875 © 2005 Materials Research Society   O4.11

X-ray diffraction characterization of suspended structures for MEMS applications

P. Goudeau[1], N. Tamura[2], B. Lavelle[3], S. Rigo[4], T. Masri[4], A. Bosseboeuf[5], T. Sarnet[5], J.-A. Petit[4], J.-M. Desmarres[6]
[1]LMP-UMR 6630 CNRS, Université de Poitiers, SP2MI, Bvd Marie et Pierre Curie, BP30179, F-86962 Futuroscope Chasseneuil Cedex
[2]Lawrence Berkeley National Laboratory, 1 Cyclotron Road, Berkeley, CA 94720, USA
[3]CEMES, UPR 8011 CNRS, 29 rue Jeanne Marvig, F-31055 Toulouse Cedex 4
[4]ENIT, avenue d'Azereix, BP 1669, F-65000 Tarbes
[5]IEF, UMR 8622 CNRS, Université Paris XI – Bât. 220, F-91405 Orsay Cedex
[6]CNES, Agence Spatiale Française, 18 avenue Edouard Belin, F-31401 Toulouse Cedex 4

ABSTRACT
Mechanical stress control is becoming one of the major challenges for the future of micro and nanotechnologies. Micro scanning X-ray diffraction is one of the promising techniques that allows stress characterization in such complex structures at sub micron scales. Two types of MEMS structure have been studied: a bilayer cantilever composed of a gold film deposited on poly-silicon and a boron doped silicon bridge. X-ray diffraction results are discussed in view of numerical simulation experiments.

Keywords: MEMS, residual stresses, failure, X-ray diffraction, micrometer scale

INTRODUCTION

Residual stresses can occur in micro devices during process and operation, giving rise to damage and even to failure. These thermo mechanical stresses are difficult to predict from empirical methods and thus experiments and numerical simulation have to be done together in order to improve our understanding of mechanical interaction at meso scales between materials present in these complex systems. Numerous mechanical testing studies have been achieved concerning MEMS structures [1-4] but only a few are dedicated to physical methods such as micro Raman [5], photoluminescence [6] or micro X-ray diffraction [7-8] for probing residual stresses or strains at a micro scale.

In this paper, micro x-ray diffraction technique (µXRD) has been used for studying residual stresses and microstructure at a micron scale in a sub micron thick gold film deposited on a polycrystalline silicon cantilever with micrometer dimensions. Changing white to monochromatic x-ray beam allows grain selection measurement in the gold layer according to their size. The obtained results are compared to macro mechanical measurements and analytical and finite element simulations performed on the same systems. In addition, the technique has been used for studying (100) Si single crystal bridges doped with boron atoms. First results concerning XRD feasibility are given.

EXPERIMENTAL

Gold coated silicon cantilever: A chip containing test structures such as suspended structures were fabricated according to a complex procedure of micro electronic technologies called CHRONOS [9]. The cantilever is composed of a 0.52 micrometer gold top layer on a poly-silicon bottom layer, 1.5 micron thick. The dimensions of the beam are 20 µm large and 100 µm long. Scanning Electron Microscopy (SEM) has been used for characterizing the structures to be analyzed and to identify markers that should be used for calibrating the x-y sample stage (fig. 1).

Si(B) bridges: The Boron doped Si bridges have been elaborated at the Institut d'électronique Fondamentale (IEF) at Orsay (France) using laser doping and selective etching [10]. The atomic percentage of boron is high (up to 3-5 %) leading then to strong elastic strains in the silicon unit cell [11]. The beam thickness is of about 200 nm; its width is between 4-6 microns and the length of these micro actuators is comprised between 50 and 500 μm (fig. 2).

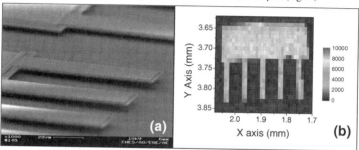

Figure 1: Array of bilayer suspended structure test: (a) SEM image and (b) florescence map of the gold top layer.

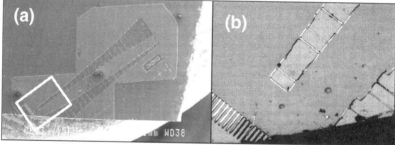

Figure 2: SEM images of the Si(B) micro actuators: figure (b) correspond to zoom of the white rectangle in figure (a). The length of the bridges varies between 50 and 500 μm.

μXRD experiments: White and monochromatic micro X-ray diffraction measurements have been done at the Advanced light source (ALS) at Berkeley (USA) on the 7.3.3 beam line [10] for measuring stress and grain orientation maps on the micro systems shown on figures 1 and 2. The diffraction patterns are recorded with a 2D CCD detector since rotations of the sample are prohibited. Silicon substrate is used for experiment calibration and the incident x-ray angle is fixed to 45° which means that x-rays entirely penetrate the cantilever and the bridge. Data are analysed using XMAS software developed by N. Tamura for extracting the average stresses.

RESULTS AND DISCUSSION

Cantilever: The suspended structure is shown on fig. 1 (a). The deflection of the free extremity due to residual stresses in the gold film is clearly visible. The array has been then imaged using gold fluorescence on fig. 1 (b). This allows for precisely calibrating the x-y sample stage.

The gold film gives a diffraction signal for both the white and monochromatic x-ray beam(fig. 3). Then, we can consider the coexistence of two grain populations in the gold film volume: large grains (size greater than 0.1 μm) visible with polychromatic x-rays and smaller grains (size down to few nanometers) giving rise to rings (isotropic texture) in the diffraction pattern.

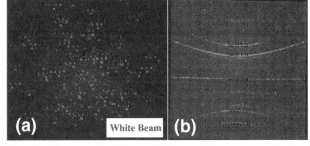

Figure 3: Diffraction patterns of gold layer: (a) indexed Laue pattern obtained with white x-ray beam and (b) diffraction rings typical of a diffracting powder. Poly –silicon contribution is only visible in figure (b) which means that all grains are smaller than 0.1 μm.

The first information that can be extracted from white beam measurements is the grain orientation map and then the pole figure. These results are given in fig. 4. We see clearly the presence of a <111> fiber texture with is not really sharp and we also observe fig. 4 (a) a progressive disinclination of the fiber axis when moving from the fixed to the free extremity of the cantilever which can be correlated with the stress induced curvature. The scan is done with a 1 μm² x-ray beam size and a step of 2 μm.

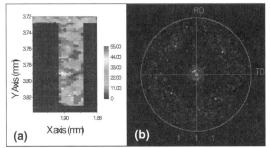

Figure 4: Grain orientation measurements in Gold layer: (a) orientation map giving the angular deviation between (111) direction and surface normal and (b) resulting (111) pole figure.

Grain orientation measurements have also been performed with a conventional XRD apparatus using an x-ray beam of 1 mm² for extracting the three pole figures corresponding to (111), (200) and (220) planes. The results are quite similar indicating a broad distribution of <111> grain fibber axis.

The stress in the region where the cantilever is maintained is approximately equal to 45 MPa (fig 5), a value close to the one determine using a simple mechanical model (Stoney equation) based on the cantilever curvature [9]. The residual stresses are mainly induced by a thermal effect rather than an intrinsic contribution which takes place during the cooling step following the gold film evaporation on the substrate [13]. The polysilicon layer is almost stress free [9]. X-ray diffraction measurements evidence also strong "grain to grain" in plane strain and texture heterogeneities. A similar scan has been done with monochromatic x-ray beam in fig. 6.

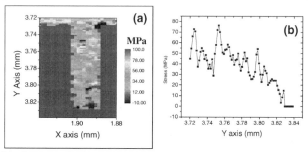

Figure 5: Residual stresses in the top gold layer determine in *larger grains* (Φ>0.1µm) with an x-ray beam size of 1µm^2: (a) principal in plane stress, (b) mean stress along the cantilever Y axis.

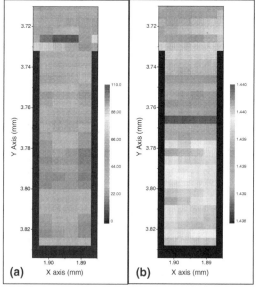

Figure 6: Analysis of the (220) diffraction ring of *small gold grains*: (a) integrated intensity (arbitrary units) and (b) d-spacing (Å). The photon energy is about 6 keV and the scan has been realized with a 5 x 3 µm^2 spot size on the sample surface and a 4 µm step size.

We observe a slight decrease of the intensity when approaching the free extremity of the cantilever and also a weak increase of the d-spacing. From classical XRD measurements, the reference or unstressed d-spacing value was found to be closed to the bulk gold (1.442 A). The calculated stress is then compressive and increases a bit at the free extremity. This is not surprising since the measured (220) planes in this experiment (45° incident angle and 6 keV

124

photon energy) are parallel to the sample surface. In plane tensile stress produces d-spacing contraction in a direction perpendicular to the sample surface.

Stresses in the cantilever have been simulated numerically by using the Finite Element Code ABAQUS in the approximation of elasticity theory since the deflexion of the beam is small (less than 2 microns) with respect to its length (100 microns). The results are given in fig. 7. The stress in the gold film is almost constant along the cantilever despite strong stresses variations close to the edges. The comparison between simulation and XRD measurements indicates that the apparent stress relaxation in large gold grains at the free extremity of the cantilever is may be not representative of the whole gold film volume. The variation in the smaller grains seems to be opposite but we do not know really the volume fraction of each grain populations.

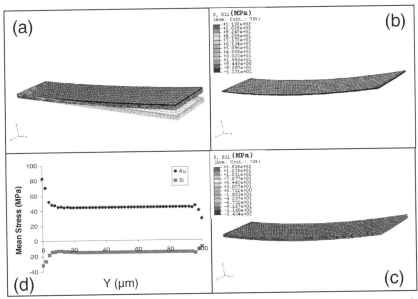

Figure 7: Numerical calculation of stresses in the bilayer suspended structure: (a) perspective view of the deformed cantilever (b) and (c) in plane stress map in the gold and silicon layers respectively, (d) stress profiles along the cantilever axis Y in the gold and silicon layers.

Bridge: The smallest bridge in fig. 2 (b) has been scanned with a 1 μm^2 white beam. The main difficulty here is that the bridge and the substrate structures are identical. However, the two signals may be differentiate on the diffraction pattern shown in fig. 8 (b) owing to the lateral curvature of the bridge which is clearly visible in fig. 8 (a). Observed distortion would result from large strains due to boron atoms which induce Si lattice parameter expansion up to 2.8 %.

CONCLUDING REMARKS

Micro XRD is a well suited technique for measuring strain/stress in complex architectures encountered in MEMS. Switching white to monochromatic x-ray allows selecting and then analyzing different grain populations of the same phase according to their size. In case of the

Si(B) bridge, the Laue signal is shifted because of the (100) plane disorientation with respect to the (100) Si substrate surface. Nevertheless, software developments are needed for extracting quantitative information which will be compared to numerical simulation for improving grain boundary conditions used in modeling.

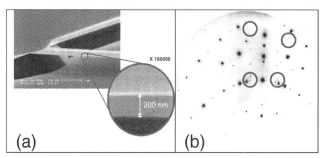

Figure 8: Si(B) bridge: (a) SEM image (b) Laue pattern obtained on the bridge. The red circles indicate the position of diffracting intensities related to the bridge.

ACKNOWLEDGMENTS

The Advanced Light Source is supported by the Director, Office of Science, Office of Basic Energy Sciences, Materials Sciences Division, of the U.S. Department of Energy under Contract No. DE-AC03-76SF00098 at Lawrence Berkeley National Laboratory.

REFERENCES
1. T. Yi and C. – J. Kim, Meas. Sci. Technol. **10**, 706 (1999).
2. J.H. Kim, J. G. Kim, J. H. Hahn, H. Y. Lee, Y. H. Kim, Nanotech **1**, 468 (2003)
3. H. D. Espion, *Microelectromechanical systems and nanomaterials: experimental and computational mechanics aspects* in Experimental Mechanics, 2003, Volume 43, N°3.
4. C. L. Muhlstein, Fatigue & Fracture of Engineering Materials &Structures **28**, 711 (2005).
5. J. C. Zingarelli, M.A. Marciniak and J. R. Foley, Mater. Res. Soc. Symp. Proc. **872**, J3.4.1 (2005)
6. A. Borowiec, D. M. Bruce, D. T. Cassidy and H. K. Haugen, Applied Physics Letters **83**, 225 (2003)
7. M.A. Phillips, R. Spolenak, N. Tamura, W.L. Brown, A.A. MacDowell, R.S. Celestre, H.A. Padmore, B.W. Batterman, E. Arzt and J.R. Patel, Microelectronics Engineering **75**, 117 (2004).
8. G.E. Ice and B.C. Larson, MRS Bulletin **29**, 170 (2004).
9. S. Rigo, P. Goudeau, J.-M. Desmarres, T. Masri, J.-A. Petit and P. Schmitt, Microelectronics Reliability **43**, 1963 (2003).
10. G. Kerrien , T. Sarnet, D. Débarre , J. Boulmer , M. Hernandez , C. Laviron and M. - N. Semeria, Thin Solid Films **453 –454** 106 (2004).
11. G. Kerrien , J. Boulmer , D. Débarre , D. Bouchier, A. Grouillet and D. Lenoble, Applied Surface Science **186**, 45 (2002).
12. N. Tamura, A. A. MacDowell, R. Spolenak, B. C. Valek, J. C. Bravman, W. L. Brown, R. S. Celestre, H. A. Padmore, B.W. Batterman and J. R. Patel, J. Synchrotron Rad. **10**, 137 (2003).
13. N.R. Moody, D.P. Adams, D. Medlin, T. Headley, N. Yang, Materials Science Forum **426-432**, 3403 (2003).

Mater. Res. Soc. Symp. Proc. Vol. 875 © 2005 Materials Research Society

Stress Analysis of Strained Superlattices

R. Peleshchak
State Pedagogical University, 24 Franko str., 82100 Drohobych, Ukraine
H. Khlyap
University of Technology, E.-Schroedinger str. 56, D-67663 Kaiserslautern, Germany, and
State Pedagogical University, 24 Franko str., 82100 Drohobych, Ukraine

ABSTRACT

The latest successful development of smart technologies, in particular, molecular-beam epitaxy technique and pulse-laser deposition method, made it possible to manufacture optoelectronic active elements based on semiconductor materials with sufficient mismatch of the lattice parameters. This problem is of special interest for preparing photosensitive devices with strained superlattices. The paper focuses on the analysis of charge carriers behavior in mechanically strained superlattices based on semiconductor materials from A^2B^6 and A^4B^6 (ZnSe, ZnTe and PbS) playing an important role in the optoelectronics design. Computational modeling is settled on the solution of one-dimensional Schroedinger equation.

INTRODUCTION

The development and improvement of novel growth technologies (in particular, molecular-beam epitaxy) have made it possible to produce extremely high-quality epitaxial interfaces, not only between lattice-matched semiconductors, but even between materials which differ in lattice constant by several percent. Such a lattice mismatch can be accommodated by uniform lattice strain in sufficiently thin layers [1]. The resulting so-called "pseudomorphic" interface is characterized by an in-plane lattice constant which remains the same throughout the structure. These strains can cause profound changes in the electronic properties, and therefore provide extra flexibility in device design. Knowledge of the discontinuities in valence and conduction bands at semiconductor interfaces is essential for the analysis of the properties of any heterojunction.

Our paper aims to analyze photoluminescent and electric field-induced properties of the heterostructures PbS/ZnSe, PbS/ZnTe taking into account strain and deformation phenomenon occurred under the growth of the structures.

EXPERIMENTAL DETAILS

The investigated PbS/ZnSe and PbS/ZnTe heterostructures were grown by the MBE technology of lead sulfide films on the (110)-oriented ZnSe and ZnTe wafers of 1 mm thickness with a substrate temperature $T_s = 540$ K (the vacuum level in the effusion cell was estimated to be about 10^{-9} Torr). The thickness of PbS films was about 1 μm. The mismatch of lattice parameters of contacting materials was estimated to be about 4%. The main parameters of the contacting materials are presented in the Table 1. Some elastic values for the materials are presented in the Table 2.

The first experimental results were obtained under the photoluminescence investigations according to the method [2] in order to clarify the dimensions of the region where the elastic and deformation effects make considerable contribution. The investigated structures (as a model the superlattice PbS/ZnSe was selected) were illuminated from the side of the substrate and from the side of the epitaxial layer (PbS). The photoluminescence in both cases was excited by a light source with $\hbar\omega = 4.42$ eV at 77 K. The dimension of the light source probe in 250 μm in plane and 100 nm in depth.

Table 1._Parameters of the components of the investigated heterostructures (all values are listed at the room temperature)_

Material	Lattice constant, Å	Band gap, eV	Dielectric constant (in units of ε_0)	Intrinsic carriers concentration, cm^{-3}
PbS	5.936	0.41	175	$2 \cdot 10^{17}$
ZnSe	5.667	2.72	9.1	$7.8 \cdot 10^{16}$ - $2.6 \cdot 10^{17}$
ZnTe	6.101	2.26	10.1	$1.7 \cdot 10^{15}$

Table 2. _Some elastic parameters of the contacting materials (all values are in MPa)_

Material	Elastic constant $c_{11}, 10^5$	Elastic constant $c_{12}, 10^5$	Elastic constant $c_{44}, 10^5$	Shear modulus $G, 10^5$	Elastic constant $D^{110}, 10^5$ (see text)
PbS	1.240	0.333	0.248	2.207	0.842
ZnSe	0.826	0.498	0.400	2.340	0.716
ZnTe	0.713	0.407	0.312	1.907	0.751

The results for the heterostructure PbS/ZnSe are shown in the Fig.1.

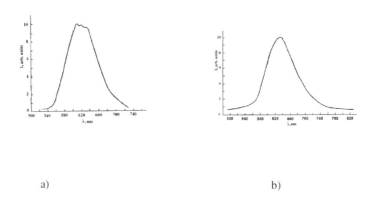

a)   b)

Fig. 1. _Photoluminescence spectra of the investigated heterostructure PbS/ZnSe at 77 K. Panel (a) shows the PL spectrum at the lead sulfide film excitation and the panel (b) demonstrates the emission spectrum at the zinc selenide substrate excitation. The spectral maximums correspond to $\lambda_{max} = 610$ nm and $\lambda_{max} = 632$ nm, respectively_

As it is shown (Fig. 1), the considerable energetic shift of the spectra maximums is occurred due to the built-in strain appeared as a result of the lattice parameters mismatch. The contribution to the strain component ε_{11} is [2]

$$\Delta \varepsilon_{11}^b = \left(\frac{a_s}{a_l} - 1 \right), \tag{1}$$

where a_s and a_l are the lattice constants of ZnSe substrate and PbS layer, respectively. The expression giving below allows to estimate a thickness of the so-called critical layer h_c where the main processes of the electron-deformation interaction take place:

$$h_c = \frac{a_l + a_s}{4\Delta \varepsilon_{11}^b}. \tag{2}$$

The calculation results in h_c = 64 Å.

So, the experimental data on the electric field-induced characteristics of the investigated structures should be analyzed as the processes occurred in the very thin region at the interface between the contacting materials.

We have estimated the strain in the structures under study according to the model-solid theory developed in [1]. Due to the sufficient lattice mismatch of the components we should take into consideration the in-plane lattice constant of the structure $a_{\parallel}$ depending on the elastic properties (see Tables 1, 2):

$$a_{\parallel} = \frac{a_1 G_1 h_1 + a_2 G_2 h_2}{G_1 h_1 + G_2 h_2}, \tag{3}$$

where a_i ($i = 1, 2$) are the equilibrium lattice parameters of the thin film and the substrate, h_j ($j = 1, 2$) are the thickness of the film and substrate, respectively, $G_{1,2}$ are the corresponding shear modules. The in-plane component of the strain tensor [1] reads

$$\varepsilon_{i\parallel} = \frac{a_{\parallel}}{a_i} - 1, \tag{4}$$

and the shear modulus G and the constant D are as follows [1]:

$$G_i = 2(c_{11}^i + 2c_{12}^i)(1 - D_i / 2), \tag{5a}$$

$$D^{110} = \frac{c_{11} + 3c_{12} - 2c_{44}}{c_{11} + c_{12} + 2c_{44}}. \tag{5b}$$

As we consider the case of the thin PbS epitaxial overlayer on the thick (up to 3 mm) monocrystalline substrate ZnSe (ZnTe), this fixes $a_{\parallel}$ = 5.65 Å for PbS/ZnSe and $a_{\parallel}$ = 6.08 Å for PbS/ZnTe structures. No strains are present in the substrates. Thus, the values of the strain are -0.049 and +0.023 for the lead sulfide film in the PbS/ZnSe and PbS/ZnTe structures, respectively. So, the epitaxial lead sulfide film is under the tensile (compress) stress and the experimental electric field-induced characteristics of our superlattices should be treated with accounting the electron-deformation phenomenon contributed in the energy spectrum of the structure [4-5].

RESULTS AND DISCUSSION

The electric field-induced characteristics of the investigated structures were studied at the room temperature. The applied electric field had not exceeded 10^4 V/m. The experimental current-voltage characteristics are shown in the Figs. 2, 3.

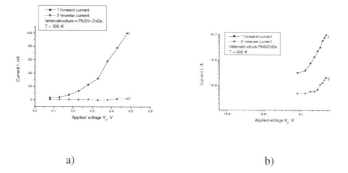

a)   b)

Fig. 2. *Current-voltage characteristics of the PbS/ZnSe structure (a) and double-log recalculation (b).*

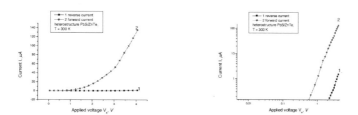

Fig.3. *Current-voltage characteristics of the PbS/ZnTe structure (a) and double-log recalculation (b).*

The forward sections of the experimental current-voltage characteristics are of exponential character. The reverse currents are at least three orders of magnitude smaller than the forward ones. To obtain more detailed information about the carriers transport in the structures the experimental current-voltage characteristics were rebuilt in double-logarithmic scale (Fig.2,b, 3,b). The numerical analysis performed according to the theories developed in [3] demonstrated the ballistic mode of the carriers transport in the superlattice PbS/ZnSe under both directions of the applied electric field and the mobility regime for the structure PbS/ZnTe also under both directions of the applied bias. The qualitative description of the experimental results for the structure PbS/ZnSe can be presented as

$$ j \sim \frac{4\varepsilon\,\varepsilon_0 A_{el}}{9L} \left(\frac{2e}{m^*}\right)^{1/2} V_a^{3/2}, \tag{6} $$

where A_{el} is the electric area of the investigated sample, L stands for the value of the critical region (in our experiments it was estimated to be about 50 Å), m^* takes care of the effective mass of the charge carriers in the PbS thin film. For the PS/ZnTe structure the experimental current is

$$j \sim \frac{9\varepsilon\,\varepsilon_0 A_{el}\,\mu}{8\,L^3}V_a^2, \tag{7}$$

where μ stands for the mobility of the charge carriers in the critical region.

To estimate the contribution of the electron-deformation potential in the electric behavior of the investigated structures [5] we should solve the one-dimensional time-independent Schrödinger equation (8)

$$\left[-\frac{\hbar^2}{2m^*}\frac{\partial^2}{\partial x^2}+V(x,L,b,n)\right]\Psi(E,x)=E\Psi(E,x), \tag{8}$$

where $V(x, L, b, n)$ is the periodical potential of the superlattice dependent on the thickness L of the narrow-gap PbS thin film, thickness of the substrate b and the carriers concentration n in the narrow-gap material:

$$V(x,L,b,n)=\Delta E_0+\Delta E_{mech}(L,b)+\Delta E_{el-d}(x,L,b,n). \tag{9}$$

The first term in the equation (9) describes the band discontinuity of the contacting materials in the non-deformed structure. The second term takes care for the change of the potential energy of the charge carriers due to mechanical distortions of the lattice appeared at the interface because of the lattice parameters mismatch, and the third term stands for the change in the potential energy of the carriers due to the local redistribution of the electron density in the vicinity of the mechanically stressed heterointerface [4 and Ref. therein]. The numerical procedure (see for details [4-5]) gives the values of the additional electron-deformation potential φ_{el-d} up to 0.1 eV.

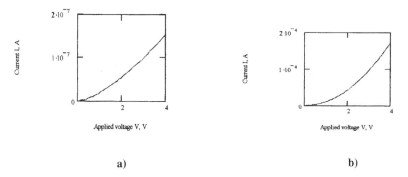

a)   b)

Fig. 4. *Numerically simulated current-voltage characteristics of the investigated structures PbS/ZnSe (a) and PbS/ZnTe (b) taking into account the contribution of the electron-deformation effects (T = 300 K).*

The numerical simulation of the experimental data performed taking into account the contribution of the electron-deformation effects in the electric characteristics of the investigated superlattices produces results illustrated by the Fig. 4. The current-voltage characteristics were calculated according to the more précised expressions:

$$I = T_{tun} \frac{4\varepsilon \varepsilon_0 A_{el}}{9L} \left(\frac{2e}{m^*} \right)^{1/2} \exp\left(\frac{e\phi_{el-d}}{k_b T} \right) V_a^{3/2},$$ (10)

for the structure PbS/ZnSe, and

$$I = T_{tun} \frac{9\varepsilon \varepsilon_0 A_{el} \mu}{8 L^3} \exp\left(\frac{e\phi_{el-d}}{k_B T} \right) V_a^2,$$ (11)

for the structure PbS/ZnTe. Here T_{tun} is the coefficient of tunneling transparency, the calculation of this value is based on the quantum mechanics consideration of the carriers transport in the barrier structures and is not under scope of this paper.

CONCLUSIONS

The strained heterostructures PbS/ZnSe and PbS/ZnTe grown by low-temperature (540 K) MBE technology of thin lead sulfide films on (110)-oriented monocrystalline substrates are investigated. Photoluminescence studies performed under 77 K revealed a so-called critical 50 Å region at the interface where the electron-deformation effects due to the lattice parameter mismatch are the most important. Experimental room-temperature current-voltage dependences measured under applied electric field up to 10^4 V/m showed ballistic and mobility components dominated in the currents flowing through the structures. The numerical simulation of the experimental data with taking into account the effects of the electron-deformation interactions demonstrated a good agreement between the measured and calculated results.

REFERENCES

1. Van de Walle C. G. Phys. Rev. **B39** 1871 (1989)
2. Tomm J. W., Griesche J., Schmidt H. et al. phys. stat. sol. (a) **114** 621 (1988)
3. Kwok Ng. The Complete Guide to Semiconductor Devices. Wiley & Sons, New York, 2002.
4. Peleshchak R., Lukiyanets B., Zegrya G. Semiconductors (Russia) **34** 1223 (2000)
5. Khlyap H., Peleshchak R. phys. stat. sol.(c) **0** 857 (2002)

Mater. Res. Soc. Symp. Proc. Vol. 875 © 2005 Materials Research Society

Experimental and Theoretical Investigations of the Magnetic Phases of Epitaxially Grown EuSe at Low Fields and Temperatures

[1] K. Rumpf, [1] P. Granitzer, [1] H. Krenn
[1] Institut für Experimentalphysik, Universität Graz, A-8010 Graz, Austria

ABSTRACT

Due to its metamagnetic behaviour the magnetic phases of EuSe are more complicated than for the other Eu-chalcogenides, e.g. EuTe. At very low temperatures (below 2K) there occurs an additional antiferromagnetic phase. This behaviour cannot be solely explained by exchange interaction of nearest neighbors and next nearest neighbors but requires more information about the NNSSNN, NSNS and NNSNNS spin arrangements. Because the nearest and next nearest neighbour exchange constants are nearly cancelled out, higher order interaction becomes important and the biquadratic exchange has also to be taken into account. Our measurements were carried out on an epitaxial grown 2.5 micrometers EuSe film on BaF_2. Three regions predicted in the phase diagram of a EuSe bulk crystal are of special interest. For low temperatures (below 1.8K) and low magnetic fields (below 0.05T) a NSNS antiferromagnetic (type II) phase occurs. For higher temperatures from 1.8K to 4.6K and B = 0.1T to 2.5T a NNSNNS ferrimagnetic order exists whereas for lower magnetic fields a further antiferromagnetic (type I) NNSSNNSS-phase is formed above a temperature of 3.6-3.7K. The experimental data show that bilinear mean field calculations taking into account only nearest (J_{NN}) and next nearest exchange (J_{NNN}) interaction fail on this system. Therefore biquadratic exchange (K) has to be included. We present a 12x12 matrix MFA-formalism to extract the proper exchange parameters, which deviate slightly from the bilinear MFA-approach: J_{NN} = 0.165 K, J_{NNN} = -0.1209 K. The biquadratic exchange constant is taken as K = -0.0458 K. We prove this theoretical implication by testing Arrott plots for a varying slope. We made a fit to the experimental data even beyond MFA by taking the corresponding critical exponents of the Heisenberg model.

INTRODUCTION

The Eu-chalcogenides (EuO, EuS, EuSe, EuTe) as magnetic semiconductors and classical Heisenberg antiferromagnets have been investigated intensively long years ago [1]. Despite the fact that the magnetic ordering temperature could not be raised up to temperatures of interest for industrial and commercial applications this substances are still of interest for studying magnetism on a very fundamental basis. Among all the Eu-chalcogenides with their fcc type rock-salt crystal structure EuSe is a special candidate concerning magnetic properties. EuSe has a metamagnetic behaviour and shows very interesting spin structures below the Neel temperature of 4.6 K. The metamagnetism of EuSe can be understood by the near-balance of the nearest (NN) and the next nearest neighbor (NNN) exchange constants, which means $J_{NN} + J_{NNN} \approx 0$ for the 12 coordinating bonds in the fcc-lattice. The nearest neighbor exchange constant has been published [2] as ferromagnetic: J_{NN} = 0.119 K, whereas the next nearest neighbour exchange constant J_{NNN} = - 0.1209 K is antiferromagnetic. We will prove in this work that these couple of bilinear exchange constants are not enough to explain simultaneously all the possible magnetic phases in EuSe. Between temperatures from 1.8 K to 2.5 K the ferrimagnetic NNSNNS phase persists. N, S denote the opposing in-plane magnetic order parameters in adjacent (111) lattice planes. The magnetic field has an in-plane direction and is swept from 0.1 T to 2.5 T. Below the Néel-transition temperature of $T_N = 4.6$ K a type-I antiferromagnetic NNSSNNSS order appears. A second NSNS antiferromagnetic arrangement (type II) is observed at temperatures below 1.8 K and at low magnetic fields not greater than 0.05 T [3].

MEANFIELD ANALYSIS INCLUDING BIQUADRATIC EXCHANGE

The Heisenberg exchange interaction J_{ij} between Eu-4f^7-spins S_i, S_j, including biquadratic exchange K_{ij}, writes as:

$$H = -\sum_{ij} J_{ij} S_i \cdot S_j - \sum_{ij} K_{ij} \left(S_i \cdot S_j \right)^2 \qquad (1)$$

From neutron diffraction it is known that for the cited type-I, II antiferromagnetic structures all spins of a certain (111) lattice plane have the same order parameter N (or S), with the meaning of a fixed magnitude and angular direction within a lattice plane. Therefore we can reduce the indices (ij) of spin sites to indices (pq) of lattice planes after renormalization by the number n of spins within a single plane:

$$\frac{H}{n} = -\sum_p J_1 S_p \cdot S_p - \sum_{p<q} J_2 S_p \cdot S_q - K \cdot \sum_p 6(S_p \cdot S_p)^2 - K \cdot \sum_{p<q} 6(S_p \cdot S_q)^2 \qquad (2)$$

In Eq.(2) $J_1 = 6J_{NN}$ denotes the total ferromagnetic exchange of a test spin within plane p to its 6 NN neighbors, $J_2 = 3(J_{NN} + J_{NNN})$ the ferromagnetic coupling to the 3 NN spins and antiferromagnetic coupling to the 3 NNN spins of the adjacent lattice plane q, respectively. In a simple approximation the biquadratic exchange constant K is not diversified for NN and NNN couplings and can be taken in front of the sum, since we anticipate (from the knowledge by neutron diffraction) that for parallel and antiparallel alignment the value in the brackets is either plus or minus, which yields a unique positive value after taking the square. In mean-field approximation (MFA) the magnetic action on a spin S_p in plane p is figured out by the so-called mean field B_{mf} from the thermal average $\langle S_{p,q} \rangle$ of all neighboring spins, which can be written as:

$$B_{mf}(S_p) = g\mu_B B + 2\sum_{p=NN} J_1 \langle S_p \rangle + 2\sum_q J_2 \langle S_q \rangle + 12K \langle S_p \rangle \underbrace{\langle S_q \rangle^2}_{=1} \qquad (3)$$

At a finite temperature T the thermal average $\langle S_p \rangle$ of spin S_p is given in MFA by the Brillouin $B_S(x)$ function [4], taking S = 7/2:

$$\langle S_p \rangle = S \cdot B_S \left(\frac{S}{k_B T} \left[g\mu_B B + 2\sum_p J_1 \langle S_p \rangle + 2\sum_q J_2 \langle S_q \rangle + 12K \langle S_p \rangle \right] \right) \approx \frac{S+1}{3}(\ldots) \qquad (4)$$

In the latter expression the linearization of $B_S(x) \cong [(S+1)/3S] \cdot x$ for small $\langle S_p \rangle$ near a phase transition has been used. Equation (4) can be mapped to a matrix ("eigenvalue") equation, with rows and colums denoting certain lattice planes. We choose 12 lattice planes by indices $p,q = 1..12$, and concatenate the top and bottom lattice plane by periodic boundary conditions. The external magnetic field is set to $B = 0$:

$$
\begin{pmatrix} \langle S_1 \rangle \\ \langle S_2 \rangle \\ : \\ \langle S_{12} \rangle \end{pmatrix} = \frac{2S(S+1)}{3k_B T} \begin{pmatrix} J_1 J_2 & 0 & .. & & .. J_2 \\ 0 J_2 & J_1 & J_2.. & & .. & 0 \\ : & : & : & : & : .. & 0 \\ J_2 & 0 & .. & & 0 J_2 J_1 \end{pmatrix} \begin{pmatrix} \langle S_1 \rangle \\ \langle S_2 \rangle \\ : \\ \langle S_{12} \rangle \end{pmatrix} + \frac{2S(S+1)}{3k_B T} 12K \begin{pmatrix} \langle S_1 \rangle \\ \langle S_2 \rangle \\ : \\ \langle S_{12} \rangle \end{pmatrix}
\tag{5}
$$

Searching for non-trivial solutions of spin-patterns $\langle S_p \rangle$, the resolving determinant has to be set to zero, which is fullfilled for a certain temperature $T = T_N$, the phase transition temperature. We find the following solutions:

Anticipated exchange parameters:

$J_{NN} = 0.165$ K; $J_{NNN} = -0.1209$ K; $J_1 = 1$ K; $J_2 = 0.1323$ K   K $= -0.0458$ K   (6)

Phase transition temperatures T_N :

NSNS → NNSNNS: $\langle S_p \rangle = (1,-1,1,-1,....,1,-1)$ $T_N = 1.805$ K   (7)
NNSNNS →NNSSNNSS: $\langle S_p \rangle = (1, 1,-1, 1,1,-1,....1,1,-1)$ $T_N = 3.625$ K   (8)
NNSSNNSS → paramagn.: $\langle S_p \rangle = (1, 1,-1, -1,1,1,....,-1,-1)$ $T_N = 4.570$ K   (9)

For experimental convenience we develop the Brillouin-function up to the order x^3, and include also a finite magnetic field $B \neq 0$:

$$
B_S(x) \approx \frac{S+1}{3S} x - \frac{2S^2 + 2S + 1}{30S^2} x^3
\tag{10}
$$

Taking for x the argument of Eq.(4), the solution $\langle S_p \rangle$ for $B \neq 0$ is after lengthy calculations and by using abbreviating constants a,b,c,d:

$$
\langle S_p \rangle^2 \approx \frac{a}{b - cKT^2} \frac{B}{\langle S_p \rangle} - \frac{d \cdot}{b - cKT^2} (T - T_N)
\tag{11}
$$

If $\langle S_p \rangle^2$ is plotted as a function of $B/\langle S_p \rangle$ for a certain temperature, one obtains the so-called Arrott-plot [4] in MFA-approximation. From Eq.(11) it is obvious, that for a non-vanishing biquadratic exchange ($K \neq 0$) the slope of the Arrott-plot depends on the temperature T: One expects for $K < 0$ an increase of the slope for decreasing temperature. Neglecting biquadratic exchange, all Arrott plots at various temperatures would exhibit the same slope with a certain isotherm $T = T_N$ bisecting the ordinate at the origin.

EXPERIMENTS

Figure 1 shows a set of magnetization curves as a function of temperature for magnetic fields up to 2.5 Tesla. The various magnetic phase regimes are denoted by I = NSNSNS and II = NNSNNS, respectively. An exact determination of the Curie-temperature from Fig. 1 is ambiguous, since for thin samples sufficient signal is only obtained for rather high magnetic fields (> 0.02T) with a concomitant rounding of the $M(T)$ characteristics. Therefore we have

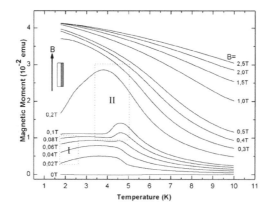

Fig. 1: Magnetization measured by SQUID for in-plane magnetic fields up to 2.5 T and 10 K

fitted Arrott-plots [4] to the data in the temperature region I, taking into account that EuSe belongs (beyond mean-field approximation) to the Heisenberg universality class. Fig. 2 shows for the temperature region I the plot $M^{1/\beta} = a(B/M)^{1/\gamma} - b(T - T_c)$ as straight lines to B-data points (B > 0.05T). It demonstrates the validity of the Heisenberg approach in such moderate magnetic fields. The line which intersects (by extrapolation) the point of coordinate origin designates the $(T = T_c)$-isotherm. We obtain $T_c = 1.8$ K for the NSNS→ NNSNNS (type-II antiferromagnetic-ferrimagnetic) phase transition. Surprisingly the Arrott plots fit the data very well in this T-range, which advises a predominance of the ferrimagnetic phase. The slope of plots from lower to higher temperature is diminished as it was predicted by our MFA model Eq.(11).
Figure 3 shows the field-dependent magnetization for temperatures in the ferrimagnetic regime ($T = 2.1 - 3.7$ K). The NNSNNS-satturation is near 1/3 of the total NNNNN-magnetization for a fully saturated sample (B > 0.3 T). Above 3.7 K the increase of magnetization is non-monotonic with temperature. For instance, at B $\leq$ 0.15 T, the magnetization at $T = 4.5$ K is *higher* than the

one at $T = 3.7$ K. This behaviour indicates the existence of a different phase (NNSSNNSS..) in this temperature regime in accordance with Eq.(8) for $T_N = 3.625$K. This transition temperature is higher than the published value $T_N = 2.8$K for bulk EuSe [1]. Recently published data [5] about strained EuSe/PbSe$_{1-x}$Te$_x$ superlattices on BaF$_2$ show the same trend of an increased T_N ($T_N = 3.25$K), even for nominally unstrained EuSe partial layers at $x = 0.21$.

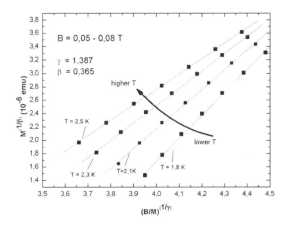

Fig. 2: Arrott plots using the Heisenberg critical exponents in the temperature range T = 1.8 – 2.5 K. The altered slope of the Arrott-plots for increasing temperature is a proof for a sizeable biquadratic exchange interaction.

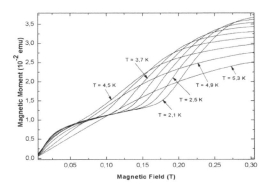

Fig. 3: The magnetization as a function of in-plane magnetic field in the ferrimagnetic (NNSNNS..) regime. The plateau magnetization is 1/3 of the total magnetization (NNNN..). Note the irregularity of the curves for 3.7 – 4.5 K.

DISCUSSION

The occurrence of various magnetic phases in metamagnetic EuSe is challenging for adopting a corresponding model which describes accurately enough the experimental findings. In the literature there exists a controversy about the question whether *bilinear* interactions beyond NN and NNN or a non-linear (*biquadratic*) exchange should be taken into account. In this work we favoured the biquadratic approach and derived a consistent set of three parameters (J_{NN}, J_{NNN}, K). The chosen 12x12 MFA-matrix formalism accounts for the commensurability of magnetic phases NSNS..(2 lattice plane period), NNSNNS..(3 lattice plane period) and NNSSNN..(4 lattice plane period). A smaller rank of the matrix could never reproduce all the observed magnetic phases. The importance of the biquadratic exchange has been also emphasized in the other Eu-chalcogenides (EuO, EuS, EuTe) in Refs. 6,7.

ACKNOWLEDGEMENTS

The support of the work by the Österr. Fonds zur Förderung der wiss. Forschung under grant No. P15397 is gratefully acknowledged, as well as the supply with samples by Dr. Springholz, University of Linz.

REFERENCES

[1] P. Wachter, Handbook on Physics and Chemistry of Rare Earths, edited by K. A. Gschneidner, Jr. and L. Eyring, North-Holland Publishing Company, 1979

[2] T. Janssen, Phys. Kond. Matt. 15, 142 (1972)

[3] J. Schoenes, J. Phys.-Cond. Matt. 15, 707 (2003)

[4] A. Aharoni, *Introduction to the Theory of Ferromagnetism*, Clarendon Press, Oxford 1996, p. 80.

[5] R.T. Lechner, G. Springholz, T.U. Schülli, J. Stangl, T. Schwarzl, and G. Bauer, Phys. Rev. Lett. 94, 157201 (2005).

[6] U. Köbler, R. Mueller, L. Smardz, D. Maier, K. Fischer, B. Olefs, W. Zinn, Z. Phys. B 100, 497 (1996).

[7] U. Köbler, I. Apfelstedt, K. Fischer, W. Zinn, E. Scheer, J. Wosnitza, H. v. Löhneysen, T. Brückel, Z. Phys. B 92, 475 (1993).

Mater. Res. Soc. Symp. Proc. Vol. 875 © 2005 Materials Research Society O4.15

Observation of Micro-Tensile Behavior of Thin Film TiN and Au using ESPI Technique

Yong-Hak Huh[1], Dong-Iel Kim[2], Jun-Hee Hahn[1], Gwang-Seok Kim[1], Chang-Doo Kee[2], Soon-Chang Yeon[3] and Yong Hyub Kim[3]
[1] Environment & Safety Research Center, Korea Research Institute of Standards and Science, P.O. Box 102, Yuseong, Daejon, 305-600, Korea, E-mail: yhhuh@kriss.re.kr
[2] Department of Mechanical Engineering, Chonnam National University
[3] School of Mechanical and Aerospace Engineering, Seoul National University

ABSTRACT

Micro-tensile properties of hard and soft thin films, TiN and Au, were evaluated by directly measuring tensile strain in film tension using the micro-ESPI(electronic Speckle Pattern Interferometry) technique. Micro-tensile stress-strain curves for these films were obtained and the properties were determined. TiN thin film 1 μm thick and Au films with two different thicknesses (t=0.5 μm and 1 μm) were deposited onto the silicon wafers, respectively, and micro-tensile specimens wide 50, 100 and 200 μm were fabricated using micromachining. In-situ measurement of the micro-tensile strain during tensile loading was carried out using the subsequent strain measurement algorithm and the ESPI system developed in this study. The micro-tensile curves showed that TiN thin film was a linear-elastic material showing no plastic deformation and Au thin film was an elastic-plastic material showing significant plastic flow. Effect of the specimen dimensions on mechanical properties was examined. It was revealed that tensile strengths for both films were slightly increased with increasing specimen width. Furthermore, variations of yielding strengths for the thin film Au with change of the dimension were investigated.

INTRODUCTION

As nano/MEMS technology is expected to be a core technology in future, micro/nano materials required for fabricating these electro-mechanical structures have been actively developed. Properties of these materials may be essential in evaluating the reliability of the system, as well as in design, manufacturing process and usage.

To measure mechanical properties of these materials, several test methods, including micro-tensile test, beam bending test, nano-indentation test and bulge test, have been suggested.[1] Strain/deformation measurement in these tests may be required to obtain more reliable properties of the materials. Instead of the conventional techniques like strain gage and extensometer, some novel non-contacting techniques, such as ISDG (Interferometric Strain Displacement Gage)[2], DIC(Digital Image Correlation)[3] and laser interferometry[4,5], etc., were suggested.

ESPI(Electronic Speckle Pattern Interferometry) technique, as a laser interferometry technique, was successively and effectively applied to measurement of the mechanical strain in micro-material.[6] Therefore, in this study, micro-tensile properties of hard and soft thin films, TiN and Au, were measured using the direct micro-tensile testing method and the micro-ESPI(electronic Speckle Pattern Interferometry) technique. Micro-tensile stress-strain curves for these films were obtained and the properties were measured using micro-sized tensile specimens, fabricated by the electromachining process, of TiN and Au thin films. Furthermore, effect of the specimen dimensions, including the specimen thickness and the width, on the mechanical

Figure 1 Typical shape of micro-tensile specimen prepared by electromachining

Figure 2 Photograph of micro-tensile testing system used

properties was examined.

EXPERIMENT

TiN and Au thin film materials, deposited on the silicon wafers by sputtering technique, were used in this study. The micro-sized specimens of 1 μm thick TiN film and Au films with two different thicknesses of 1 and 0.5 μm were fabricated by electromachining process. Figure 1 shows a typical shape of the specimens which were prepared in length of 2 mm and three different widths of 50, 100 and 200 μm.

Micro-tensile tests were carried out using the micro-tensile testing system developed in this study. This system consisted of a micro-tensile loading system and an in-plane tensile strain measuring system, i.e. micro-ESPI system. The load cell installed in the micro-tensile loading system had a capacity of 500 mN and the actuator had a resolution of 4.5 nm in stroke. Figure 2 shows the loading system used. Using the micro-ESPI system, the in-plane tensile strain during micro-tensile testing was continuously measured. Dual laser beams, which were generated from He-Ne laser and were symmetric with respect to the observation direction, with the same intensity were exposed on the surface of the tensile specimen. The speckle pattern images produced by the simultaneous illumination were focused on the screen of the CCD camera. The images were sequentially taken in real time through a frame grabber and digitally stored. Using the subsequent strain measurement algorithm[6], the in-plane tensile strain values corresponding to each deformed states of the specimen can be determined, and these values are calculated, counting the number of fringes over the test-section or measuring the spacing between fringes, according to the following:

$$\varepsilon = \frac{\lambda}{2D_p \cos\theta} \tag{1}$$

, where λ and θ represent the wave length of laser light source (632.8 nm for He-Ne laser) and incident angle of dual illumination lights, respectively, and D_p means spacing between two fringes. The resolution of the average strain, which could be determined by counting the number

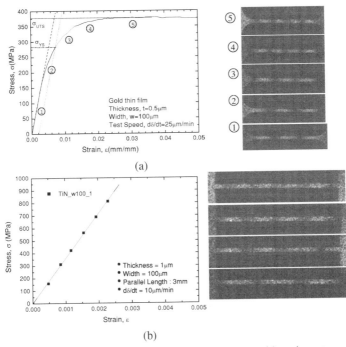

(a)

(b)

Figure 3 Typical stress-strain curves and ESPI fringe patterns with various stress levels (a) for Au thin film and (b) TiN thin film

of the fringes over the gage section of the specimen 2mm long, was $\lambda/4\cos\theta$. The resolution can be improved by introducing the phase shift techniques.

Using the cameras installed in micro-tensile loading system, the specimen was aligned and installed on the grips with UV adhesives. Tensile test were carried out in the displacement control, where 10 and 25 μm /min were selected for TiN and Au thin film, respectively.

RESULTS AND DISCUSSION

Micro-tensile stress-strain curves

Using micro-tensile specimens with various dimensions, micro-tensile tests were carried out with the in-plane tensile strain measurement by micro-ESPI technique. Figure 3 shows an example of micro-tensile stress-strain curves and fringe patterns, corresponding to various stress levels, for hard TiN and soft Au thin films. As shown in figure 3, fringe patterns normal to the loading axis appear clearly to be parallel with increment of tensile stress. This means that the tensile stress is uniformly distributed over the cross-section normal to the loading axis. The parallel fringe patterns like these patterns were observed for all thin films tested in this study.

From the strain calculated from these fringe patterns and the stress corresponding to the strain, stress-strain curves for the two thin film materials could be determined. These curves shown in

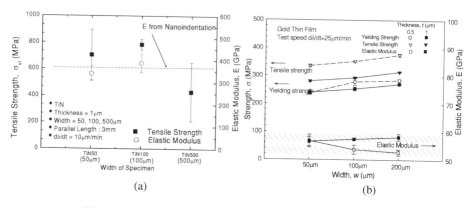

(a)   (b)

Figure 4 Micro-tensile properties measured in different widths (a) for TiN thin film and (b) Au thin film

figure 3 indicates that TiN film is a linear elastic material showing no plastic deformation, while Au thin film is an elastic-plastic material showing significant plastic flow. Therefore, only tensile strength and elastic modulus can be defined in the tensile test for TiN film and, for Au film, yielding strength (0.2% offset), tensile strength and elastic modulus can be defined.

Micro-tensile properties : Dependency of width and thickness of the specimen

From the stress-strain curves obtained using micro-ESPI technique, micro-tensile properties for TiN and Au thin films can be determined as shown in figure 4. Furthermore, figure 4 shows the dependency of the mechanical properties on the specimen width. In this study, for TiN film 1 μm thick, micro-tensile specimens with three different widths, i.e. 50, 100, 500 μm, were prepared and, for Au thin film, those with three different widths, i.e., 50, 100, 200 μm, and two different thicknesses, i.e. 0.5 and 1 μm , were used. For the TiN film 1 μm thick, elastic modulus and tensile strength for 50 μm wide specimens were obtained as 338 GPa and 705 MPa, respectively, and those for 100 μm wide specimens were 387 GPa and 784 MPa, respectively. From these results, it is revealed that mean values of respective tensile strength and elastic modulus are increased slightly with increasing specimen width, but respective values are nearly identical if the error bound of those values are considered. However, the width dependency for the relatively soft thin film, Au, could be more clearly observed in figure 4(b). For Au thin film 0.5 μm thick and 50 μm wide, values of the yielding strength and tensile strength were obtained as 243 MPa and 334 MPa, respectively, while those values for the wider Au film specimens were slightly greater than those for the specimen 50 μm wide. Similar trend in the yielding strength and tensile strength was also shown for the thicker Au thin film. Therefore, it can be said that the yielding strength and tensile strength for TiN and Au thin films show a dependency of the specimen width. However, the variation of those values with the specimen width can be compared to the result obtained from membrane deflection tests[7], where these strengths decreased with increasing width.

Values of elastic modulus for Au thin film 0.5 μm thick and 50 μm wide ranged from 51 to 59

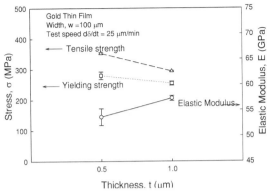

Figure 5 Micro-tensile properties measured in different thickness of the specimen for Au thin film

GPa for all the specimens tested in this experiment. These values are nearly identical to 53 ~ 55 GPa obtained by Espinosa and Pronrok[7], and are within the range of values, 30 ~78 GPa, reported by Nix.[8], while these are quite lower than elastic modulus of 78 GPa for bulk Au. As shown in figure 4 (b), the respective values of the elastic modulus for the specimen 100 and 200 μm wide are not greatly different, while those are a little lower than that for the specimen 50 μm wide. However, for Au thin film 1 μm thick, significant variation of elastic modulus with increment of the specimen width is not observed.

Effect of the specimen thickness on the mechanical properties is presented in figure 5, where Au thin films with two different thicknesses of 0.5 and 1 μm were used. Figure 5 shows the typical thickness dependency for Au film 100 μm wide, where the micro-tensile tests were carried out under the displacement control of 25 μm/min. As shown in figure 5, the yielding strength and tensile strength decreased with increasing thickness. This trend is quite similar to the result shown by Espinosa and Pronrok [7]. In general, the thickness of the deposited layer may be related to the grain size of the layer. Therefore, it can be said that decrease of the strength with increase of the thickness may be closely related to the coarsened grain size of the layer. Contrary to the trend in strength, the elastic modulus is found to be increased slightly with increasing thickness.

SUMMARY AND CONCLUSIONS

For hard and soft thin films, TiN 1 μm thick and 1 and 0.5 μm thick Au films, micro-tensile tests were carried out using the micro-tensile loading system and the micro-ESPI(electronic Speckle Pattern Interferometry) system developed in this study. Micro-tensile specimens wide 50, 100 and 200 μm for these films were fabricated using micromachining. The subsequent strain measurement algorithm and the ESPI system developed in this study were used in in-situ measurement of the micro-tensile strain during tensile loading. From these measurements, micro-tensile stress-strain curves for these films were obtained and the properties were measured. The micro-tensile curves showed that TiN thin film was a linear elastic material showing no plastic deformation and Au thin film was an elastic-plastic material showing significant plastic flow. It was revealed that 0.2% yielding strength and tensile strength for both films were slightly increased with increasing specimen width, while these strengths decrease slightly with increasing specimen thickness. Contrary to this variation, the elastic modulus for Au thin film was found to be increased with increasing specimen thickness.

ACKNOWLEDGMENTS

The authors thank the support of 21C Frontier R&D Programs, *Development o, Nanostructured Materials Technology*, and R&D program of KRISS.

REFERENCES

1. Mohamed Gad-el-Hak: *The MEMS Handbook* (CRC Press, 2002)
2. W.N., Sharper, Jr., B. Yuan, and R.L., Edwards, *J. microelectromechnical systems* 6, 193 (1997).
3. M.A. Sutton, W.J. Wolters, W.H. Peters, and S.R. W.F., Image Vision Computing 1, 133 (1983)
4. D.T., Read, *Mesaurement Science and Technology*. 9, 676-685 (1998).
5. H.D. Espinosa, and B.C. Prorok, M. Fisher, in *Proc. of the SEM Annual Conf.* (2001) p. 446-449.
6. Y.-H. Huh, D.I. Kim, D.J. Kim, P. Park, C.D. Kee, and J.H. Park , *Key Eng. Materials* 279, 744 (2004)
7. Espinosa, H.D.and Pronrok, B.C, *J. of Mater. Sci.*, 38, 4125-4128 (2003)
8. Nix, W. D., *Met. Trans. A 20A*, 2217 (1989)

Mater. Res. Soc. Symp. Proc. Vol. 875 © 2005 Materials Research Society   O4.18

Analysis of Film Residual Stress on a of 4-point Bend Test for Thin Film Adhesion

Sassan Roham, Timothy Hight
Dept. of Mechanical Engineering
Santa Clara University, Santa Clara, CA.

ABSTRACT

The four-point bend (4PB) test has emerged as a method of choice in semiconductor industry for obtaining bimaterial interface adhesion data. When measuring the interface adhesion using 4PB test, it is essential to obtain a crack through the interface of interest. The deposited films, however, posses intrinsic and extrinsic stresses which affect the ratio between energy release rates for interface cracking and crack penetration. Crack penetration and deflection at a bimaterial interface and the role of residual stress has been broadly studied before. However, the results are based on asymptotic analysis regarding interface between two semi-infinite half spaces, where the results do not directly account for boundary conditions and finite size effects of an actual test specimen. In this paper, we look at the role these residual stresses play on the competition between deflection and penetration energy release rates of a bimaterial interface and the extent of which the previous assumption of two semi-infinite media can be accepted.

INTRODUCTION

In current IC devices, interfacial adhesion of a bimaterial interface is a key reliability issue of the system. During manufacturing, operation or deposition, intrinsic and extrinsic stresses imposed on multiplayer films can cause degradation, delamination and ultimately failure of the device. Many tests have been developed that measure the adhesion energy of a bimaterial interface. Several articles have reviewed these test methods and detailed discussions are presented regarding mechanics of thin films, interface adhesion and advantages and disadvantages of each test. The most recent article is by Volinsky et al., [1], where they review adhesion tests based on their classifications and discuss some theoretical models. Bagchi & Evans, [2], also review several of these test methods, stating their respective advantages and disadvantages.

In this paper, we first discuss the theoretical framework for calculating the stress intensity factors for this test configuration. This is followed by a discussion on the role of residual stress on interface crack penetration and crack deflection at a bimaterial interface. Finally, we present the results from the literature and put our work in a suitable viewpoint with concluding remarks.

BACKGROUND

A typical 4PB specimen is shown in Figure 1a. In a 4PB test configuration a vertical pre-crack is notched on the top substrate. During initial loading of the specimen, the notch breaks placing the vertical crack past the top substrate. A desired crack configuration is when the vertical crack deflects along the interface of the film stack. Figure 1b) shows a typical 4PB film stack, which is the subject of the study here. The interface critical energy release rate, G_c, is the energy per unit increase in crack area required to fracture the interface thereby creating two free surfaces. According to Rice [3] the critical energy release rate can be related to remote stresses through the J-integral:

$$G = J = \oint_{\Gamma} (w\,dy - p_i \frac{\partial u_i}{\partial x} ds) \qquad (1)$$

where u is the displacement vector, y is the distance along the direction normal to the plane of the crack, s is the arc length along the contour, p is the traction vector and w is the strain energy density. For any material characterized by linear or nonlinear elastic behavior J is independent of the path Γ taken to compute the integral.

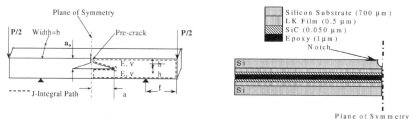

Figure 1. a) 4-Point bend test specimen. b) Cross-section of sandwich film stack.

The framework for the analysis for short cracks and the applicability of the classical formulas that ignore film thickness in computation of adhesive fracture energy from 4PB test data has been studied in [4]. Also discussed in [4], are critical studies and theoretical validation of results for near singular crack tip fields. Furthermore, [5] explored both singly deflected as well as doubly deflected crack configurations and the role of film thickness in obtaining a successful test. We further extend the work in [5] to include the effect of residual stress.

Dimensionless Residual Stress & Stress Intensity Factors

For a bimaterial interface, the residual stresses can be shown as in Figure 2. He-Hutchinson [6] define two nondimensional residual stress values:

$$\eta_n = \frac{\sigma_n a_d^\lambda}{k_1} \text{ , and } \eta_t = \frac{\sigma_t a_p^\lambda}{k_1} \tag{2}$$

where a is the crack length at either the interface, a_d, or through the adjacent material, a_p. λ is the stress singularity exponent for the main crack and k_1 is a "stress intensity like" factor proportional to the applied far field stresses. Knowledge of k_1 requires solution for the main crack problem at the tip of biomaterial interface with the prescribed boundary conditions.

The main crack is anticipated to advance in one of the ways shown in Figure 2. It either penetrates through the interface or deflects along the interface as a single or double crack.

Figure 2. Crack propagation or Deflection and residual stresses.

In case of crack penetrating through the interface the loading is pure mode I. Per He-Hutchinson [6], the stress intensity factor is given as:

$$K_I = c(\alpha)k_1 a_p^{1/2-\lambda} + h(\alpha)\sigma_t a_p^{1/2} \tag{3}$$

where c and h are dimensionless functions of Dundurs Parameters [7], α and β. For deflected cracks, the loading is combined mode I and II. Again per He-Hutchinson [6], the stress intensity factor is a complex number, which is given as:

$$K_I + iK_{II} = d(\alpha)k_1 a_d^{1/2-\lambda} + g(\alpha)\sigma_n a_d^{1/2} \tag{4}$$

where d and g are dimensionless complex functions of α and β. Tabulated values of c, h, d and g are given in He-Hutchinson [6]. Experience has suggested that from the two parameters, α and β, the dependence on α is the strongest one. Therefore, we take $\beta=0$ for all our calculations. The residual stress, σ_n has no effect on K_I, since it acts parallel to the advancing crack. Similarly, the residual stress σ_t has no effect on K_I and K_{II} since it acts parallel to the advancing crack.

Stress Intensity Factors and Mode Mixity Derived from Interaction Integral

As seen in equations (3) and (4), obtaining either stress intensity factor requires the knowledge of k_I, or vice versa. Charalambides, et al. [8] developed the so called "mixed-mode specimen" shown in Figure 1a, where the top and bottom plates are different materials, to measure the fracture resistance at a bimaterial interface. They developed a finite element approach to portray trends in stress intensity factors, energy release rates based on specimen dimensions, elastic properties, and crack length.

Using Green's functions derived for the plane elastostatics problem of a dislocation in a bimaterial strip, Ballarini and Luo [9] have published stress intensity factors for several combinations of parameters, describing various loading, geometry and elastic mismatch conditions. Of particular interest they have a 4PB specimen with different upper and lower plates and with a vertical crack resting at the bimaterial interface.

In our research, the 4PB specimen has the same material in both upper and lower plates. The thickness of the film stack that is enclosed by these upper and lower plates is significantly smaller. We can derive stress intensity factors (and Mode Mixity) from the use of the interaction integral. Note that in a 4PB test, if the deflected crack is large, K_I and K_{II} are present simultaneously.

For a general mixed-mode stress state, Irwin's equation gives a relation between the J-integral and the stress intensity factors:

$$J = \frac{(1-v^2)K_I^2}{E} + \frac{(1-v^2)K_{II}^2}{E} \tag{5}$$

Deriving J-Integrals per Equation (1) using superposition of loads and the Interaction Integral for the Figure 1a loading conditions yields the stress intensity factors:

$$K_I = \frac{M}{2}\sqrt{\frac{12}{h^3}}, \; K_{II} = \frac{3M}{2}\sqrt{\frac{1}{h^3}} \tag{6}$$

and the phase angle, ψ, becomes

$$\Psi = \tan^{-1}\left(\frac{K_{II}}{K_I}\right) = \frac{3}{\sqrt{12}} = 40.9° \tag{7}$$

Equations 6 – 7 give stress intensities and the phase angle for the four-point bend configuration of Figure 1a provided that the horizontal crack is long, i.e., on the order of specimen thickness.

In the 4PB configuration of Figure 1a when the notch breaks and the vertical crack rests on the interface of interest, i.e., low-k/SiC in this case, the competition between deflection and penetration energy release rates start. In case of deflection, the initial crack size is small $(a_d << h)$ and the loading state is still primarily mode I. Therefore, in calculating our dimensionless residual stress values of equation (2) we only need K_I, which can be used to obtain k_I, and subsequently η_t and η_n. K_I is calculated from published data for a standard beam in 4PB configuration, Tada, Paris & Irwin [10].

FINITE ELEMENT MODELING

The purpose of this paper is to study the effect of film residual stress on the competition between the penetrating energy release rate, G_p, and deflecting energy release rate G_d and how this ratio relates to material properties, namely the crack will tend to grow along the interface (deflect) if

$$G_d/G_p > G_{dc}/G_{pc}$$
(8)

Otherwise the crack will tend to grow through the interface into the adjacent material (cohesive failure). The approach here is to compare the energy release rates for a virtual crack is allowed to grow along the interface with the energy from a crack growing into the adjacent material. For both of these conditions the governing equation is given by Irwin [11]:

$$G=dU/da$$
(9)

where dU is the change in strain energy of the body when the crack is allowed to grow a distance da. According to fracture mechanics theory, the crack will tend to deflect along the interface if condition of equation (8) is met.

For the actual modeling of the films and their adhesive and cohesive behavior, 2-D plane strain models of several sandwich film stacks were analyzed using ABAQUS solver [12].

Figure 1b shows the ½–symmetric model of the film stack that was analyzed. The interface of interest is between the low k film and the barrier. The crack is assumed to have traveled vertically through the top silicon wafer to the interface of interest through the application of constant applied load equivalent to the test specimen. The interface fracture is associated with a sharp drop in load that is characteristic of rapid fracture. Once rapid fracture occurs, a dynamic equilibrium is reached meaning that the rate of delamination is maintained at the rate of displacement. Because of initial damage due to rapid fracture, it is important to initiate delamination at the lowest possible load. Deeper notches accomplish this.

He-Hutchinson [6] shows that the critical ratio of the adhesive and cohesive energy release rate for the case of two half spaces bonded together is given as follows:

$$\frac{G_d}{G_p} = \frac{|d|^2 + \eta_n(d\overline{g} + \overline{d}g) + \eta_n|g|^2}{[c^2 + 2\eta_t ch + \eta_t^2 h^2](1-\alpha)}(\frac{a_d}{a_p})^{1-2\lambda}$$
(10)

where c, and h are dimensionless functions of α and β and $d= d_R + id_I$ and $g = g_R + ig_I$ are dimensionless complex functions of α and β. Furthermore, η_t and η_n are non-dimensional residual stresses in tangent and normal directions to the interface that are defined by equation (2) and shown in Figure 2. For the film stack of Figure 1b, the crack tip at the interface of interest is meshed with square elements along the interface and through the adjacent film, thus eliminating the dependence of G_d/G_p on crack (or flaw) size ratio (a_d/a_p) in equation (10). Virtual crack growth method is achieved by releasing nodes one element length at a time to open the crack in vertical and horizontal directions. Energy release rates of several crack sizes are calculated per equation (9).

148

Effect Of Residual Stress

The energy release ratio of G_d/G_p has been computed for a range of the elastic mismatch parameter, α, for $a_d=a_p$, as a function of dimensionless residual stress, η_t. Essentially, η_t is the dimensionless residual stress due to tension or compression in the film. The parameter η_n has not been computed here due to the nature of our problem. Thin films develop tensile or compressive stresses due to reasons discussed earlier. Unlike He-Hutchinson where two semi-infinite materials' interface, the development of normal stresses in thin film is considered negligible compared to tension or compression.

Figures 3 a – c show the G_d/G_p vs. ratio of crack size to adjacent film thickness, for film stacks where $\alpha<0$. The dimensionless residual stress values has been varied from $\eta_t=-0.1$ to $\eta_t=+0.1$. The actual stress values in the films that correspond to this range depend on the elastic mismatch, α. For example for $\alpha =0.8$, $\eta_t=-0.1$ to $\eta_t=+0.1$ produce a stress of -455 MPa to $+343$ MPa in the SiC film. Although this residual stress range represents a sizeable range, the G_d/G_p ratio does not change a great deal.

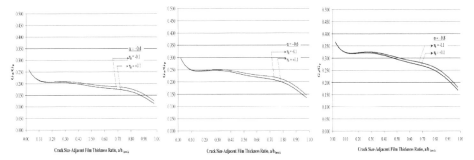

Figure 3. Residual Stress effect on G_d/G_p, a) $\alpha = -0.4$, b) $\alpha = -0.6$, c) $\alpha = -0.8$.

Figures 4 a – c show the G_d/G_p vs. crack size-to-film thickness ratio for film stacks of Figure 1 where $\alpha>0$. The film stack is similar to Figure 1 with barrier-Low-k places reversed. Figures 4 a – c repeat the trend seen for $\alpha<0$. Even though the residual stresses represent a significant range, the effect on G_d/G_p is negligible.

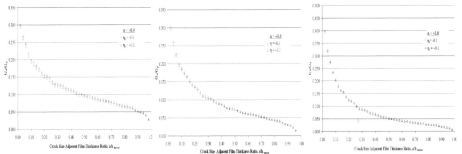

Figure 4. Residual Stress effect on G_d/G_p, a) $\alpha = +0.4$, b) $\alpha = +0.6$, c) $\alpha = +0.8$.

149

SUMMARY AND CONCLUSIONS

Comparing these results to He-Hutchinson shows a similar trend. As η_t becomes negative, G_d/G_p increases indicating better chance of crack deflection at the interface. This is because the film being penetrated is in compression, which impedes crack growth. If the film is in tension it facilitates crack growth. Therefore, films in residual compression, produce a higher G_d/G_p ratio, indicating better change of success, i.e., crack deflection. However, our results show that this trend is more obvious for larger flaw size, while He-Hutchinson did not have crack size (flaw size) dependence on G_d/G_p ratio.

Residual stresses in films represent a form of energy density in the specimen. Since the thickness of the film studies here is extremely small compared to the dimension of the specimen, residual stress in the films do not have significant effect on the G_d/G_p ratio. It is possible that thicker films can store more energy therefore, residual stress within them produce significant strain energy to affect the G_d/G_p ratio. To study the effect of film thickness on adhesive and/or cohesive crack growth more numerical and analytical studies are necessary.

REFERENCES

1. Volinsky, A.A., Moody, N.R., Gerberich, W.W., Acta Materialia Vol. 50, pp. 441–466, 2002.
2. Bagchi, A, Evans, A.G., Interface Science Vol. 3, pp. 169 – 193, 1996.
3. Rice, J.R., J. App. Mech., Vol. 35, pp. 379 – 386, 1968.
4. Roham, S., Hardikar, K., Woytowitz, P., Material Research Society Symposium Proceeding, Vol. 778, pp. 73 – 78, ©2003 Materials Research Society.
5. Roham, S., Hardikar, K., Woytowitz, P., Journal of Materials Research, Vol. 19, No. 10, pp. 3019 – 3027, 2004.
6. Hutchinson, J.W., He, M.Y., Evans, A.G., Int. Journal of Solids Structures, Vol. 31, No. 24, pp. 3443-3455, 1994.
7. Dundurs, J., J. Appl. Mech., Vol. 36, pp. 650 – 652, 1969.
8. Charalambides, P.G., Lund, J., Evans, A. G., McMeeking, R.M., J. Appl. Mech., Vol. 56, pp. 77-82, 1989.
9. Ballarini, R., Luo, H.A., International Journal of Fracture, Vol. 50, pp. 239 – 262, 1991.
10. Tada, H., Paris, P.C., Irwin, G.R., Third Edition, ASME, N.Y., N.Y., ©2000
11. Irwin, G.R., "*Fracture Dynamics,*" in Fracturing of Metals, edited by F. Jonassen et al., American Soc. Of Metals, Cleveland, OH, pp. 147 – 166, 1948.
12. ABAQUS Structural solver, Hibbit, Karlsson & Sorensen, Inc., Pawtucket, RI 02860.

Mater. Res. Soc. Symp. Proc. Vol. 875 © 2005 Materials Research Society   O4.19

Practical Work of Adhesion of Polymer Coatings Studied by Laser Induced Delamination

A. Fedorov, J. Th. M. De Hosson, R. van Tijum, W.-P. Vellinga
Department of Applied Physics, Materials Science Centre and the Netherlands Institute for
Metals Research, University of Groningen, Nijenborgh 4, 9747 AG Groningen, The Netherlands

ABSTRACT

Laser Induced Delamination is a novel technique aimed at measuring the practical work of adhesion of thin polymer coatings on metal substrates. In this technique a laser pulse is used to create initial blisters which initiate further delamination of the film under the blister pressure. A simple elastic model is developed to describe the formation of the blisters. The model predicts the values for the blister height and pressure, which are in fair agreement with the experimental results. In order to account for possible plastic deformations, simulations using a finite element model with a mixed mode cohesive zone were carried out. The calculated stress fields are in agreement with those predicted by the elastic model suggesting that the contribution of plastic deformation to the measured work of fracture is rather limited.

Measurements are carried out on a number of samples, presenting ECCS steel substrate covered with 35 μm thick polyethylene terephthalate (PET) film. The tensile stresses created in the film at the interface required for delamination are estimated at 7-8 MPa, which corresponds to the practical work of adhesion $G = (0.6\pm0.1)$ J/m^2.

INTRODUCTION

Adhesion properties of polymer coatings on steel are of great interest in various industrial applications. However there are not so many experimental techniques present which are capable of providing quantitative characterization of the strength of the polymer-steel interface [1]. Especially problems arise if a coating has a good adhesion and during the testing procedure considerable plastic deformations are introduced [2]. In the present work polymer coatings on steel were studied with the Laser Induced Delamination technique [3-4]. In this technique a laser pulse is used to create initial blisters which initiate further delamination of the film under the blister pressure. A simple elastic model is developed to calculate the stresses in the film, blister pressure and practical work of adhesion. In order to validate the model the stress fields in the film were simulated with a finite element model (FEM) with a mixed mode cohesive zone. The comparison of two approaches is presented in this work.

EXPERIMENTAL

In the Laser Induced Delamination technique a coating under study is subjected to a series of laser pulses with stepwise increase of intensity. Every shot is carried out through a mask resulting in a formation of two parallel cylindrically shaped blisters as shown on the left in Figure 1. The strip of a film between the blisters is not exposed to the laser irradiation and is attached to the substrate in the beginning of each series. Upon increasing the laser pulse intensity, the pressure inside the blisters reaches the critical value, resulting in further delamination of the middle strip as shown on the right in Figure 1. By measuring the blister

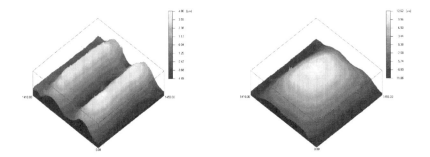

Figure 1. Blisters observed in the confocal microscope. Left: two cylindrical blisters before the delamination. The central, unexposed to the laser region is attached to the substrate. Right: after the delamination takes place and two blisters merge together.

profile and fitting it to the model presented below, stresses in the film and the blister pressure are calculated.

In the experiments we used the infrared Nd:YAG laser NL303HT from EXPLA with the maximum pulse energy of 800 mJ and 5 nanosecond pulse duration. The blister profiles were measured with the stylus Perthometer S2 from Mahr. Samples under study were obtained from CORUS and presented 35 μm layer of PET on ECCS steel substrate.

THIN PLATE MODEL AND FEM CODE

In order to calculate the stresses in the film, blister pressure and other parameters a linear elastic model based on the Kirchhoff assumptions [5] is developed. The model is described in details elsewhere [4] and therefore only the most important equations are summarized below.

A blister of cylindrical symmetry aligned along the x-axis has the following profile:

$$w(y) = 16H\left(\left(\frac{y}{b}\right)^2 - \frac{1}{4}\right)^2, \tag{1}$$

where H is the blister height and b is the blister width. The blister is clammed at the boundaries $y = -b/2$ and $y = b/2$. Blister height depends on the blister overpressure p and the flexural rigidity D:

$$H = \frac{pb^4}{24D}\frac{1}{16}, \tag{2}$$

where $D = Et^3/12(1-\nu^2)$, E is the modulus of elasticity, ν is the Poisson's ratio and t is the film thickness. The stresses σ_x, τ_{xy}, τ_{xz} are equal to zero because of symmetry reasons. The non-zero stresses obtained from the Hooke's law and by integrating the equations of equilibrium are given by the following expressions:

152

$$\sigma_y = -\frac{Ez}{1-v^2}\frac{pb^2}{6D}\left(3\left(\frac{y}{b}\right)^2 - \frac{1}{4}\right),$$ (3)

$$\sigma_z = -\frac{E}{2(1-v^2)}\left(\frac{z^3}{3} - \frac{t^2}{4}z - \frac{t^3}{12}\right)\frac{p}{D} - p,$$ (4)

$$\tau_{yz} = \frac{E}{2(1-v^2)}\left(z^2 - \frac{t^2}{4}\right)\frac{py}{D}.$$ (5)

Upon increasing the blister pressure, the blister can grow through further delamination of the film. The work produced by the gas at constant temperature is given by the change in the Helmholtz free energy $dF=-pdV$. Delamination also causes relaxation of the strain energy of the blister cap U. Thus the condition for delamination is $-(dF+dU)\geq GdS$, where G is the practical work of adhesion or the fracture energy and dS is the elementary change of the blister area. The final expression for the practical work of adhesion is as follows:

$$G = \frac{b^4 p^2}{288D}\frac{(5p_{atm}+2p)}{p_{atm}+2p},$$ (6)

where p_{atm} is the atmospheric pressure. It should be mentioned that delamination which takes place in the Laser Induced blister tests is essentially a crack propagation process. This process is neither reversible and nor likely to proceed through a sequence of equilibrium states. Therefore, the thermodynamic description presented above has to be taken with caution.

The FEM code used in the present calculations is described elsewhere [6]. The polymer-steel interface interaction is described by a rate independent mixed mode cohesive zone model which uses the following form of potential [7]:

$$\phi = \phi_n + \phi_n \exp\left(-\frac{\Delta_n}{\delta_n}\right)\left\{\left[1-r+\frac{\Delta_n}{\delta_n}\right]\frac{1-q}{r-1} - \left[q+\left(\frac{r-q}{r-1}\right)\frac{\Delta_n}{\delta_n}\right]\exp\left(-\frac{\Delta_t^2}{\delta_t^2}\right)\right\},$$ (7)

where Δ_n and Δ_t are the normal and tangential displacement components, respectively. q is the ratio of the normal (ϕ_n) and tangential (ϕ_t) work of separation: $q = \phi_t/\phi_n$. In the simulations the following values were used: $\phi_n = 1$ J/m^2 and $q = 0.95$. $\delta_n = 30$ nm and $\delta_t = 42.4$ nm, are the characteristic lengths and $r = 0.95$ is a parameter that governs the coupling between the normal and tangential separations. The corresponding traction components T_n and T_t are calculated as follows: $T_\alpha = \partial\phi/\partial\Delta_\alpha$ ($\alpha = n, t$). Other material parameters used both in the linear elastic model and FEM simulations are: $E = 2\times10^9$ Pa and $v = 0.33$.

Comparison between the calculations carried out with the linear elastic model and the FEM code is presented in Figure 2. On the left figure the profiles of the top and the bottom surfaces of the film are shown. The simulations with the FEM code are carried out for the blister overpressure of $p = 0.20$ bar. In order to obtain the best fit the blister pressure used in the thin plate model calculations was taken as $p = 0.16$ bar. On the right figure, stresses σ_y are calculated

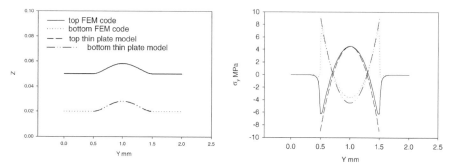

Figure 2. Comparison between the results obtained with the elastic model and the FEM code simulations. Left: profiles of the top and bottom surfaces of the blister. Film thickness used in the simulations is 30 µm. Right: y-component of the stress field calculated for the top and bottom surfaces of the film.

for the top and the bottom surfaces ($z = t/2$ and $z = -t/2$, respectively). At the bottom surface the stress σ_y is compressive at the center of the blister, and tensile at the boundaries. Opposite behavior is observed for the stress at the top surface of the film. The maximum stress in the film is tensile and is achieved at the boundaries ($y = -b/2$ and $y = b/2$), at the interface with the substrate ($z = -t/2$).

Reasonable agreement between the calculations with the thin plate model and the FEM code is observed. As expected, major discrepancy is seen at the boundaries: while the thin plate solution abruptly switches from the maximum value to zero at the clammed boundary, the FEM code provides a smooth solution, with stresses penetrating beyond the blister boundaries.

In Figure 3 the stress fields σ_y and τ_{yz} calculated with the FEM code are presented. As

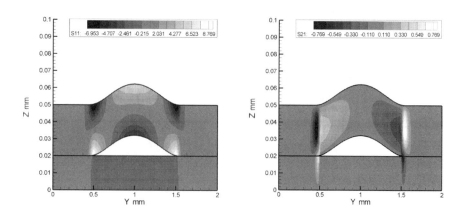

Figure 3. Stress fields σ_y (left) and τ_{yz} (right) calculated with the FEM code.

predicted with the thin plate model the maximum stress in the film is tensile and is achieved at the boundaries of the blister. The main features of the share stress filed τ_{yz} are in line with Equation 5: the field is anti-symmetric with y, and has a quadratic-like behavior in the z direction.

EXPERIMENTAL RESULTS

Two profiles measured with the stylus profiler, one before the delamination and another after, are shown in Figure 4. The profiles are fitted to Equation 1. The agreement between the measurements and the fit proves that the polymer exhibit pure elastic behavior, and the presented elastic model adequately describes the blisters formation process.

Figure 5 presents the evolution of the blister width (left) and the maximum tensile stress σ_y^{max} (right) as a function of the laser pulse intensity. Before delamination the width is calculated as a sum of the widths of both blisters. After the delamination the width of the resulting blister is taken. A steep increase in the blister(s) width indicates the moment of delamination. Just before the delamination stress σ_y^{max} reaches a maximum value of 7-8 MPa. This value corresponds to the stress at which the crack stops to propagate and the blister takes the final shape. By using Equation 6 the practical work of adhesion is estimated: $G = (0.6\pm0.1)$ J/m^2. This value is close to the normal (and tangential) work of separation used in FEM simulations.

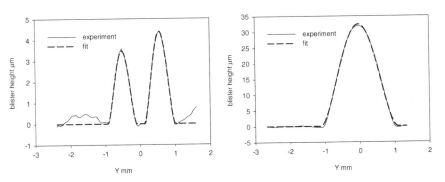

Figure 4. Blister profiles measured with the stylus profiler and fit with Equation 1. Two blisters before the delamination (left) merge together in one blister (right).

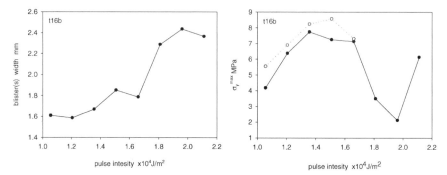

Figure 5. Left: Blister width, before the delamination is calculated as a sum of the widths of two blisters, after the delamination the width of the resulting blister is taken. Right: Evolution of the maximum stress at the blister boundary, σ_y^{max}, with the laser pulse intensity. Before the delamination both blisters are measured, resulting in two points for every laser shot. After the delamination only one blister is present, therefore only one data point per shot is depicted.

CONCLUSIONS

Blister profiles calculated with a linear elasticity model are in fair agreement with those obtained experimentally. The calculations of the stress fields carried out with the thin plate model and a finite element model with a mixed mode cohesive zone are compared. Although the FEM model includes plastic response of the polymer, the stresses created in the blister wall are much lower than the yield stress of PET of 50-60 MPa. That explains a fair agreement between the FEM and thin plate model calculations.

Samples presenting PET film on steel were measured and the tensile stresses created in the film at the interface required for delamination are estimated as 7-8 MPa, which corresponds to the practical work of adhesion $G = (0.6 \pm 0.1)$ J/m^2.

REFERENCES

[1] A.A. Volinsky, N.R. Moody, W.W. Gerberich, Acta Materialia **50**, 441-466 (2002).
[2] I. Georgiou, H. Hadavinia, A. Ivankovic, A.J. Kinloch, V. Tropsa, J.G.Williams, The Journal of Adhesion **79**, 239-265 (2003).
[3] A. Fedorov, A. van Veen, R. van Tijum, J. Th. M. De Hosson, Mat. Res. Soc. Symp. Proc. **795**, U8.61 (2004 Material Reasearch Society).
[4] A. Fedorov and J. Th. M. De Hosson, J. of Appl. Physics, (2005) (in press).
[5] E. Fentsel, Th. Krauthammer, *Thin plates and Shells* (Marcel Dekker 2001).
[6] E. Vander Giessen, Eur. J. Mech., A/Solids **16**,87-106 (1997).
[7] X.-P. Xu, and A. Needleman, Model. Simul. Mater. Sci. Eng. **1**, 111-132 (1993).

Mater. Res. Soc. Symp. Proc. Vol. 875 © 2005 Materials Research Society   O4.21

Strain Relaxation in Si$_{1-x}$Ge$_x$ Thin Films on Si (100) Substrates: Modeling and Comparisons with Experiments

Kedarnath Kolluri[1], Luis A. Zepeda-Ruiz[2], Cheruvu S. Murthy[3], and Dimitrios Maroudas[1]
[1]Department of Chemical Engineering, University of Massachusetts, Amherst, MA 01003, U.S.A.
[2]Chemistry & Material Sciences Directorate, Lawrence Livermore National Laboratory, Livermore, CA 94550, U.S.A.
[3]IBM Semiconductor Research & Development Center, Hopewell Junction, NY 12533, U.S.A.

ABSTRACT

Strained semiconductor thin films grown epitaxially on semiconductor substrates of different composition, such as Si$_{1-x}$Ge$_x$/Si, are becoming increasingly important in modern microelectronic technologies. In this paper, we report a hierarchical computational approach for analysis of dislocation formation, glide motion, multiplication, and annihilation in Si$_{1-x}$Ge$_x$ epitaxial thin films on Si substrates. Specifically, a condition is developed for determining the critical film thickness with respect to misfit dislocation generation as a function of overall film composition, film compositional grading, and (compliant) substrate thickness. In addition, the kinetics of strain relaxation in the epitaxial film during growth or thermal annealing (including post-implantation annealing) is analyzed using a properly parameterized dislocation mean-field theoretical model, which describes plastic deformation dynamics due to threading dislocation propagation. The theoretical results for Si$_{1-x}$Ge$_x$ epitaxial thin films grown on Si (100) substrates are compared with experimental measurements and are used to discuss film growth and thermal processing protocols toward optimizing the mechanical response of the epitaxial film.

INTRODUCTION

Strained Si devices on Si$_{1-x}$Ge$_x$ virtual substrates enhance electron and hole mobility compared to unstrained substrates of the same material [1]. When an alloyed Si$_{1-x}$Ge$_x$ layer is grown on a Si substrate, biaxial strain develops due to lattice mismatch between the substrate and the grown film. Possible mechanisms of strain relaxation include misfit dislocation generation at the film/substrate interface [2] beyond a critical film thickness, as well as film surface morphological transitions [3]. In practice, large numbers of threading dislocations are nucleated which, after gliding a short distance, become immobilized, resulting in a high dislocation density in the film. Device-quality materials, however, need to have a high degree of strain relaxation, low threading dislocation densities, and smooth surfaces. Recently, it has been reported that He ion implantation and subsequent annealing at temperatures (T) over the range 1023 K $\leq T \leq$ 1123 K can result in thin Si$_{1-x}$Ge$_x$ layers possessing a high degree of strain relaxation, as well as relatively low densities of threading dislocations [4].

In this paper, we report a hierarchical approach for computational analysis of the mechanical response of Si$_{1-x}$Ge$_x$ films on Si substrates. We use continuum elasticity and dislocation theory to study the critical thickness of Si$_{1-x}$Ge$_x$ films as a function of the alloy composition in an Si$_{1-x}$Ge$_x$/Si(100) heteroepitaxial system. Subsequently, we employ a phenomenological model to examine the kinetics of strain relaxation during thermal annealing that follows a typical ion implantation process used for the post-growth treatment of heteroepitaxial films. Finally, we examine the role of atomistic simulations in parameterizing consistently the continuum models for critical film thickness calculation and strain relaxation kinetic analysis.

HIERARCHICAL COMPUTATIONAL APPROACH

Our study focuses on theoretical analysis of the deformation mechanics, interfacial stability, strain relaxation kinetics, and surface morphology of strained-layer $Si_{1-x}Ge_x/Si$ heteroepitaxial systems. Toward this end, we have developed a hierarchical approach combining continuum elasticity and dislocation theory with atomistic simulations of structural and compositional relaxation, within a reliable empirical description of interatomic interactions. Special emphasis is placed on the case of heteroepitaxial growth on compliant substrates of finite thickness. The analysis aims at rigorous parameterization of continuum theoretical models for the mechanical response of strained-layer heteroepitaxial systems. A diagrammatic outline of our hierarchical computational approach is given in figure 1.

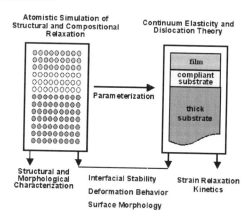

Figure 1. Diagrammatic outline of hierarchical computational approach to study mechanical response of strained-layer heteroepitaxial systems.

CALCULATION OF CRITICAL FILM THICKNESS

First, we consider the case where the epitaxial film of thickness h_f and the substrate of (generally finite) thickness h_s are coherently elastically strained. Taking the equilibrium lattice parameter of the film to be less than that of the substrate, the lattice mismatch results in compressive strain, ε_f, for the film, and tensile strain, ε_s, for the substrate. Assuming uniform deformation, $\varepsilon_{xx,i} = \varepsilon_{yy,i} = \varepsilon_i$ and $\sigma_{xx,i} = \sigma_{yy,i} = M_i\varepsilon_i$, where $i = f, s$ and M_f and M_s are the corresponding biaxial moduli defined as $M \equiv 2\mu(1+\nu)/(1-\nu)$ with μ and ν being the shear modulus and Poisson's ratio, respectively. The condition of zero net force on any atomic plane perpendicular to the interface requires that $\sigma_f h_f + \sigma_s h_s = 0$, which yields $M_s\varepsilon_s h_s + M_f\varepsilon_f h_f = 0$. In addition, the compatibility condition for perfect interfacial coherence requires $\varepsilon_f - \varepsilon_s = \varepsilon_m$, where ε_m is the mismatch strain in the film in the limit $h_s \to \infty$. Solving the mechanical equilibrium and compatibility conditions for the elastic strain in the absence of dislocations, we obtain $\varepsilon_f = \varepsilon_m/[1+\Lambda(h_f/h_s)]$ and $\varepsilon_s = -\varepsilon_m\Lambda(h_f/h_s)/[1+\Lambda(h_f/h_s)]$, where $\Lambda \equiv M_f/M_s$ is the ratio of the moduli of the film and the substrate.

Next, we consider a dislocation with Burgers vector $\{b_x, b_y, b_z\}$ that is introduced within the substrate directed at the film/substrate interface. The work of the stress done on the dislocation can be expressed as $W_1 = M_s \varepsilon_s h_s b_x$ and the self-energy of dislocation, W_2, according to Ref. [5]. The critical condition for the introduction of a misfit dislocation at the interface requires that $W_1 + W_2 = 0$, which yields

$$\frac{[b_x^2 + b_y^2 + (1-\nu_s)b_z^2]}{8\pi(1+\nu_s)b_x} \frac{(h_s + \Lambda h_{f,c})}{\Lambda h_s h_{f,c}} \ln\left[\frac{4h_s h_{f,c}}{b(h_s + h_{f,c})}\right] - \varepsilon_m = 0 \; , \tag{1}$$

an equation that can be solved for the critical film thickness, $h_{f,c}$. The work of stress in the film done on the dislocation is $W = M_f \varepsilon_l h_f b_x - W_2$, which can be rewritten as $W = \tau_{eff} h_f b_x$ to define the effective stress, τ_{eff}, on the dislocation. The resulting expression for the effective stress is

$$\tau_{eff} = \frac{M_f \varepsilon_m}{1 + \Lambda(h_f / h_s)} - \frac{M_s[b_x^2 + b_y^2 + (1-\nu_s)b_z^2]}{8\pi(1+\nu_s)b_x h_f} \ln\left[\frac{4h_s h_f}{b(h_s + h_f)}\right] . \tag{2}$$

The above analysis is a simple extension of the Freund & Nix theory for compliant-substrate heteroepitaxial systems [5], which yields the well-known Matthews-Blakeslee result [6] for infinitely thick substrates. Figure 2 shows the results of the analysis for the dependence on the substrate thickness of the critical thickness of $Si_{1-x}Ge_x$ films on Si substrates for various film compositions, x. The results of figure 2 were obtained by solving the nonlinear algebraic equation for $h_{f,c}$, Eq. (1), numerically for given h_s and x and carrying out a parametric study over a broad range of h_s and x. The beneficial effects of using thin compliant substrates are evident in figure 2, considering the critical film thickness as a typical metric for strain relaxation. To generate the results of figure 2, we have assumed, as a first approximation, that the film properties (lattice parameter and elastic moduli) vary linearly with the Ge composition, analogous to Vegard's law. A more rigorous dependence of the film properties on the Ge composition can be obtained using atomistic simulations (as described below) for coherently strained systems.

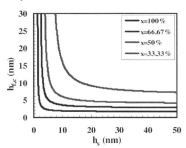

Figure 2. Dependence of the critical film thickness, $h_{f,c}$, on the substrate thickness, h_s, and the film composition, x, for heteroepitaxial $Si_{1-x}Ge_x/Si(100)$ systems.

MODELING OF STRAIN RELAXATION KINETICS

To study the strain relaxation kinetics after the onset of dislocation generation, we have adopted a variant of the phenomenological model proposed by Alexander and Haasen to describe plastic deformation dynamics in semiconductor crystals [7,8,9,10]. In our formulation, the speed of a gliding dislocation, $V(t)$, is given by

$$V(t) = V_o \left[\frac{\tau_{eff}}{\mu_f}\right]^m \exp\left(-\frac{Q_v}{k_B T}\right) , \tag{3}$$

where V_0 is a mobility pre-exponential factor, τ_{eff} is the effective stress, Q_v is the Peierls activation barrier [11], k_B is Boltzmann's constant, T is temperature, and the exponent $m = 2$ for $Si_{1-x}Ge_x/Si$ [9]. The rate of dislocation generation is given by

$$\frac{dN(t)}{dt} = BN(t) \left[\frac{\tau_{eff}}{\mu_f} \right]^n \exp\left(-\frac{Q_n}{k_B T} \right), \tag{4}$$

where B and n are material constants, N is the dislocation density, and Q_n is the activation barrier for dislocation nucleation. The strain relaxation rate in the film is given by Orowan's equation

$$\frac{d\varepsilon_f(t)}{dt} = N(t)V(t)\cos\lambda \tag{5}$$

where λ is the angle between the Burgers vector and the corresponding strain relaxation direction.

The effective stress appearing in Eqs. (3) and (4) is reduced further from the expression of Eq. (2) by subtracting the term $\alpha\mu_f bN^{1/2}$ to take dislocation-dislocation interactions into account; α is a case-dependent numerical constant that can be used as a fitting/adjustable parameter and b is the magnitude of the Burgers vector. In addition, we assume that the deformation remains biaxial as strain relaxation occurs through dislocation formation. The validity of this assumption depends on the surface orientation and is guaranteed for $Si_{1-x}Ge_x/Si(100)$. Finally, in simulating growth experiments we take $dh_f/dt = V_g$, where V_g is the growth velocity and is assumed to be constant.

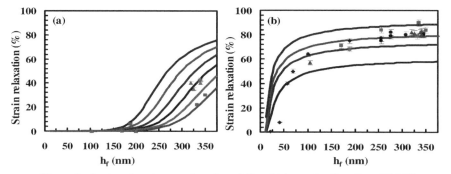

Figure 3. Strain relaxation as a function of film thickness for $Si_{0.80}Ge_{0.20}/Si(100)$ samples annealed after epitaxial growth unimplanted (a) or after He ion implantation (b). The solid curves correspond to the modeling results, while the discrete points correspond to experimental data; different symbols correspond to different experimental conditions [from Ref. 4].

We have integrated Eqs. (3)-(5) to model the thermal annealing experiments reported by Cai, et al. [4], where $Si_{0.80}Ge_{0.20}/Si(100)$ samples were annealed after epitaxial growth either unimplanted or following post-growth He ion implantation. For quantitative predictions, we have used dislocation parameters from the literature [9] and taken the experimental conditions carefully into account. During growth, we have used the elastic equations discussed above to model the deformation mechanics for film thicknesses less than critical. Adjusting the parameter α, we maintained a low dislocation density at the end of the growth process ($N < 10^3$ cm^{-2}). For modeling the annealing of unimplanted samples, the initial values of the dislocation density and film strain were taken equal to the corresponding final values from the kinetic modeling of epitaxial growth. Ion implantation causes substantial dislocation nucleation; therefore, for modeling the annealing of the He implanted samples, we used initial values for the dislocation density higher by orders of magnitude than those obtained from the kinetic modeling of growth. The modeling results and comparisons with the experimental data are shown in figure 3 for annealing of unimplanted samples, figure 3(a), and annealing after sample implantation with He ions, figure 3(b). The agreement of the modeling results with the experimental data is good and comparable to the agreement reported between the data and state-of-the-art discrete dislocation-dynamics simulations for the case of post-implantation annealing [12].

ATOMIC–SCALE ANALYSIS

Atomic-scale simulations based on accurate many-body interatomic potentials can be used to carry out detailed structural and compositional relaxation and calculate the energy and strain in the relaxed state of epitaxially grown films on thin or thick substrates. These calculations can be used to parameterize phenomenological models of mechanical behavior, which can then be used to evaluate compositional grading schemes in order to grow higher-quality films. The key ingredient of our atomistic simulation procedure is a variant of the Monte Carlo (MC) method originally suggested by Foiles [13,14]. The MC simulation is preceded and followed by energy minimization based on a conjugate-gradient (CG) scheme to account for local structural relaxations. In $Si_{1-x}Ge_x$ systems, the combined MC/CG approach minimizes the system energy by distributing the Si and Ge atoms and relaxing the atomic coordinates after the compositional distribution; therefore, it generates the equilibrium configurations resulting from solute segregation at lattice defects, surfaces, and interfaces. Our MC method employs a three-step sequence: (i) one "compositional" MC sweep over all Ge atoms, where each MC step consists of a trial to exchange a Ge atom with a randomly chosen Si atom (chemical identity switching); (ii) many (typically 50) MC sweeps over all atoms, where each MC step consists of a continuous-space atomic displacement trial for "structural" relaxation; and (iii) one MC step for cell-size ("strain") relaxation, consisting of a trial to adjust the cell dimensions in the principal directions normal to the film's free surface. In all three steps, trials are accepted or rejected according to the Metropolis criterion.

We have implemented the above atomistic simulation method to model the relaxation of a prototypical system, consisting of a $Si_{0.50}Ge_{0.50}$ slab with (100) free surfaces. The interatomic interactions were described according to Tersoff's many-body potential [15]. The results are shown in figure 4 for the evolution of the system energy during relaxation (inset to figure 4(a)), the equilibrated Ge distribution in the slab (figure 4(a)), and the final relaxed atomic configuration (figure 4(b)). The results of figure 4 demonstrate the capabilities of the method to capture the Ge segregation at the slab's surfaces.

SUMMARY AND CONCLUSIONS

We have analyzed the energetics of dislocation formation and the strain relaxation kinetics in $Si_{1-x}Ge_x$ epitaxial films grown layer-by-layer on Si(100) substrates. The critical film thickness for the onset of misfit dislocation formation has been calculated as a function of the substrate thickness and the film composition. A phenomenological model that describes dislocation kinetics during growth and annealing also has been discussed and used successfully to interpret experimental data for strain relaxation in $Si_{0.80}Ge_{0.20}$/Si(100) systems. Finally, an atomic-scale method for modeling structural and compositional relaxation in $Si_{1-x}Ge_x$/Si systems

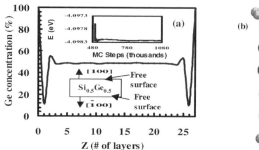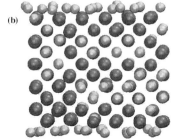

Figure 4. (a) Ge distribution in relaxed configuration of $Si_{0.50}Ge_{0.50}(100)$ slab starting from a random distribution of Ge atoms. The inset gives the slab energy evolution during relaxation. (b) Atomic configuration representative of the slab's equilibrated state. Gold and silver spheres denote Ge and Si atoms, respectively.

toward parameterizing coarse-grained mechanical-behavior models has been presented and demonstrated through a prototypical system consisting of a $Si_{0.50}Ge_{0.50}$ slab with (100) free surfaces; the atomistic simulation has captured the Ge segregation at the slab surfaces.

ACKNOWLEDGEMENTS

Fruitful discussions with K. W. Schwarz and J. O. Chu of IBM's T. J. Watson Research Center, M. R. Gungor, and M. S. Valipa are gratefully acknowledged. This work was supported in part by the National Science Foundation (Award Nos. CMS-0201319 and CMS-0302226).

REFERENCES

[1] Y. J. Mii, Y. H. Xie, E. A. Fitzgerald, D. Monrow, F. A. Thiel, B. E. Weir, and L. C. Feldman, *Appl. Phys. Lett.* **59**, 1611 (1991).

[2] J. H. van der Merwe, *J. Appl. Phys.* **34**, 117 (1963).

[3] D. J. Srolovitz, *Acta. Metall.* **37**, 621 (1989).

[4] J. Cai, P. M. Mooney, S. H. Christiansen, H. Chen, J. O. Chu, and J. A. Ott, *J. Appl. Phys.* **95**, 5347 (2004).

[5] L. B. Freund and W. D. Nix, *Appl. Phys. Lett.* **69**, 173 (1996).

[6] J. W. Matthews and A. E. Blakeslee, *J. Crystal Growth* **27**, 118 (1974).

[7] H. Alexander and P. Haasen, in: F. Seitz, D. Turnbull, H. Ehrenreich (Eds.), *Solid State Physics* vol. 22, Academic Press, New York, 1968, p. 27.

[8] H. Alexander, in: F.R.N. Nabarro (Ed.), *Dislocations in Solids* vol. 7, North Holland, Amsterdam, 1986, p. 113.

[9] D. C. Houghton, *J. Appl. Phys.* **70**, 2136 (1991).

[10] L. A. Zepeda-Ruiz, B. Z. Nosho, R. I. Pelzel, W. H. Weinberg, and D. Maroudas, *Surf. Sci.* **441**, L911 (1999).

[11] J. P. Hirth and J. Lothe, *Theory of Dislocations* (Wiley, New York, 1982).

[12] K. W. Schwarz, J. Cai, and P. M. Mooney, *Appl. Phys. Lett.* **85**, 2238 (2004).

[13] S. M. Foiles, *Phys Rev. B* **32**, 7685 (1985).

[14] P. C. Kelires and J. Tersoff, *Phys. Rev. Lett.* **63**, 1164 (1989).

[15] J. Tersoff, *Phys. Rev. B* **39**, 5566 (1989).

Comparison Between *In-situ* Annealing and External Annealing For Barium Ferrite Thin Films Made by RF Magnetron Sputtering

A. R. Abuzir and W. J. Yeh
Dept. of Physics, University of Idaho
Moscow, Idaho 83844, U.S.A.

ABSTRACT

Due to their large magnetic anisotropy perpendicular to the film plane, barium ferrite thick films ($BaFe_{12}O_{19}$, or BaM) with c-axis orientation are attractive candidates for microwave applications [1,2]. Barium ferrite thin films on silicon substrates without under layer have been deposited under various conditions by RF magnetron sputtering. The structure of the as-grown films is amorphous. External annealing in air has been done at $950^{\circ}C$ for ten minutes to crystallize the films. C-axis oriented thin films with squareness of about 0.87 and coercivity of about 3.8KOe are obtained.

Thick BaM films with c-axis orientation are difficult to achieve with one single deposition. Multilayer technique looks promising to grow thick films [3]. The external annealing process is difficult to incorporate with the multilayer procedure. An *in-situ* annealing procedure has been developed to obtain films, which can be used as the basic component for future multilayer deposition. Barium ferrites are first magnetron sputtered on bare silicon substrates in $Ar + O_2$ atmosphere at substrate temperature of $500-600^{\circ}C$, the deposition pressure was kept about 0.008 torr. After the deposition, the temperature of the substrate is immediately increased to about $860^{\circ}C$ for ten minutes in 140 torr of argon (80%) and oxygen (20%) mixture of gas, which was introduced into the chamber without breaking the vacuum. With the *in-situ* process, c-axis oriented thin films of 0.88 squareness and coercivity value of about 4.3KOe are obtained.

Both annealing methods seem to have the similar effect on the perpendicular squareness and coercivity at various film thicknesses. The average value of the saturation magnetization M_s obtained from the *in-situ* annealing using multilayer technique is higher than that of the external one. We have grown films up to 1.0 micron thickness using the multilayer technique, in which three layers of $0.3 \mu m$ thickness each are deposited until the final thickness is reached. After the deposition of each layer, it was *in-situ* annealed before starting the deposition of the next layer. With the multilayer technique, coercivity of about 3.5 KOe and average value of the saturation magnetization M_s of about 4.0 K Gauss is obtained.

I. INTRODUCTION

Because of their large uniaxial anisotropy, $BaFe_{12}O_{19}$ thin films with c-axis oriented along the film normal have become an attractive candidate for ultra-high-density recording media applications, their high mechanical hardness allow closer head-disk separation while their good chemical stability maintains adequate medium lifetime. Microwave applications have emphasized growth of thick (above $1.0 \mu m$), c-axis oriented; nearly single crystal films for the construction of microwave filters and other devices [4].

RF sputtering [5], pulse laser ablation deposition (PLD) [6], liquid-phase epitaxy (LPE) [7], sol-gel [8], and metalorganic chemical vapor deposition (MOCVD), have been used over the last decade by many research groups to deposit high quality hexaferrite films. Radio frequency (rf) magnetron sputtering is the most popular method to grow BaM films. The difficulty to grow BaM thick films raise from the deterioration of c-axis orientation as the film thickness increase [9]. Many efforts have been employed to improve the c-axis orientation by using underlayers of different kinds and thickness, or by using different doping elements substituted for the Ba or the Fe atoms in the $BaFe_{12}O_{19}$ formula, however the c-axis orientation does not improve much [10,11].

With the rf magnetron sputtering method, the structure of the as-grown films is amorphous, external annealing and *in-situ* annealing can be used to crystallize the films. We have used the rf magnetron sputtering method to grow films of about 1.0µm by using multilayer technique. Both external and *in-situ* annealing has been used to crystallize each layer of the films. In this paper, we will compare between the two methods and study their magnetic and structural properties at different film thickness.

II. EXPERIMENT

Barium rich targets have been made by the standard solid-state reaction after mixing 10% pure $BaCO_3$ with 90% pure $BaFe_{12}O_{19}$ commercial powder. Barium ferrite films were deposited on 10mm×17mm silicon (100) substrates using a rf magnetron sputtering system. A magnetically turbo molecular pump was used to reach 3×10^{-6} torr. The gas pressure during deposition was fixed to about 8×10^{-3} torr using 20% pure oxygen and 80% pure argon. The target was 2.0 inch diameter disk and placed at 6.3 cm from the substrate. The rf power was fixed at 50W, our deposition rate was about 0.1-0.15 µm/h according to VB-250 vase elliposemeter thickness measurement. During the deposition, the substrate temperature was kept between 550-600°C.

Multilayer technique was used to reach the desired thickness, after each thin layer, either external or *in-situ* annealing take place, in the former, the film was heated in air at about 950°C for 10 minutes using an electric furnace. In the later, the films were heated inside the chamber without breaking the vacuum, typically after deposition, the pump was shut down, 140 torr of argon (80%) and oxygen (20%) was introduced into the chamber, then immediately increased the substrate temperature up to 850-900°C for 10 minutes. This procedure continued after each thin layer of BaM until the final thickness is achieved.

III. RESULTS AND DISCUSSION

A. *In-situ* film measurements

The hysteresis loop in perpendicular and parallel directions of the 0.3µm *in-situ* thin film is shown in figure 1 which was grown at 550°C on a silicon substrate with *in-situ* annealing; the magnetic properties of the BaM films were measured by a Vibrating Sample Magnetometer (VSM). The data for the perpendicular loop have been compensated for demagnetizing fields using a demagnetizing factor of 4π. Coercivity was about 4300Oe for the perpendicular and 3700Oe for the parallel direction. Squareness was about 0.87 for the out of plane direction and the average value of the saturation magnetization M_s is about 2400 Gauss.

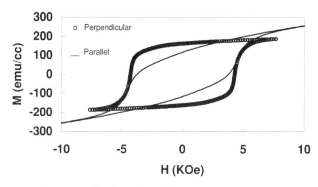

Figure 1. The perpendicular and parallel VSM hysteresis loops for the 0.3 μm in-situ thin film.

We have used x-ray diffraction diagrams XRD to verify the structure of the 0.3 μm *in-situ* thin films which is shown in figure 2. Relatively strong 006 and 008 peaks are observed which indicates good perpendicular c-axis orientation. . The coercivity in the plane direction is probably due to the small (107 and 114) peaks.

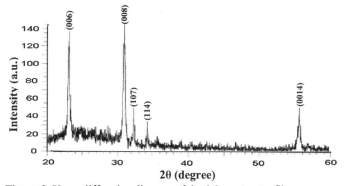

Figure 2. X-ray diffraction diagram of the 0.3 μm *in-situ* film.

In our attempt to grow 1.0 μm films, we have employed the multilayer technique and have used three layers of 0.3 μm thickness each to reach the final value. We have used a superconducing quantum intereference device (SQUID) to study the magnetic properties of the BaM films, which is shown in figure 3. We have noticed that as the thickness increases, the c-axis orientation of the films starts to deteriorate due to the in-plane orientation. The coercivity of perpendicular orientation was decreased from 4300Oe for the 0.3 μm film to 3000Oe for the 1.0 μm ones; perpendicular squareness was also decreased to 0.65, however we have noticed a significant increase in the average saturation magnetization Ms that reached 4000 Gauss for the 1.0 μm films, the. From the value of the average saturation magnetization M_s the *in-situ* method could be good choice to grow thin films, and a promising technique to fabricate thick films by using multilayer technique.

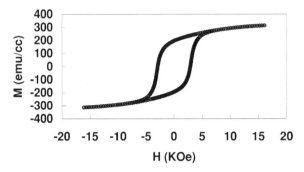

Figure 3. The perpendicular SQUID hysteresis loops for the 1.0 μm *in-situ* multilayered films.

B. External annealing measurements

The magnetization measured as function of the external field H by VSM is shown in figure 4, H was applied perpendicular to and in the plane of the external annealed 0.3 μm film. The Film was grown at similar condition to those of the *in-situ* ones but heated externally in air at about 950°C for 10 minutes. The data in figure 4 were also compensated for demagnetizing field by using a demagnetizing factor of 4π.

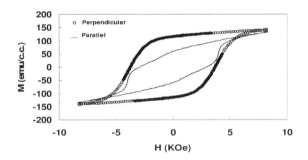

Figure 4. The VSM hysteresis loops for the 0.3 μm external annealed thin films.

The X-ray diffraction for the external annealed film in figure 5 shows slightly better c-axis orientation than the *in-situ* ones. To minimize the in plane peaks (107 and 114) a seed layer could be helpful [12, 13].

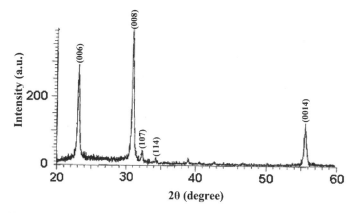

Figure 5. X-ray diffraction diagram of the 0.3 μm external annealed film.

Multilayer technique was employed again to grow thick film of 1.0 μm consisting of three layers of 0.3 μm each, after each layer was deposited; it was heated externally in air at about 950°C for 10 minutes prior to the deposition of the next layer. SQUID has been used to measure the hysteresis loop for the multilayered external annealed film in the perpendicular direction, which is shown in figure 6 below. The perpendicular squareness was about 0.66; coercivity remained about the same around 3000Oe. The average saturation magnetization M_s increased from 1800 Gauss for the 0.3 μm films to about 2600 Gauss for the 1.0μm external annealed multilayered films. In comparing both *in-situ* and external annealing methods, it seems that the average saturation magnetization M_s of the *in-situ* method is better than that of the external annealed one by about 25% at the 0.3 μm thickness, and about 35% higher at the 1.0 μm thickness.

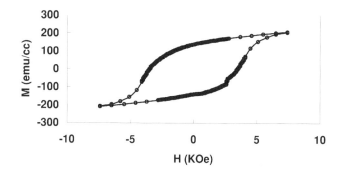

Figure 6. The perpendicular SQUID hysteresis loop for the 1.0μm external annealed multilayered films.

CONCLUSION

We have developed an *in-situ* method to grow films up to 1.0 μm by using multilayered technique. The multilayer average saturation magnetization value M_s was as high as 4000 Gauss, which qualify the multilayer technique to be a possible way to grow thick films using the *in-situ* method. The perpendicular squareness and Coercivity was about the same in both *in-situ* and external annealing methods. Better c-axis orientation in the normal plane was achieved using the external annealing. As the thickness of the film increases in either method, the squareness and the coercivity value decreases.

ACKNOWLEDGEMENT

The authors would like to thank Prof. Y. K. Hong and Mr. S .H. Gee for the VSM measurements. This research was supported by the U.S. NSF grant EPS-0132626 and ONR grant N000140410272.

REFERENCES

1. Sergey V. Lebedev, Carl E. Patton, and Michael A. Wittenauer, J. Appl. Phys. 91, 4426 (2002).
2. S. D. Yoon, and C. Vittoria, S. A. Oliver, J. Appl. Phys. 93, 4023 (2003).
3. S. H. Gee, Y. K. Hong, D. W. Erickson, T. Tanaka, and M. H. Park, J. Appl. Phys. 93, 7507 (2003).
4. I. Zaquine, H. Benazizi, and J. C. Mage, J. Appl. Phys. 64, 5822 (1988).
5. Z. Zhuang, M. Rao, R. M. White, D. E. Laughlin, and M. H. Kryder, J. Appl. Phys. 87, 6370 (2000).
6. S. A. Oliver, S. D. Yoon, I. Kozulin, M. L. Chen, and C. Vittoria, Appl. Phys. Lett. 76, 3612 (2000).
7. S. D. Yoon, and C. Vittoria, J. Appl. Phys. 96, 2131 (2004).
8. M. N. Kamansanan, S. Chandra, P. C. Joshi, and A. Mansingh, Appl. Phys. Lett. 59, 3547 (1991).
9. T. S. Cho, S. J. Doh, J. H. Je, J. Appl. Phys. 86, 1958 (1999).
10. Y. Hoshi, Y. Kubota, and H. I. Kawa, J. Appl. Phys. 81, 4677 (1997).
11. J. Feng, N. Matsushita, K. Watanabe, S. Nakagawa, and M. Naoe, J. Appl. Phys. 85, 6139 (1999).
12. S. G. Wang, S. D. Yoon, and C. Vittoria, J. Appl. Phys. 92, 6728 (2002).
13. Z. Zhuang, M. Rao, D. E. Laughlin, and M. H. Kryder, J. Appl. Phys. 85, 6142 (1999).

Correlation Between Elastic Constants and Magnetic Anisotropy in Co/Pt Superlattice Thin Films

Nobutomo Nakamura[1], Hirotsugu Ogi[1], Teruo Ono[2], Masahiko Hirao[1], Takeshi Yasui[1], and Osamu Matsuda[3]

[1]Graduate School of Engineering Science, Osaka University, Machikaneyama 1-3, Toyonaka, Osaka 560-8531, Japan.
[2]Institute for Chemical Research, Kyoto University, Gokasho, Uji, Kyoto 611-0011, Japan.
[3]Faculty of Engineering, Hokkaido University, Kita 13 Nishi 8, Kita-ku, Sapporo 060-8628, Japan.

ABSTRACT

In this paper, we study contribution of the magnetoelastic anisotropy effect on perpendicular magnetic anisotropy of Co/Pt superlattice thin films. The effective-magnetic anisotropy energy was measured for Co/Pt thin films with a constant Co-layer thickness (4 Å) deposited on silicon substrate by the ultrahigh-vacuum deposition method. As the Pt-layer thickness increases, the effective-magnetic anisotropy energy increases and decreases after showing a maximum at the Pt-layer thickness of 12 Å. This behavior can be explained only by the magnetoelastic anisotropy effect, that is, elastic-strain energy released through the magnetostriction effect with the strain-dependent elastic constants.

INTRODUCTION

Magnetization direction of a ferromagnetic thin film is usually parallel to the film surface because of strong contribution of the magnetic shape-anisotropy effect. However, it can be perpendicular to the film surface for such a system that consists of magnetic-material/noble-metal multilayer thin film, when the thickness of the magnetic-material layer is of the order of a few angstroms. This magnetic property is called perpendicular magnetic anisotropy (PMA) [1,2]. PMA will appear when the total energy of the system takes a minimum with the perpendicular magnetization. The total energy of the system consists of magnetocrystalline anisotropy energy, magnetic shape-anisotropy energy, magnetic interfacial-anisotropy energy, and magnetoelastic anisotropy energy. Considering the magnetocrystalline anisotropy and magnetic shape-anisotropy energies, the magnetization direction of a thin film of Co always lies in the film plane. However, the magnetization direction of Co/Pt superlattice thin films can be perpendicular to the film surface because of the magnetic interfacial-anisotropy energy and magnetoelastic anisotropy energy. PMA can be candidates for ultrahigh-density-magnetic-recording media, and clarification of its mechanism is a long running problem.

For deep understating of the cause of PMA in the Co/Pt system, previous works studied dependence of PMA on the Co-layer thickness [1-5]. However, changing the Co-layer thickness affects both the magnetic interfacial and magnetoelastic anisotropy energies per unit volume of Co, and their contributions to PMA cannot be separately evaluated. In this paper, we study contribution of the magnetoelastic anisotropy energy to PMA by measuring PMA with various Pt-layer thickness, remaining the Co-layer thickness unchanged, where only the magnetoelastic anisotropy energy is changed, depending on the Pt-layer thickness. Therefore, if PMA was independent of the Pt-layer thickness, the magnetoelastic anisotropy energy would hardly affects PMA. However, the degree of PMA changed with the Pt-layer thickness and showed a maximum at the Pt-layer thickness of 12 Å. This behavior is explained by taking account only of the magnetoelastic anisotropy energy with the strain-dependent elastic constants.

MAGNETIC ANISOTROPY ENERGIES

The degree of PMA is evaluated by the effective-magnetic anisotropy energy K_{eff}, which is difference between the total energy of the system per unit volume of Co with the in-plane magnetization and that with the out-of-plane magnetization ($K_{eff}=K^{//}-K^{\perp}$). The total energy K of the system consists of the magnetocrystalline anisotropy energy K_c, magnetic shape-anisotropy energy K_s, magnetic interfacial-anisotropy energy K_i, and magnetoelastic anisotropy energy K_e.

K_c depends on crystal structure and magnetization direction of magnetic material. When Co/Pt superlattice thin film shows PMA, closed-packed planes of Co and Pt layers are parallel to the film surface. Co exhibits hcp structure at room temperature. Then, the c axis of hcp Co is perpendicular to the film surface. The hcp Co shows uniaxial magnetic anisotropy [6], and K_c takes a minimum when the magnetization direction is perpendicular to the film surface.

K_s is related to shape of the magnetic material. When the magnetic material shows an elliptical shape, K_s is calculable [6]. We can assume thin films to be an ellipsoid with infinite aspect ratio. Then, K_s takes a minimum with the in-plane magnetization.

Considering the sum of K_c and K_s, the magnetization direction of Co thin film always lies along the in-plane direction, because K_c+K_s shows minimum with the in-plane magnetization. However, for Co/Pt superlattice thin films, the magnetization direction can be perpendicular to the film surface because of K_i and K_e.

Direction of magnetic spin of an atom is generally determined by interactions with surrounding magnetic spins. The magnetic interfacial-anisotropy originates from the magnetic anisotropy caused by atoms near the interfaces because of the reduced symmetry. K_i basically depends on the number of the interface per unit volume of Co. However, quantitative evaluation of K_i is not straightforward. CoPt intermetallic alloy could occur at the interfaces to contribute to PMA [5]. This contribution also depends on the number of the interfaces and we include this effect in the magnetic interfacial anisotropy.

When Co/Pt superlattice thin film shows PMA, Co and Pt layers are epitaxially bonded on their closed-packed planes; (111) of fcc Pt and (0001) of hcp Co are parallel to the film surface. Then, Co/Pt superlattice thin film possesses as large as 10.7 % lattice misfit at the interface, and significantly large elastic strains occur, which reach in the order of 10^{-2}. K_e is the elastic strain energy in the system per unit volume of Co and it is given by

$$K_e = \frac{1}{d_{Co}} \left(\frac{d_{Co}}{2} S_i^{Co} C_{ij}^{Co} S_j^{Co} + \frac{d_{Pt}}{2} S_i^{Pt} C_{ij}^{Pt} S_j^{Pt} \right). \tag{1}$$

Here, C_{ij} are the elastic constants and S_i the engineering strain. d_{Co} and d_{Pt} denote the thickness of Co and Pt layers, respectively. In bulk Co, an interatomic distance in (0001) plane becomes larger with the magnetization perpendicular to the (0001) plane than that with the magnetization along the (0001) plane because of the magnetostriction effect [6]. Therefore, the perpendicular magnetization decreases the lattice misfit and the elastic strain energy.

DEPENDENCE OF PMA ON PLATINUM-LAYER THICKNESS

We study the effect of the magnetoelastic anisotropy on PMA by measuring dependence of K_{eff} on Pt-layer thickness, keeping Co-layer thickness unchanged. Because K_c, K_s, and K_i depend on the structure, shape of Co layer, and the number of the interface, respectively, they hardly change depending on the Pt-layer thickness. Therefore, the dependence of K_{eff} on the Pt-layer thickness reflects the dependence of K_e.

Co/Pt superlattice thin films were deposited on (001) plane of monocrystal silicon substrates by the ultrahigh-vacuum-evaporation method. The background pressure was of the order of 10^{-10} Torr. First, Pt-buffer layer of 16 Å thickness was deposited. Then, Co and Pt were alternately deposited; the number of each layer was 40 and the superlattices included 80 Co-Pt interfaces. The thickness of the Co layer was 4 Å for all specimens and the thickness of the Pt layer was 2, 4, 8, 12, 16, or 20 Å (Si/Pt$_{buffer}$(16 Å)/[Co(4 Å)/Pt(d_{Pt} Å)]$_{40}$). Bilayer thickness ($d_{Co}+d_{Pt}$) was determined from X-ray diffraction spectra in low-angle regions.

Magnetic hysteresis loops were measured using a superconducting quantum interference device (SQUID). External magnetic field in the range of -4 to 4×10^6 A/m was applied parallel or perpendicular to the film surface at 300 K. K_{eff} is determined from the magnetization curves with external magnetic fields parallel and perpendicular to the film surface. The difference of areas surrounded by the initial-magnetization curves of the parallel and perpendicular external fields equals the difference of the energies needed to saturate the magnetization along the in-plane and out-of-plane directions. Thus, this energy is equal to K_{eff} and $K_{eff}>0$ means that the film exhibits PMA.

RESULTS

Figure 1 shows magnetic hysteresis loops when $d_{Pt}=2$ and 16 Å. The vertical axis is magnetization normalized by the saturation magnetization I_s. The hysteresis loops indicate that K_{eff} depends on d_{Pt}. Figure 2 shows correlation between the Pt-layer thickness and K_{eff} normalized by the K_s. When the Pt-layer thickness is smaller than the Co-layer thickness, the Co/Pt superlattice hardly shows PMA. As the Pt-layer thickness increases, K_{eff} increases and takes a maximum near $d_{Pt}=12$ Å.

DISCUSSION

As described above, K_c, K_s, and K_i will be independent of the Pt-layer thickness because of the constant Co-layer thickness. Here, we discuss the contribution of K_e on PMA by calculating K_e, taking account of lattice anharmonicity.

In the calculation of K_e, we must consider the *strain-dependent* elastic constants due to lattice

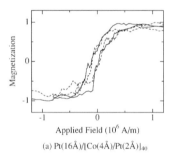
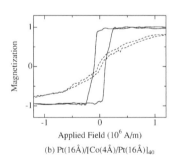

(a) Pt(16Å)/[Co(4Å)/Pt(2Å)]$_{40}$ (b) Pt(16Å)/[Co(4Å)/Pt(16Å)]$_{40}$

Figure 1   Magnetic hysteresis loops of Co/Pt superlattice thin films of (a) Pt(16 Å)/[Co(4 Å)/Pt(2 Å)]$_{40}$ and (b) Pt(16 Å)/[Co(4 Å)/Pt(16 Å)]$_{40}$. The external magnetic field is applied perpendicular (solid lines) and parallel (dashed lines) to the film surface. Magnetization is normalized by the saturation magnetization.

anharmonicity. Elastic constants change depending on the interatomic distance: compressive strain enhances the elastic constants and tensile strain decreases them. The calculation of the strain-dependent elastic constants follows three steps:

(i)   Calculate the elastic strain.
(ii)   Calculate change of bulk modulus ΔB using the averaged internal energy.
(iii) Derive ΔC_{ij} from ΔB and their temperature derivatives.

First, the elastic strain caused by the lattice misfit is calculated considering the multilayer thin films as layered rectangular-parallelepiped plates. A two-dimensional stress field without shear stresses and with $\sigma_3=0$ are assumed (plane stress condition).

Among independent components of the elastic constants, the bulk modulus B is closely related with the internal energy [7]. In this study, we use an embedded-atom-method (EAM) potential [8], and calculate ΔB with the elastic strain calculated in (i).

ΔB is converted into ΔC_{ij} using temperature derivatives of elastic constants. We assume that the ratio of the elastic-constant change to the bulk-modulus change $\Delta C_{ij}/\Delta B$ caused by the internal strain is the same as that caused by temperature change, because the temperature change also changes the elastic constants by changing the interatomic distance (due to thermal expansion). We referred the temperature derivatives of Co given by Fisher and Dever [9] and

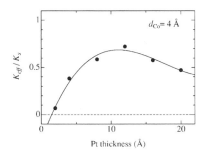

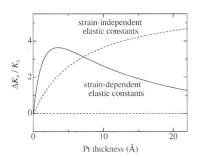

Figure 2 Correlation between K_{eff} and Pt-layer thickness. K_{eff} is normalized by the magnetic shape-anisotropy energy K_s. The Co-layer thickness d_{Co} is kept to be 4 Å.

Figure 3 Dependence of ΔK_e on the Pt-layer thickness. Solid and dashed lines denote ΔK_e calculated by considering the strain-dependent and strain-independent elastic constants, respectively.

those of Pt given by Macfarlane [10] for this calculation.

Difference of K_e thus calculated between the in-plane magnetization and the out-of-plane magnetization ($\Delta K_e = K_e{}^{//} - K_e{}^{\perp}$) is shown in Fig. 3. ΔK_e denotes the energy released by reduction of the lattice misfit due to the magnetostriction effect, being equivalent to contribution of K_e on K_{eff}. In the calculation of K_{eff}, we assumed complete bonds at the Co-Pt interfaces and used magnetostriction coefficients of bulk Co. For comparison, we calculated ΔK_e assuming the strain-independent elastic constants, which is unrealistic. In this case, however, ΔK_e increases monotonously as the thickness of Pt layer increases and decrease of K_{eff} never occurs. When elastic constants depend on the elastic strain, ΔK_{eff} shows a maximum. This trend is the same as the experimental result.

SUMMARY

PMA of Co/Pt superlattice thin film showed dependence on the Pt-layer thickness despite of the constant Co thickness. Considering that the magnetocrystalline anisotropy, magnetic shape-anisotropy, and magnetic interfacial-anisotropy energy are hardly affected by the Pt-layer thickness, this result indicates that the magnetoelastic anisotropy is dominant factor on PMA. This view was supported by the calculation of ΔK_e, and the strain-dependent elastic constants played an important role.

REFERENCES

1. P. F. Carcia, A. D. Meinhaldt, and A. Suna, *Appl. Phys. Lett.* **47**, 178 (1985).
2. P. F. Carcia, *J. Appl. Phys.* **63**, 5066 (1988).
3. S. Hashimoto, Y. Ochiai, and K. Aso, *J. Appl. Phys.* **66**, 4909 (1989).
4. S. J. Greaves, P. J. Grundy, and R. J. Pollard, *J. Magn. Magn. Mater.* **121**, 532 (1993).
5. H. Takahashi, S. Tsunashima, S. Iwata and S. Uchiyama, *J. Magn. Magn. Mater.* **126**, 282 (1993).
6. S. Chikazumi, *Physics of Magnetizm* (Wiley, New York, 1964).
7. D. Wallace, *Thermodynamics of crystals* (John Wiley & Sons, New York, 1972).
8. X. W. Zhou, H. N. G. Wadley, R. A. Johnson, D. J. Larson, N. Tabat, A. Cerezo, A. K. Petford-Long, G. D. W. Smith, P. H. Clifton, R. L. Martens, and T. F. Kelly, *Acta Materialia* **49**, 4005 (2001).
9. E. S. Fisher and D. Dever, *Trans. Metal. Soc. AIME* **239**, 48 (1967).
10. R. E. Macfarlane, J. A. Rayne, and C. K. Jones, *Physics Letters* **18**, 91 (1965).

Thin Film Plasticity—
Size Effects

Hillock formation and thermal stresses in thin Au films on Si substrates

Linda Sauter, T. John Balk[1], Gerhard Dehm[2], Julie A. Nucci and Eduard Arzt
Max Planck Institute for Metals Research and Institut für Metallkunde, University of Stuttgart, 70569 Stuttgart, Germany
[1]Now at: University of Kentucky, Department of Chemical and Materials Engineering, Lexington, KY 40506, USA
[2]Now at: Erich Schmid Institut für Materialwissenschaft, Österreichische Akademie der Wissenschaften, and Department für Materialphysik, Montanuniversität Leoben, 8700 Leoben, Austria

ABSTRACT

The wafer curvature technique was used to analyze stresses in fine-grained, 50 nm to 2 µm thick Au films on silicon substrates between room temperature and 500 °C. The microstructural evolution was analyzed by scanning electron microscopy (SEM), focused ion beam (FIB) microscopy and transmission electron microscopy (TEM). *In situ* heating experiments inside a scanning electron microscope provided a comparison between the morphological development and the stress-temperature behavior of the film. Hillock formation was observed, but it can only partially account for the stress relaxation measured by the wafer curvature technique.

INTRODUCTION

Thin metal films on rigid substrates often sustain high stresses during thermal cycling and annealing. Compressive stresses at elevated temperatures can enhance grain growth, grain boundary diffusion, or hillock formation. Au films are typically deposited onto metallic adhesion layers including V, Ti, Cu, Ni, Sn, In, and Cr [1-3]. These elements diffuse into the Au film, segregate to grain boundaries and/or the surface, and may form a surface oxide. The microstructural evolution, especially hillock and void formation, has been found to depend crucially upon the behavior of these interlayers [2]. Hillock growth was also reported in Au films with an interlayer of immiscible Mo [4]. The lateral diameter of these hillocks was predicted as a function of isothermal annealing time, in reasonable agreement with the experimental data. Zhang *et al.* [5] found that a 40 nm alumina passivation layer on Au/Si bilayer beams with a 20 nm Cr interlayer prevented grain coarsening and hillock formation. Thus, since the stress and morphological evolution in thin Au films is complicated by interaction with the surrounding layers, a clear picture of the origins and consequences of hillocks in Au films is not yet available.

In the present study we deposited Au films directly onto silicon nitride-coated silicon wafers. A thin metallic interlayer was not necessary, as the Au adhered to the silicon nitride throughout the temperature cycling. This avoided the complications imposed by an interlayer and allowed more accurate film stress measurements over a broad film thickness range (50 nm to 2 µm). Hillock formation and stress evolution in these films were studied systematically.

EXPERIMENTAL

Nine different Au film thicknesses between 50 nm and 2 μm were DC magnetron sputtered at room temperature under ultrahigh vacuum conditions (10^{-9} mbar) onto silicon nitride-coated (100)-oriented silicon wafers. The 50 nm amorphous silicon nitride coating was deposited onto both sides of a single-side polished wafer that had previously been coated with 50 nm of thermal silicon oxide. The samples were annealed in the sputtering chamber at 500 °C for 30 min immediately after Au deposition to stabilize the microstructure. Film texture was determined from θ-2θ X-ray scans and (111) pole figures recorded with a Phillips diffractometer.

The thermomechanical behavior of these Au films was studied by the wafer curvature method [6], where the film stress was measured between 50 and 500 °C at heating and cooling rates of 4-6 K/min for one to three thermal cycles. The film surface topography was investigated before and after thermal cycling using a LEO 1530 VP field emission scanning electron microscope (SEM) and *in situ* observations were obtained with a CamScan SEM equipped with a heating stage. The heating rates were between 2 and 5 K/min. The film and hillock microstructure were analyzed using focused ion beam (FIB) microscopy and transmission electron microscopy (TEM). An FEI 200xP FIB was employed to image the grain structure and to prepare hillock cross sections. Plan-view and cross-sectional TEM investigations of the film microstructure were carried out in a JEOL 2000 FX operated at 200 kV. For plan-view characterization of films thinner than 100 nm, the silicon was mechanically polished to approximately 80 microns, dimpled, and chemically etched with an $HF/HNO_3/CH_3COOH$ mixture at room temperature. Ion milling was necessary to produce electron transparency in Au films thicker than 100 nm. For cross-sectional evaluation, two 3 x 10 mm² samples, glued together with the Au films facing each other, were sandwiched between two pieces of silicon for mechanical stability, such that the total stack height was about 3 mm. This sample was sliced, mechanically ground, polished, dimpled, and finally ion milled (2 h at 3.3 kV and 20 min at 1.8 kV) to obtain an electron transparent film.

Sets of micrographs from plan-view TEM samples recorded at different tilt angles and magnifications were taken for grain area evaluation. The grain boundaries were traced on a transparency and a scanned image was evaluated using commercial software (Quantimet Q500/W, Leica). At least 250 grains were measured for each sample.

The mean hillock density for the different film thicknesses after three cycles was determined by counting at least 26 hillocks per film. The mean hillock area was determined by tracing the hillock contour onto a transparency and was evaluated using the same software. By assuming the hillocks were hemispherical caps, their mean volume and the stress they relaxed were estimated.

RESULTS AND DISCUSSION

All films are (111) fiber textured with a full width half maximum of $(3.7 \pm 0.2)°$, independent of film thickness. Despite the *in situ* ultrahigh vacuum anneal at 500 °C for 30 min, the grain size stayed nearly constant at about 100 nm for all film thicknesses. The 100 nm films were columnar (Fig. 1a), while the 300 nm film was only partially columnar (Fig. 1b). All thicker films were non-columnar, for example, the 800 nm thick film is shown in Fig. 1c.

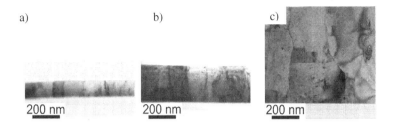

a)　　　　　　　b)　　　　　　　c)

200 nm　　　200 nm　　　200 nm

Figure 1. TEM micrographs of 100, 300 and 800 nm thick films. (a) The 100 nm film shows a columnar grain structure, whereas (b) the 300 and (c) the 800 nm films exhibit twins and grain boundaries parallel to the film/substrate interface.

Wafer curvature results for the first and second temperature cycles of a 1 µm film are shown in Fig. 2. The room temperature stress before the first cycle was tensile and followed the thermoelastic slope upon heating until it became compressive at about 200 °C. It then exhibited a compressive stress plateau starting at about 250 °C. At 350 °C the stress started to increase again until another stress relaxation occurred between 430 and 500 °C. The heating portion of the second cycle was also thermoelastic up to 250 °C, above which stress increased only slightly with temperature; a second stress plateau was not observed in this case. The 1 µm film was subjected to additional thermal cycling in an SEM in order to observe surface changes. Grain boundary grooving occurred between 200 and 250 °C during the first cycle, indicating that surface and grain boundary diffusion set in at this temperature. Voids formed upon cooling.

Au films of different thickness exhibited some common features in their stress-temperature behavior. The stresses decreased thermoelastically during heating until they became compressive between 200 and 250 °C. In compression the stress was either constant or increased slightly, as for the 1 µm film, but no systematic trend in the compressive regime was observed with respect to film thickness. Direct comparison of the first cycles is difficult, since the initial room temperature stress decreased as the time between deposition and testing increased. The initial stress directly after deposition and the time-dependent stress relaxation was not systematically studied. However, it is assumed that the stress after deposition is the same as the stress at the end of the first cycle, since the films were annealed in the sputtering chamber at 500 °C, which was also the maximum temperature during thermal cycling. For 600 and 800 nm thick films this behavior was observed. The stress after the first thermal cycle increased with decreasing film thickness. Therefore, taking the above assumption into consideration, the initial stress in the films directly after deposition likely increased with decreasing film thickness.

A comparison of the second thermal cycles for three film thicknesses (100 nm, 300 nm and 1 µm) is given in Fig. 3. Stresses became compressive at 250 to 350 °C, compared to 200 to 250 °C for the first cycle, since the room temperature stresses at the start of the second cycle were higher. However, the stress evolution in compression changed as a function of film thickness: The films with thickness 100 nm and below exhibited increasing compressive stresses up to 500 °C, whereas the stress in the thicker films remained constant or decreased slightly at high temperatures. For the 1 µm Au film, the wafer curvature data can be compared to an *in situ* SEM cycle. Hillocks were observed to form (Fig. 4a, 4b) and grow (Fig. 4c) between 350 and 400 °C, when the film was in compression according to the stress-temperature measurements. They did not disappear during cooling (Fig. 4d), although voids were seen to nucleate and grow.

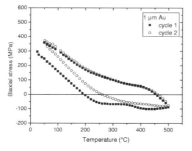
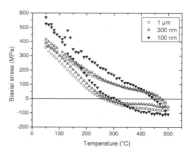

Figure 2. First and second temperature cycle of a 1 μm thick Au film.

Figure 3. The second thermal cycle for 100 nm, 300 nm and 1 μm thick Au films.

The stress-temperature curve for a third cycle was identical to the second cycle. *In situ* microscopy revealed the formation and growth of new hillocks between 300 and 400 °C, as well as void nucleation and growth upon cooling. SEM investigations of all films after three thermal cycles showed that only non-columnar films with thicknesses of 300 nm or above formed hillocks, whereas hillock formation was not observed in thinner, columnar films.

Evaluation of SEM images revealed that both the hillock density and radius increased with increasing film thickness, except for the 2 μm thick film, which did not follow this trend. The stresses relaxed by hillock formation were estimated from calculations of the hillock volume, described in the experimental section. The compressive stress relaxed by a hillock of volume V_{hill} is given by [7]

$$\Delta\sigma = M_{111}\frac{V_{hill}\rho_{hill}}{2h_f} \tag{1}$$

where $M_{111} = 178$ GPa is the biaxial modulus of the film, ρ_{hill} is the hillock areal density and h_f is the film thickness. A summary of the hillock statistics is found in Table I. The standard deviation of the hillock radii is an order of magnitude lower than the radii.

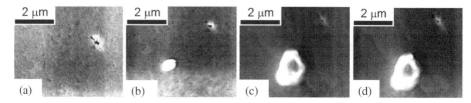

Figure 4. Micrographs of a 1 μm thick film during the second *in situ* thermal cycle in an SEM: (a) at 350 °C, (b) first hillock observed at 400 °, (c) hillock growth up to 450 °C, (d) hillock after cooling to 74 °C. The void shown in these figures formed during the first thermal cycle.

Table I. Hillock statistics and stress relaxation after 3 thermal cycles.

film thickness (nm)	hillock density ($10^{-3}/\mu m^2$)	hillock radius (nm)	Calculated stress relaxed by hillocks (MPa)	Total stress relaxed between 50 and 500 °C from wafer curvature (MPa)
50	0	-	-	237
100	0	-	-	245
300	1.7	344	39	480
400	0.5	591	41	481
500	1.4	681	160	518
600	3.4	748	418	596
2000	8.4	636	189	646

The stress relaxation data were obtained from wafer curvature measurements by extrapolating the thermoelastic slope of -2.09 MPa/K from 50 to 500 °C and then subtracting the measured 500 °C value. Table I shows that the actual stress relaxation significantly exceeds the value estimated from hillock formation. With the exception of the 600 nm thick film, which showed an unusually high hillock volume, the stress relaxation by hillock formation generally increased with increasing film thickness. Hillocks contribute to, but do not govern, the stress relaxation for all film thicknesses. In contrast, Kim *et al.* [7] found that hillock formation was the dominant stress relaxation mechanism in a 1 μm thick Al film.

The cross sections in Fig. 5 show that hillocks are composed of several grains that are each much larger than the average grain size. The FIB images suggest that hillocks form at the original film surface, since they are largest in this plane, and simultaneously grow both towards the substrate and out of the film plane. Fig. 5a shows the onset of hillock growth at the film surface. The hillock in Fig.5b reaches farther into the film and exhibits a more spherical shape. Finally, the hillock in Fig. 5c extends completely through the Au film. Due to surface topography, electron back scattering diffraction (EBSD) was not possible for hillocks such as in Fig. 5c. Smaller, flatter hillocks (e.g. Fig. 5a, 5b) measured by EBSD showed a (111) out-of-plane orientation, consistent with the original film texture.

Hillock formation was also observed *in situ* in the second and third cycle. Its onset between 350 and 400 °C was in good agreement with the deviation from elastic behavior in the compressive region of the stress-temperature plots. At these temperatures, surface and grain boundary diffusion are active, as indicated by the onset of grain boundary grooving during the first cycle. An important observation is that hillocks form only in non-columnar films, whereas grain boundary grooves are observed for all film thicknesses. We speculate that grain boundaries parallel to the surface in non-columnar films provide the diffusive pathways and mass transport needed to form hillocks. In addition, it can be further speculated that local variations in grain boundary misorientation angles, and therefore diffusivities, lead to preferred sites for hillock formation at certain triple junctions.

Both the Au and Al hillocks [8] consisted of grains much larger than the average grain size. However, the Au hillocks grew from a different interface than the hillocks in the fine-grained, self-passivated Al film. Since the Al/Al_2O_3 oxide interface is a low diffusivity pathway, the Al hillocks grew instead from the underlying Al/SiO_2 interface. While grain boundary diffusion is likely a significant factor in hillock formation in both the Au and Al films, it is clearly not the only factor. Comparison between these two studies shows that *both* the interfacial/surface diffusion and the grain boundary diffusion determine where hillocks most easily nucleate.

Figure 5. Different hillocks in a 1µm thick film: a) initial state with large lateral expansion, but little penetration into the film, b) hillock extending halfway into the film, c) hillock penetrating fully to the film/substrate interface.

CONCLUSION

Hillock formation in fine-grained Au films depends critically on the film microstructure. Columnar Au films less than 200 nm thick exhibited no hillock formation during thermal cycling to 500 °C. However, non-columnar Au films with thicknesses between 300 nm and 2 µm formed hillocks during the same temperature cycling. Hillock nucleation sites are likely determined by the local grain boundary misorientation angles at triple points. FIB analysis revealed that hillocks nucleated at the free Au surface and grew both toward the underlying interface and out of the film plane. Hillock growth presumably requires long range material transport along a continuous network of grain boundary pathways found only in the non-columnar films. Based on comparison with a previous study in Al films, it is concluded that the relative diffusivities of the underlying and overlying interfaces must also be considered to fully predict hillock formation behavior.

ACKNOWLEDGMENTS

The authors thank the thin film laboratory of the MPI for Metals Research for preparing the Au films, as well as N. Sauer for TEM sample preparation. L. Sauter acknowledges funding by the International Max Planck Research School for Advanced Materials.

REFERENCES

1. H. Hieber, Thin Solid Films **37** (3), 335 (1976).
2. J. Y. Kim and R. E. Hummel, Physica Status Solidi A **122** (1), 255 (1990).
3. D. C. Miller, C. F. Herrmann, H. J. Maier, S. M. George, C. R. Stoldt, and K. Gall, Scripta Materialia **52**, 873 (2005).
4. W. B. Pennebaker, Journal of Applied Physics **40** (1), 394 (1969).
5. Yanhang Zhang, M. L. Dunn, K. Gall, J. W. Elam, and S. M. George, Journal of Applied Physics **95** (12), 8216 (2004).
6. W. D. Nix, Metall. Trans. A **20A**, 2217-45 (1989).
7. K. Kim, B. Heiland, W. D. Nix, E. Arzt, M.D. Deal, and J. D. Plummer , Thin Solid Films **371** (1-2), 278 (2000).
8. D. K. Kim, W. D. Nix, M. D. Deal, and J. D. Plummer, Journal of Materials Research **15** (8), 1709 (2000).

Mater. Res. Soc. Symp. Proc. Vol. 875 © 2005 Materials Research Society   O5.5

How Stretchable Can We Make Thin Metal Films?

Candice Tsay[1], Stephanie P. Lacour[1], Sigurd Wagner[1], Teng Li[2], Zhigang Suo[2]
[1]Department of Electrical Engineering, Princeton University, Princeton, NJ, USA
[2]Division of Engineering and Applied Sciences, Harvard University, Cambridge, MA, USA

ABSTRACT

Thin metal films deposited on elastomeric substrates can remain electrically conducting at tensile strains up to ~100%. We recently used finite-element simulation to explore the rupture process of a metal film on an elastomer. The simulation predicted the highest stretchability on stiff elastomeric substrates [1]. We now report experiments designed to verify this prediction. A ~15-μm thick silicone elastomer layer with Young's modulus E ~ 160 MPa is deposited on a 1mm thick membrane of polydimethylsiloxane (PDMS), a silicone elastomer with E ~ 3 MPa. Metal stripes consisting of 25-nm thick gold (Au) film sandwiched between two 5-nm thick chromium (Cr) adhesion layers are fabricated either on top of the stiff layer spun onto the soft membrane substrate, or are encapsulated at the interface between the two elastomers. Encapsulated gold films remain electrically conducting beyond 40% strain. But conductors deposited on top of stiff elastomer lose conduction at strains of 3-8%. These results suggest that, in addition to the stiffness of the elastomeric substrate, the initial microstructure of the metal film plays a role in determining its stretchability.

INTRODUCTION

Conformable or skin-like electronics must withstand repeated and large deformations. For example, a stretchable sensor array wrapped around the elbow-joint of a prosthetic arm may experience tensile strains of 10% or more. However, thin film electronic materials fracture at lower strains than this. Free-standing gold films, for example, rupture at 1-2% strain [2].

We have demonstrated experimentally that a thin gold film bonded to an elastomeric substrate with Young's modulus E ~ 3 MPa can stretch and remain conducting to 100% strain [3, 4]. This is a step toward fabricating elastic electronic circuits on elastomeric substrates using arrays of rigid device islands connected by stretchable metallization [5]. We used finite-element simulations to model the rupture process for these stretchable conductors. The results show that, while freestanding metal films rupture by deformation localization, an elastomeric substrate suppresses the localization and allows the metal film to elongate without immediate rupture [1, 6]. Furthermore, the simulation predicts greater stretchability for metal films on stiffer elastomeric substrates. This paper describes experiments designed to test the simulation prediction and also presents our observations on the effect of an encapsulating silicone layer on the conductor's stretchability.

Four types of samples are prepared, summarized in Figure 1a. Type A, in which the gold conductor is fabricated on top of the compliant silicone (E ~ 3 MPa) substrate, is the configuration tested many times before [3, 4], and is used here as a basis for comparison. For type B, we introduce a moderately stiff silicone material (E ~ 160 MPa) as the substrate. The ~15-μm thick stiff silicone layer is spun onto the 1-mm thick compliant silicone substrate for easier handling. Given the large compliance of the 1-mm substrate, the metal film behaves mechanically as if it were attached to the stiff elastomer alone. For the other two configuration

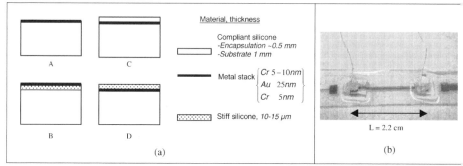

Figure 1. (a) Cross-section views. Type A: compliant silicone (E ~ 3 MPa) as substrate; type B: stiff silicone (E ~ 160 MPa) bottom layer; type C: compliant silicone encapsulation; type D: stiff silicone encapsulation. (b) Top view photograph of a type C conductor 600-μm wide with embedded gold wires and compliant PDMS on top.

types, the gold conductor is sandwiched between the compliant silicone substrate and a encapsulating silicone film. For type C, the compliant silicone is used as the top encapsulatin material. Figure 1b shows a conductor encapsulated by the clear, compliant silicone and lea wires to the contact pads. For type D, the stiff silicone is used as the top encapsulating material.

EXPERIMENTAL DETAILS

Sample preparation

The compliant silicone substrate material (Sylgard 184[®], Dow Corning) is cast in plastic Petri dish and oven-cured at 60°C to a 1-mm thick PDMS membrane. Electron-bean evaporation is used to deposit the metal through a polyimide shadowmask [3]. The metal film consists of a 25-nm thick Au layer between two 5-nm thick Cr layers. The thin Cr layers ar needed for the adhesion of Au to the adjoining PDMS. After the deposition, the shadowmask i peeled away, resulting in conductors that are 2-cm long and 600-μm wide.

The stiff silicone (WL-5150 photo-patternable silicone, Dow Corning) is spin-coate either on top of the compliant PDMS membrane (for type B) or on top of the metallic conducto (for type D), then oven-cured at 80°C for 10 hours. The film is typically 10 to 15-μm thick. Fo the conductors with the stiff silicone underneath, the silicone is spun on and cured before meta deposition.

When the stiff silicone is spun directly onto the conductors, the substrate, i.e. the compliant PDMS membrane, is mounted on a 0.5-mm thick aluminum sheet prior to meta deposition. The aluminum backing provides stability and rigidity during the subsequent spinnin and baking steps. After metal deposition, the stiff silicone is spun on top of the conductors to 1 to 15-μm thickness and is then patterned to open contact holes to the conductor pads. Th patterning process is similar to a photoresist process: a UV exposure, a 150°C baking step fo polymerization, and a solvent development [9]. After the silicone film is cured, the aluminur backing is removed in an HCl etching solution.

The symmetrically encapsulated conductors (type C) are topped with Sylgard 184 PDM after electrical connections are established. To ensure contact, 100-μm diameter gold wires ar first attached to the conductor pads with uncured epoxy-based conductive paste. The PDMS i.

184

then hand-deposited on the areas surrounding the pads. After a 60°C oven-cure, the gold wires are effectively embedded in the PDMS (Fig. 1b).

Sample characterization

For electromechanical measurements, compliant electrical contacts are made to the conductors using the same method as described above for the symmetric encapsulated conductors. Electrical resistance is recorded with a Keithley 4140 source-meter.

The uni-axial stretching is done with an automated stretching device [3]. Data acquisition and control of the stepper motor is done through a Labview program. In all experiments, the strain is increased in steps of 0.1% every 12 seconds.

For Scanning Electron Microscopy observation, the samples are coated with a 5-nm thick iridium layer.

RESULTS

Type A and type B. No encapsulating layer.

We compare the stretchability of type A and type B samples. The electrical resistance is measured as the sample is stretched with increasing applied tensile strain and plotted in Figures 2a and 2b. Surprisingly, the conductors on the stiff silicone do not retain electrical conduction at as high strains as the conductors on the compliant silicone. The electrical resistance of the conductor on complaint silicone (type A) increases steadily with the applied strain. At 50% strain, its resistance has increased by over 1200%, but the sample does not fail electrically and no macroscopic cracks appear in the film. Instead, as shown in Fig. 2c, the metal usually buckles, by Poisson compression, in the direction perpendicular to the applied strain [7]. After relaxation

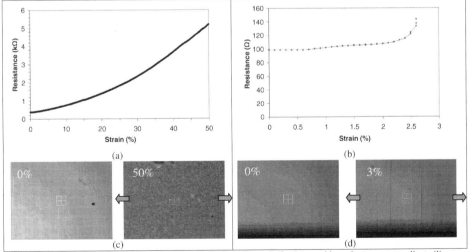

Figure 2. Electrical resistance as a function of applied tensile strain for (a) a gold conductor on compliant silicone; (b) a gold conductor on stiff silicone. 320-μm x 240-μm optical micrographs of (c) top surface of metal of sample in (a) during stretch at 0% and 50% strain; and (d) top surface of metal of sample in (b) during stretch at 0% and 3% strain. Arrows indicate stretching direction.

back to 0% strain, the resistance value recovers to a value slightly higher than the pre-stretch initial resistance value.

In contrast, electrical failure occurs at 3% strain for the conductor on the stiff silicone. Multiple small transversal cracks form at the edges of the conductor and gradually traverse across the width of the line, as seen in Figure 2d. Loss of conduction corresponds to the moment in which cracks from opposite edges meet to form a line of discontinuity across the width of the metal. After relaxation to 0% strain, electrical conduction returns, but parallel cracks are visible in the metal layer. All 600-μm wide samples of this type display this behavior, and none retain conduction above 8% strain.

In addition, although the two sample types are prepared in the same metallization batch, the samples on stiff silicone have four times lower initial resistance than the samples on compliant silicone. This disparity may be explained by considering surface roughness of the silicone materials. The surface roughness of gold on the compliant silicone is twice that of gold on the stiff silicone [8]. The difference is clear in the SEM micrographs shown in Figure 3.

Type C and type D. With encapsulating layer.

All encapsulated conductors are highly stretchable, regardless of the coating layers' stiffness. Figure 4a shows the normalized resistance ($R/R_{initial}$) plotted against tensile strain for the two encapsulated configurations as well as the non-encapsulated conductor on compliant silicone. Both the resistance values and the metal morphology follow closely those of a non-covered conductor. The samples can be stretched beyond 30% strain, but this particular compliant silicone encapsulated sample slipped from the stretching apparatus during the test. Encapsulated samples and non-encapsulated samples fabricated in the same metallization batch

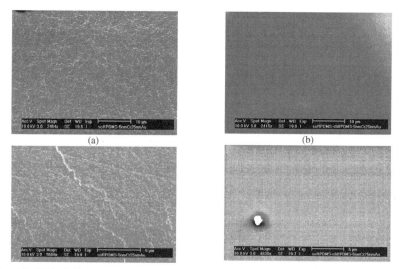

(a)   (b)

Figure 3. SEM micrographs of (a, top) a bare compliant silicone surface; (a, bottom) 5-nm Cr and 25-nm Au on compliant silicone; (b, top) a bare stiff silicone surface; (b, bottom) 5 nm-Cr and 25-nm Au on stiff silicone. Samples have been coated with 5-nm iridium.

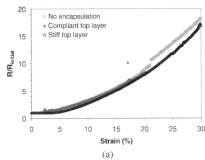
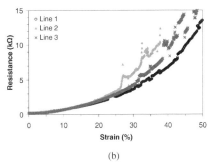

| (a) | (b) |

Figure 4. Electrical resistance measured with increasing applied strain. (a) Normalized resistance of conductors with stiff silicone encapsulation, compliant PDMS encapsulation, and no encapsulation; (b) Three conductors with stiff encapsulation.

all fail electrically near the same critical strain. Figure 4b shows the electrical resistance values measured with increasing strain for three stiff silicone-covered conductors of the same batch.

Two different conductor widths, 350-μm and 600-μm, are also tested for each configuration. Wider conductors have lower initial resistance in every case and generally retain conduction at higher strains. The effect is more pronounced for the conductors on stiff silicone.

CONCLUSIONS

The present experiments show that using the stiff elastomer as the substrate material does not increase stretchability of the metal film, although the simulation predicted highest stretchability on stiff elastomeric substrates. This discrepancy in metal film stretchability is explained as follows. Rupture of thin metal film results from localized plastic deformation [2] such as necking or shear-off. A polymer substrate delocalizes the deformation in a metal film under tension, thus carrying the film to large strain before rupture [6]. In the finite-element simulation, good adhesion between the metal film and the elastomer substrate is assumed, thus excluding delamination along the film/substrate interface. Under such an assumption, a stiffer elastomer substrate more effectively delocalizes the deformation in the metal film, thus leading to better stretchability of the film. In practice, the metal/elastomer interface is never perfect. When such a structure is stretched, debonding may occur, so that the metal film becomes freestanding and ruptures at a small strain. The adhesion between the metal film and the elastomer substrate is critical to the stretchability of the metal film [10]. The formation of multiple transverse cracks in the metal conductor (as in Fig. 2d) has also been observed in metal films on polymer substrates by other researchers [10]-[12]. This rupture behavior of the metal conductor on a stiff elastomer suggests weak adhesion between the film and the substrate, which leads to poor stretchability of the metal conductor.

Another type of rupture behavior is observed in gold films deposited directly on the compliant silicone substrate. While we have previously observed two types of gold film on compliant silicone – buckled and smooth, or flat and microstructured [7] - the samples tested in the current experiment appear to be the latter. These gold films display a complex microstructure that results from the surface roughness of the underlying substrate. This microstructure is not reflected in the finite-element simulation model. Experimentally, these samples retain electrical conduction at very high strains without rupturing. This suggests that the gold film's initial microstructure plays an important role in determining its stretchability.

This study also finds that depositing another elastomer layer on top of a stretchable conductor does not significantly change the electrical behavior of the metal film during stretching. Cycling experiments can be done to see if the encapsulation improves the fatigue life of the conductors.

ACKNOWLEDGMENT

This research was supported by DARPA-funded AFRL-managed Macroelectronics Program Contract FA8650-04-C-7101, and by the New Jersey Commission on Science and Technology.

REFERENCES

1. T. Li, Z. Huang, Z. Suo, S. P. Lacour, S. Wagner, Appl. Phys. Lett. **85**, 3435 (2004).
2. D.W. Pashley, Proc. Roy. Soc. Lond. A **255**, 218 (1960).
3. S.P. Lacour, S. Wagner, Z. Huang, Z. Suo, Appl. Phys. Lett. **82**, 2404 (2003).
4. S.P. Lacour, J. Jones, Z. Suo, S. Wagner, IEEE Elec. Dev. Lett. **25**, 179 (2004).
5. S.P. Lacour, C. Tsay, S. Wagner, IEEE Elec. Dev. Lett. **25**, 792 (2004).
6. T. Li, Z. Y. Huang, Z. C. Xi, S. P. Lacour, S. Wagner, and Z. Suo, Mechanics of Materials **37**, 261 (2005).
7. C. Chambers, S.P. Lacour, S. Wagner, Z. Suo, Z. Huang, Mat. Res. Soc. Symp. Proc. **769** Apr. 2003, pp. H10.3.1-6.
8. W. Zhang, J.P. Labukas, S. Tatic-Lucic, L. Larson, T. Bannuru, G.S. Ferguson, Technical Digest Eurosensors XVIII (Rome, 2004), 552.
9. H. Meynen, M. Vanden Bulcke, M. Gonzalez, B. Harkness, G.Gardner, J. Sudbury-Holtschlag, B. Vandevelde, C. Winters, E. Beyne, Microelec. Eng. **76**, 212 (2004).
10. Y. Xiang, T. Li, Z. Suo, J.J. Vlassak, to be pubished.
11. S. L. Chiu, J. Leu and P. S. Ho, J. Appl. Phys., **76**, 5136 (1994).
12. B. E. Alaca, M. T. A. Saif and H. Sehitoglu, Acta Mater., **50**, 1197 (2002).

Mater. Res. Soc. Symp. Proc. Vol. 875 © 2005 Materials Research Society   O5.6

An Investigation of Film Thickness Effect on Mechanical Properties of Au Films Using Nanoindentation Techniques

Yifang Cao, Zong Zong, and Wole Soboyejo
Princeton Institute of Materials Science and Engineering(PRISM)
and Department of Mechanical and Aerospace Engineering
Princeton University
Princeton, NJ 08544

ABSTRACT

This paper presents the results of nanoindentation experimental studies of Au thin films with different thicknesses. The effects of film thickness and microstructure on the hardnesses of electron-beam deposited Au films were studied in terms of Hall-Petch relationship. The effects of different thicknesses on indentation size effects (ISE) are explained within the framework of mechanism-based strain gradient (MSG) theory using the concept of microstructural length scale.

I. INTRODUCTION

In recent years, significant efforts have been made to develop micro-electronics and micro-electro-mechanical systems (MEMS) structures that include metallic contacts at the micro- and nano-scales [1–3]. In most cases, gold has been used due to its exceptional combination of oxidation resistance and electrical conductivity [4]. Within this context, the physics of contact-induced deformation of Au films needs to be fully investigated. In addition, the mechanical properties of the small structures may be significantly different from those of bulk materials [5]. Therefore, there is a need to obtain measurements of the mechanical properties at the appropriate scales for Au films.

Several prior experimental efforts have been made to measure the nano/micro-scale mechanical properties of polycrystalline Au using nanoindentation [6,7], uniaxial tensile testing [8], membrane deflection experiment [9], and bimaterial microcantilever testing [10]. However, a complete understanding of mechanical properties of Au films in terms of film thickness effect, microstructure effect, substrate effect is yet to emerge.

This paper presents the results of an experimental study of film thickness effect on the mechanical properties of Au films using nanoindentation techniques. Following a brief description of sample preparation and surface microstructure, film mechanical properties (hardnesses) are characterized using nanoindentation techniques. The thickness effect on yield strength of Au films is discussed in terms of Hall-Petch relationship. The effects of different thicknesses on hardness indentation size effects (ISE) are analyzed within the framework of the mechanism-based strain gradient (MSG) theory. Electron-beam deposited thin films Au with thicknesses of 100 nm, 500 nm, 1000 nm and 2000 nm on Si substrates are examined in this study.

II. ANALYSIS THEORIES

Doerner and Nix [11], and later on Oliver and Pharr [12,13] developed a most comprehensive method for determining the hardness and modulus from depth sensing indentation (DSI) load-displacement data. In the theory, the Meyer's definition of hardness, H was adopted. This is given by:

$$H = \frac{P_{max}}{A} \tag{1}$$

where P_{max} is the maximum load, A is the projected contact area, which can be calculated using the method shown in Ref. [12]. Usually tip rounding effects [14] and indentation pile-up [15] are needed to be taken account for the estimation of A.

The indentation size effects, the phenomenon that the hardness increases as the indentation size decreases, were observed by many researchers [14, 16, 17]. Recent mechanism-based strain gradient

(MSG) plasticity theory, established by Nix and Gao [16], leads to the following characteristic form of the depth dependence of hardness:

$$\frac{H}{H_0} = \sqrt{1 + \frac{h^*}{h}}$$ (2)

where h is the depth of indentation. H_0 is the hardness in the limit of infinite depth, and h^* is a characteristic length scale that depends on the shape of indenter, the shear modulus, and H_0. This model was used to derive the following law for strain gradient plasticity:

$$(\frac{\sigma}{\sigma_0})^2 = 1 + \hat{l}\chi$$ (3)

where σ is the effective flow stress in the presence of a gradient, σ_0 is the flow stress in the absence of a gradient, χ is the effective strain gradient and $\hat{l}$ is a characteristic material length scale. $\hat{l}$ can be expressed in terms of the Burger's vector b and shear modulus μ [14]. This gives:

$$\hat{l} = \frac{b}{2}(\frac{\mu}{\sigma_0})^2$$ (4)

This length scale can be thought of as a formalism that enables the strain contributions from strain gradients to be modeled within a continuum theory framework. For the case of pure FCC metals, this scale positively scales with the mean spacing between statistically stored dislocations (SSD) L_s through the following expression [16]:

$$\hat{l} = \frac{4}{3}\frac{L_s^2}{b}$$ (5)

The plastic flow stress of polycrystalline materials σ_0, is effected by the grain size. The well-known Hall Petch relationship [18], which describes the grain size effect on the flow stress, is given by the following:

$$\sigma_0 = \sigma_{0i} + \frac{k_y}{\sqrt{d}}$$ (6)

where d is the grain size, σ_{0i} is the intrinsic flow stress in the absence of grain size effects, k_y is a constant coefficient.

III. MATERIALS

A. Sample Preparation

Polycrystalline Au thin films with different thickness were deposited onto substrates of Si. The Si substrates were obtained from Silicon Quest International Company, Santa Clara, CA. Prior to deposition, the silicon wafers were prepared by submerging it in a sulfuric acid and hydrogen peroxide solution in order to clean the wafer. They were then rinsed in hydrofluoric acid to remove the remaining oxide layer and then rinsed with water to remove any residual impurities. Subsequently, Au films were deposited on the silicon wafers using a Denton/DV-502A E-Beam Evaporator (Denton Vacuum, Moorestown, NJ) that was operated at pressures below 10^{-6} Torr, and substrate temperature below 40 ^{o}C. A deposition rate of 0.5 nm/sec was used. In this way, films with thicknesses of 100 nm, 500 nm, 1000 nm and 2000 nm were produced.

B. Microstructure and Topology

The microstructure and surface topographies of the deposited films were examined using a Dimension 3100 atomic force microscope (AFM) (Veeco Instruments, Woodbury, NY) operated in the tapping mode. Typical AFM images of Au films (with thicknesses of 100 nm, 500 nm, 1000 nm and 2000 nm) deposited on silicon substrates are presented in Figure 1 as well as a typical scanning electron microscopy (SEM) image of a 1000-nm-thick Au film. It is shown an AFM image has similar grain size distribution to the SEM image for the same Au film thickness. The increase in the grain size with increasing film thickness is quite evident in the micrographs. The root mean square (RMS) roughness of the Si substrate is 0.5 nm. The RMS roughness and grain size that were obtained for the Au films were summarized in Table I. The results show an increase in the RMS roughness, increasing from $\sim$1 nm, for 100 nm thick Au film, to $\sim$4 nm for 2000 nm thick Au film. The average grain size of the Au films also increased with increasing film thickness, increasing from approximately 20 nm, for the

Table I. Root mean square (RMS) roughnesses and grain sizes obtained for the Au films using AFM scans

Film	RMS (nm)	Grain Size (nm)
100 nm Au	0.8	21
500 nm Au	2.2	39
1000 nm Au	2.3	76
2000 nm Au	4.0	126

100 nm thick film, to ~130 nm, for the 2000 nm thick film. The results are, therefore, consistent with prior studies of film microstructure [19–21], which show that grain size increases with increased film thickness.

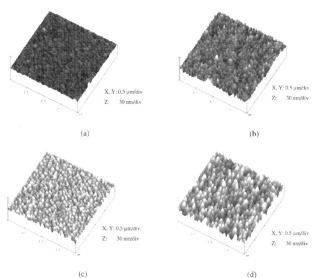

(a)　　　　　　　　　　　　　　(b)

(c)　　　　　　　　　　　　　　(d)

Figure 1. AFM images of surface topography of Au films on silicon substrates prepared by e-beam evaporation with film thickness 100 nm for (a), 500 nm for (b), 1000 nm for (c), and 2000 nm for (d).

IV. EXPERIMENTAL TECHNIQUES

The nanoindentation measurements were performed using a TriboScope (Hysitron Inc., Minneapolis, MN) Nanomechanical Testing System integrated with a DI Dimension 3100 AFM frame. A Berkovich (three-sided pyramid tip, 142.3°) diamond indenter was used in this study. The tests were conducted under load control, using peak loads in the range between 100 μN and 11,000 μN. Only regions with relatively low roughness values were chosen for performing the indentations. Hence, the possible effects of rough surfaces (on the hardness and modulus measurements) were minimized. Furthermore, in an effort to decrease the possible interactions between adjacent indents, all the indents were separated by 5-10 μm. During the test, contact mode AFM scans were obtained before and after each indentation. A load-displacement curve was also recorded for each indent.

Significant material pile-up was found to occur in the polycrystalline Au thin films. The material pile-up information recorded by AFM scan is obtained by AFM image post-processing. One typical image showing pile-up as well as the relevant section analysis is presented in Figure 2(a) and Figure 2(b). The pile-up is due to the severe constrained plastic deformation in the films. Its sizes were found to

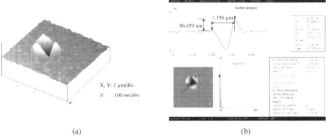

(a)   (b)

Figure 2. (a) A typical AFM image taken after a nanoindentation test on a Au film on Si substrate shows significant pile-up. (b) Typical section analysis performed for a pile-up AFM image taken after a nanoindentation test.

monotonically increase with indentation depths. The pile-up deformation of Al and LIGA Ni films were prior studied in Ref. [15] and Ref. [14], respectively. Here we used the approach in Ref. [14] to account for the effects of material pile-up in the analysis of the data. The loading profile included three segments: loading to a peak load; holding at the peak load; unloading back to the zero load. Loading and unloading rates of 400 μN/sec were applied. A holding period of 3 sec was applied to allow time dependent effects to diminish. In this way, the loading curves obtained for various peak loads were found to exhibit good reproducibility. For consistency, only load-displacement profiles that mapped onto the same master curve were accepted as valid measurements. The invalid profiles have a fraction less than 5% for all the indentations we made.

V. RESULTS AND DISCUSSION

The hardness values obtained for Au films of different thickness (versus normalized indentation depth) are presented in Figure 3. The usual indentation size effects are observed for indents with depth to film thickness ratios of up to ~0.2, which is attributed largely to the effects of geometrically necessary dislocations at small scales [16]. This critical ratio 0.2, below which the indentation size effect can be well captured by the effects of geometrically necessary dislocations [16], matches well with the critical ratio 0.18-0.22 measured in [6] for the Au film on Ni substrate system. Note for the thinnest film (Au film thickness of 100 nm), the measured hardness values increase with increasing indentation depth for indents with depth to film thickness ratios of up to ~1. This is due to the significant effects of film substrate modulus mismatch.

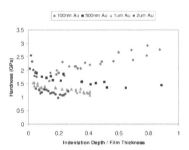

Figure 3. Film thickness effect on hardnesses of Au films.

The depth independent hardnesses for different film thickness can be calculated using the indents with depth to film thickness ratios of beyond 0.2. They can also be taken as the hardnesses in the

Table II. Comparison between H_0 and the average hardness values calculated using the indents with depth to film thickness ratios of beyond 0.2

Film	H_0	Average H (h/t>0.2))
500 nm Au	1.58	1.53±0.15
1000 nm Au	1.27	1.21+0.07
2000 nm Au	1.06	1.11±0.07

limit of infinite depth H_0, which can be obtained by fitting Eq. (2) to the experimental data. Those hardness values obtained uisng H_0 are very close to those calculated using the indents with depth to film thickness ratios of beyond 0.2, as shown in Table II. It is noticed that the hardnesses decreased with increasing Au film thickness, from approximately 1.5 GPa to 1.1 GPa, for 500 nm thick film and 2000 nm thick film. This shows some evidences of material strengthening. The strengthening is possibly due to the increased effective grain boundary area associated with grain size reduction as the film thickness decreases. Figure 4 shows the experimental data of H versus $d^{-1/2}$, which can be well described by a linear fitting with the correlation coefficient of 0.984. The Hall Petch relationship (Eq. (6)) can be used to characterize the grain size effect in Figure 4 since hardnesses can be estimated to be roughly three times of the yield strengthes for FCC metals (Tabor's relation) [16]. Note the slope of the linear fitting in Figure 4 is three times of the Hall Petch coefficient k_y.

The obtained k_y for the Au films in this study is ~ 2 GPa$(nm)^{-0.5}$, which is comparable to the reported Hall-Petch coefficients of ~ 3 GPa$(nm)^{-0.5}$ for Au films [22]. The difference could be due to the significant statistical scatter of the determination of both the yield strength and grain size, shown in the literatures [22] and our study. In fact, the Hall Petch coefficient is strain, temperature, and strain rate dependent [18].

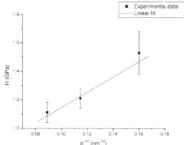

Figure 4.　The experimental data of H versus $d^{-1/2}$ and the corresponding linear fitting.

Table III. Microstructural length scales for different samples

	Sample Specification		b (nm) [a]	μ (GPa) [b]	H_0 (GPa)	$\hat{l}$ (μm)
Au	Thin	500 nm	0.2885	30.4	1.58	0.48
	Film	1000 nm	0.2885	30.4	1.27	0.74
	Thickness	2000 nm	0.2885	30.4	1.06	1.07

[a] [b]: Ref. [23]

The material microstructure length scale $\hat{l}$ can be thought of as a formalism that enables the strain contributions from strain gradients to be modeled within a continuum theory framework. Table III lists the material length scales obtained for Au films. Note that the material length scale is comparable to the film thickness, for the range of film thicknesses that were examined in the current work. $\hat{l}$ values are on the order of 1 μm, which are close to those reported by Begley and Hutchinson [24] with $\sim$1 μm for a range of single crystals (Cu, Ag and W). Hence, the measured values of $\hat{l}$ are consistent with prior values reported in the literature for other materials. The material length scales seem to increase with increasing film thicknesses. $\hat{l}$ increases from slightly less than 0.5 μm for 500 nm thick film, to 1.1 μm for 2000 nm thick film. Based on Eq. (5), bigger $\hat{l}$ values in thicker films indicate bigger spacings of statistically stored dislocations, which could lead to weaker dislocation interaction and lower strengthening, and suggest that thicker films have lower intrinsic hardnesses. This is consistent with our experimental results that thicker films show significantly lower intrinsic hardness values.

VI. CONCLUDING REMARKS

This paper presents the results of an experimental study of mechanical properties of polycrystalline Au thin films using nanoindentation techniques. The salient conclusions arising from this study are summarized below: (a) The grain sizes of the Au films increase with increasing film thickness from 100 nm to 2000 nm. The hardnesses of the Au films decrease with increasing film thickness. The strengthening with increasing film thickness can be reasonably characterized by the Hall Petch relation. (b) The measured length scales were shown to increase with increasing film thickness. This indicates bigger dislocation spacings and therefore lower hardnesses in thicker films.

Acknowledgements This work was supported by the National Science Foundation (Grant No: DMR 0213706 and Grant No: DMR 0231418). Appreciation is extended to the Program Managers (Dr. Ulrich Strom and Dr. Carmen Huber) for their encouragement and support. The authors would also thank Dr. Seyed Allameh for helping on e-beam depositions and Mr. Derek Nankivil for assistance on nanoindentation.

REFERENCES

[1] S. Majumder, N. E. McGruer, G. G. Adams, P. M. Zavracky, R. H. Morrison, and J. Krim, *Sensors and Actuators A - Physical*, **93**, 19, 2001.

[2] P. Zavracky, N. McGruer, and S. Majumder, *J. Micro-electromech. Systems*, **6**, 3, 1997.

[3] J. Plummer, M. Deal, and P. Griffin, *Silicon VLSI Technology: Fundamentals, Practice and Modeling*. Upper Saddle River, NJ: Prentice Hall, 2000.

[4] R. Holliday and P. Goodman, *IEE Review*, **48**, 15, 2002.

[5] M. GadElHak, *The MEMS Handbook*. New York: CRC Press, 2002.

[6] Z. Xu and D. Rowcliffe, *Surface and Coatings Technology*, **157**, 231, 2002.

[7] J. F. Smith and S. Zheng, *Surface Engineering*, **16**, 143, 2000.

[8] R. Emery and G. Povirk, *Acta Materialia*, **51**, 2067, 2003.

[9] H. D. Espinosa and B. Prorok, *Journal of Materials Science*, **38**, 4125, 2003.

[10] K. Gall, N. West, K. Spark, M. L. Dunn, and D. S. Finch, *Acta Materialia*, **52**, 2133, 2004.

[11] M. F. Doerner and W. D. Nix, *J. Mater. Res.*, **1**, 601, 1986.

[12] W. C. Oliver and G. M. Pharr, *J. Mater. Res*, **7**, 1564, 1992.

[13] G. M. Pharr and F. B. W. C. Oliver, *J. Mater. Res*, **7**, 613, 1992.

[14] J. Lou, P. Shrotriya, T. Buchheit, D. Yang, and W. O. Soboyejo, *Journal of Materials Research*, **18**, 719, 2003.

[15] R. Saha and W. D. Nix, *Acta Materialia*, **50**, 23, 2002.

[16] W. D. Nix and H. Gao, *J. Mech. Phys. Solids*, **46**, 411, 1998.

[17] Q. Ma and D. R. Clark, *J. Mater. Res.*, **10**, 853, 1995.

[18] J. Aldazabal and J. G. Sevillano, *Materials Science and Engineering A*, **365**, 186, 2004.

[19] Y. Xiang, X. Chen, and J. J. Vlassak, *Mat. Res. Soc. Symp. Proc*, **695**, L.4.9.1–6, 2002.

[20] J. F. Chang, H. H. Kuo, I. Leu, and M. H. Hon, *Sensors and Actuators B: Chemical*, **84**, 258, 2002.

[21] M. Aguilar, P. Quintana, and A. I. Oliva, *Materials and Manufacturing Processes*, **17**, 57, 2002.

[22] S. Sakai, H. Tanimoto, and H. Mizubayashi, *Acta Materialia*, **47**, 211, 1999.

[23] T. Trimble, R. Cammarata, and K. Sieradzki, *Surface Sicence*, **531**, 8, 2003.

[24] M. Begley and J. Hutchinson, *J. Mech. Phys. Solids.*, **46**, 2049, 1998.

Thin Film Plasticity—
Creep

Mater. Res. Soc. Symp. Proc. Vol. 875 © 2005 Materials Research Society

Fabrication and Composition Control of NiTi Shape Memory Thin Films for Microactuators

David J. Getchel and Richard N. Savage
Materials Engineering Department, California Polytechnic State University,
San Luis Obispo, CA 93407, U.S.A.

ABSTRACT

Microactuators fabricated with NiTi thin films take advantage of the shape memory effect's large energy density ($\sim$5-10 joules/cm^3) and high strain recovery ($\sim$8%). Microelectromechanical Systems (MEMS) designed with these actuators can serve as biosensors, micro-fluidic pumps or optical switches. However, the fundamental mechanical properties of these shape memory NiTi films have not been fully characterized with micro-scale test structures. Equiatomic NiTi thin films were deposited by co-sputtering NiTi and Ti targets with the intension of fabricating such test structures. Dual cathodes allowed direct control of the film composition by adjusting the Ti cathode power. Energy Dispersive Spectroscopy (EDS) quantified the film composition relative to pure standards. A thin ($\sim$50 nm) chromium film on a pure silicon substrate created excellent film adhesion. Oxidized Si wafers did not bond with the Cr and NiTi films. This deposition method enabled control of film composition and the necessary adhesion.

INTRODUCTION

The shape memory effect (SME) of NiTi outperforms traditional electrostatic and thermal MEMS actuators when large range of motion and high actuation forces are desired [1]. NiTi is a challenging actuator material to use because the shape memory behavior is sensitive to deposition parameters and post deposition annealing [2]. The realization of NiTi MEMS actuators requires performance data from test structures similar in size to current MEMS devices. Previous tensile and diaphragm NiTi testing provide valuable film properties, but not on the micro-scale [3, 4]. Smaller test structures may expose how factors such as surface oxides or compositional variation in thickness [5] affect the mechanical and shape memory properties. Fabricating and testing a series of micro-scale NiTi test structures to examine how deposition parameters affect actuator performance motivated this study.

An ideal MEMS test structure would use scanning probe microscopy (SPM) instruments to deflect miniature thin film cantilevers. Nanoindenters are the ideal instruments for measuring force vs. displacement data from cantilevers because they are capable of measuring both tip displacement and tip force continuously during measurements [6]. Triangular cantilevers with optimized strain profiles [7] and pure tensile fixed beams [8] have also been fabricated for nanoindenter testing. A profilometer is a less expensive alternative with comparable displacement measurement precision, and is capable of measuring stylus displacement at fixed tip forces. Force displacement measurements of Ni cantilevers over silicon trenches have already been taken with a profilometer [9]. NiTi cantilevers have been made and tested with tiny weights [10], but they were made with very thick rolled sheet (50-95 microns). Bending micro-scale NiTi beams with a profilometer will confirm SMA actuator abilities for MEMS.

EXPERIMENTAL DETAILS

NiTi films were deposited with DC magnetron co-sputtering onto un-heated 100 mm <100> silicon wafers. Independently controlled 2" diameter cathodes of equiatomic NiTi and Ti were chosen to create equiatomic NiTi thin films. Pure Argon gas (99.999%) at 3.5 mTorr supplied the ions for sputtering. The pure Ti cathode compensated for the imbalanced sputtering rate of Ni and Ti. Substrate rotation during the deposition increased compositional uniformity. Piranha etchant (H_2SO_4 (96%): H_2O_2 (30%), 9:1 mixture by volume, at 70°C for 10 minutes) removed organic material from the wafers, and a buffered oxide etchant (1 minute at 25°C) stripped the native oxide prior to sputtering. Pre-sputtering both cathodes at 100W for 30 minutes with a shutter covering the substrate removed target surface oxidation. Pre-sputtering also replicated a sputter-ion pump, and reduced the system base pressure below 2 x 10^{-7} Torr. A 100 nm chromium film on top of the clean silicon served as an adhesion promoter, and will later serve as a mask layer for etching the silicon substrate. Varying the Ti cathode power through a series of NiTi film depositions changed the resulting film composition. Energy dispersive x-ray spectroscopy (EDS) analysis with pure metal calibration standards determined the film composition. Profilometer measurements of film thickness across the entire wafer determined the deposition rate to be 12Å per second.

DISCUSSION

The deposition process is capable of producing equiatomic NiTi thin film at a reasonable deposition rate. Table I. lists the values for Ti-rich and Ni-rich thin films and the power settings. Single watt adjustability of the Ti cathode power allows for precise composition control since a 2 at. % shift in titanium content occurs over a 25 W range. Similar deposition parameters of 250 W to a NiTi target and 20 W to a Ti target were used by Shih [11] to obtain equiatomic films. The use of 3" targets by Shih, instead of 2" targets, most likely reduced Ti depletion and the need for high Ti cathode power. The NiTi film adhesion to silicon with a Cr adhesion layer is excellent. Figure 1(a), shows the torn silicon surface after removing the NiTi/Cr film. The adhesion between the metal layers and the pure silicon was stronger than the internal silicon bonds. Figure 1(b) shows how the poor adhesion between NiTi/Cr and SiO_2 allowed easy removal of the film without altering the substrate. Future processing of test structures should be robust because of the good composition control and the excellent adhesion.

Table I. NiTi thin film composition measurements at different Ti target power settings

NiTi Power (W)	Ti Power (W)	At. % Ti	At. % Ni
250	50	48	52
250	75	50	50
250	100	52	48

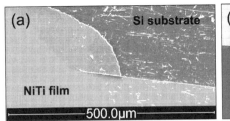 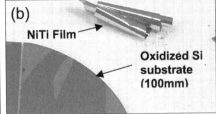

Figure 1. (a) Rough Si substrate created by removing NiTi film during wafer cleaving, and (b) easily removed NiTi film from oxidized Si wafer.

CONCLUSIONS

Co-sputtering of NiTi thin films with excellent deposition rate is possible with the described system and sputtering procedure. Pre-sputtering with a substrate shield is essential to eliminate oxygen content from the cathode surface and the vacuum chamber before deposition. The native oxide on silicon substrate must be removed to ensure good adhesion.

Future experimental work will include post deposition annealing and shape memory evaluation of the thin films. Differential Scanning Calorimetry (DSC) analysis will determine the start and finish transition temperatures for the low and high temperature crystalline phases. Grazing incidence XRD in conjunction with a temperature stage will verify the presence of each phase. Finally, measuring the displacement of micro-scale NiTi cantilevers while applying known loads at various temperatures will yield the power capabilities and fundamental mechanical properties. Evaluating the shape memory and mechanical behavior, as functions of temperature, should yield the efficiency of micro-scale NiTi actuators, and more importantly optimize film processing.

REFERENCES

1. P. Krulevitch, A. Lee, P. Ramsey, J. Trevino, J. Hamilton, M. Northrup, J. MEMS. Vol. 5 No. 4, 274 (1996)
2. Y. Fu, H. Du, S. Zhang, Surface and Coating Technology 167, 120 (2003)
3. A. Ishida, A. Takei, M. Sato, S. Miyazaki, Thin Solid Films 281-282, 337 (1996)
4. Y. Fu, H. Du, W. Huang, S. Zhang, M. Hu, Sen. & Act. A 112, 397 (2004)
5. A. Ishida, M. Sato, Acta Materialia 51, 5578 (2003)
6. D. Son, J. Jeong, D. Kwon, Thin Solid Films 437, 183 (2003)
7. J. Florando, W. Nix, J. Mech. and Phys. of Solids 53, 621 (2005)
8. H. Espinosa, B. Prorok, M. Fischer, J. Mech. and Phys. of Solids 51, 49 (2003)
9. J. Luo, A. Flewitta, S. Spearingb, N. Flecka, W. Milnea, Mate. Lett. 58, 2307 (2004)
10. M. Kohl, D. Allen, T. Chen, S. Miyazaki, M. Schwörer, Mat. Sci. & Eng. A 270, 146 (1999)
11. C. Shih, B. Lai, H. Kahn, S. Philips, A. Heuer, J. MEMS VOL. 10 NO. 1, 73 (2001)

Mater. Res. Soc. Symp. Proc. Vol. 875 © 2005 Materials Research Society   O6.5

Influence of Gas Atmosphere on the Plasticity of Metal Thin Films

T. Wübben[1], G. Dehm[3], E. Arzt[1,2]

[1]Max-Planck Institute for Metals Research, Stuttgart, Germany
[2]Institut für Metallkunde, University of Stuttgart, Stuttgart, Germany
[3]now at Erich Schmid Institut für Materialwissenschaft, Österreichische Akademie der Wissenschaften, and Department für Materialphysik, Montanuniversität Leoben, Leoben, Austria

ABSTRACT

Stresses in thin films are routinely measured by the so-called substrate curvature technique. These experiments are usually carried out in air or under a protective gas atmosphere. In this contribution we describe a new set-up capable of performing substrate curvature measurements under ultra-high vacuum conditions. The advantages are the absence of possible artifacts due to gas/film interactions, better control of gas composition, and the possibility to measure chemical effects on mechanical properties in a controlled way. We present first results that indicate an unexpected sensitivity even of polycrystalline Cu films to the gas environment.

INTRODUCTION

The substrate curvature technique is a widely used method for investigating the elastic and plastic properties of metal thin films. It makes use of the stress that is induced in the film due to thermal mismatch during temperature changes. The film stress exerts a bending moment on the substrate, whose curvature can be measured by different techniques.

Substrate curvature experiments can in principle be carried out on every kind of material. However, most of the equipment in use today performs the required thermal cycles in high temperature furnaces without control of the atmosphere. To prevent the samples from oxidation, a flow of inert gas, e.g. nitrogen or argon, is usually sustained and a possible interaction of gas and sample is commonly neglected.

In view of the high specific surface area of thin films, the surface state may strongly influence the experimental results. It is for instance known that a passivation such as an oxide layer on the sample surface can lead to an increased flow stress (1-4). The presence of an oxide layer during a temperature cycle, however, cannot be controlled under gas flow conditions. A controlled atmosphere during thermal cycling is also desirable for experiments investigating the stress development in a metal under catalytic conditions. This is of special interest for chemical applications, e.g. when metals are used as catalysts in the steam reformation or oxidation of methanol (CH_3OH) in production lines for formaldehyde (CH_2O). The influence of stress on the catalytic performance is currently under investigation in this community (5).

In view of these limitations, we have developed a substrate curvature apparatus capable of operating under ultra high vacuum (UHV). Several design problems had to be solved to reach this goal. The new equipment now allows experiments under controlled ambient conditions.

EXPERIMENTAL

A schematic 3D representation of the system is shown in figure 1. The specimen is heated and cooled inside a vacuum chamber while a laser beam is directed through a viewport onto the substrate to measure its curvature (for details on this method, see e. g. (6, 7)). In the following the components are described in more detail.

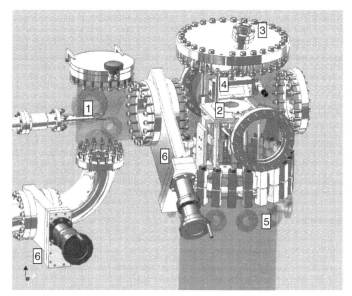

Figure 1: New system for substrate curvature measurements under UHV conditions; (1) sample transfer, (2) heater and sample, (3) viewport for pyrometer measurement, (4) mirror, (5) feedthroughs for external supplies, (6) gate valves. The pumping system and the optical measurement devices are not visible here.

Vacuum system

The central component is a self-built UHV vacuum chamber, which is evacuated by a pumping system consisting of an oil diffusion pump and a turbo molecular pump (200 liters/s). A vacuum quality of the order of 10^{-7} mbar (10^{-5} Pa) can be achieved. To maintain the vacuum or the gas atmosphere during sample exchange, a manual gate valve (6) is used to separate sample chamber (left in figure 1) and furnace chamber. For optical access, both chambers are equipped with several viewports.

For the control of the gas atmosphere in the UHV chamber, a leak valve mounted near the sample position is used (not visible in figure 1). In this way the pressure inside the chamber can be controlled within the accuracy of the pressure measurement. The leak valve can be connected to various gas supplies, depending on the requirements of the planned experiment. This includes an argon supply for purging the chamber or sustaining an inert atmosphere as well as evaporators used for example to provide a methanol-rich ambient. In addition, the chamber is equipped with a quadrupole mass spectrometer to analyze the partial pressures of the gas constituents.

Experiments can be carried out under flow or non-flow conditions. However the maximum pressure cannot exceed approximately 1 mbar in the flow case due to the limitations of the pumping system.

Heating system

To provide direct and efficient heating of the sample, a ceramic heater element was chosen (2). The maximum temperature achievable by this device is above 1500°C. However, due to limited temperature capability of other parts, sample temperatures are currently restricted to 900°C. A direct measurement and control of the sample temperature is ensured by the use of a pyrometer (through viewport (3)).

Sample heating rates above 20K/min can easily be achieved in the present set-up. Cooling limits the rate of the measurements as cooling rates in vacuum are usually low. In the present equipment, the cooling rate was found to be better than 2K/min, even near room temperature. Above 200°C cooling rates are in the range of 6K/min.

Sample transfer and curvature measurement

A manually driven transfer bar (1) is used for transferring wafers up to 4" diameter. Upon transfer into the chamber, the sample is lifted from the holder by means of a stepper motor that moves three tips mounted in plane with the heater element (2). Before starting a measurement, the holder is removed and the sample is lowered towards the heater element. The distance between heater and sample can be adjusted and is usually about 1mm.

The curvature of the film-substrate composite is measured by a laser optical system. A laser beam is directed to the sample via an arrangement of two mirrors outside and one inside (4 in fig. 1) the UHV system. All mirrors can be adjusted separately by micrometer screws. The beam scans once across the sample diameter and its reflection is then guided back to its origin, where a position sensitive detector is mounted. Approximately 100 data points are recorded per scan. From the measured signal the curvature of the sample is then calculated by a computer program and for each data point measurement parameters such as time are saved. To obtain the initial curvature change induced during thermal annealing, reference data of the bare substrate recorded before film deposition can be subtracted. In this way absolute film stresses can be measured.

For the calibration of the system, two mirrors with defined curvatures of 20.098 and 100.8 m, respectively, were used. After at least 10 calibration runs the measurements reproduce these values with a relative standard deviation of less than 0.2%. For radii above 100 m but below 500 m, the accuracy is still better than 1% (relative standard deviation) of the measured value. The maximum detectable radius of curvature is of the order of 2 km.

FIRST RESULTS

Accuracy and reproducibility of temperature cycles

To evaluate the reproducibility of the measurements as a major quality criterion, copper films on silicon were used as test samples. This material is known from earlier investigations (8, 9) to show a high reproducibility of temperature cycles.

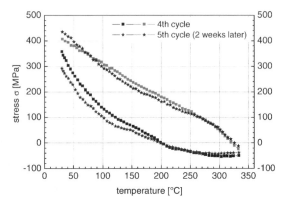

Figure 2: Two temperature cycles of a polycrystalline 200 nm thick Cu film in vacuum under identical conditions.

In figure 2 we show two temperature cycles of 200 nm thick polycrystalline copper on a silicon wafer. The copper film was magnetron sputtered and annealed for 10 min. at 600°C under UHV conditions. Since the first cycle after deposition usually differs significantly from subsequent measurements, the fourth and the fifth cycle, taken with a time difference of about 2 weeks, are shown as examples.

As can be seen from figure 2, both cycles show a very similar behavior. Especially the flow stresses in compression and at 50°C in the cooling cycle are almost identical within about 10 MPa. Only at the beginning of the heating cycles can differences be observed. The lower stress in the second cycle is probably due to relaxation effects that have taken place during the

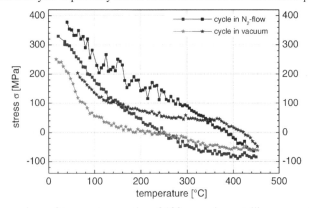

Figure 3: Comparison of temperature cycles of 100 nm polycrystalline copper thin films in vacuum (better than 10^{-6} mbar) and under nitrogen flow; the N_2-cycle was carried out less than 4 hours after the vacuum cycle. The noise in the nitrogen cycle is due to external disturbances.

storage of the sample. The high reproducibility could be shown also for other samples with different film thicknesses and thus confirms the reliability of the new system.

Comparison with measurements under nitrogen flow conditions

The influence of the gas ambient on substrate curvature measurements has, to our knowledge, not been explored previously. It is to be expected that the behavior of oxidation-resistant metals should not be affected by different oxygen partial pressures. This has been confirmed by first measurements at palladium films. By contrast, copper samples showed different behavior when cycled under nitrogen flow compared to vacuum conditions. In figure 3 two cycles of a 100 nm thick polycrystalline copper sample, under nitrogen flow and vacuum conditions (better than 10^{-6} mbar), respectively, are shown. To avoid relaxation effects, the nitrogen cycle was carried out a short time after the vacuum cycle.

As can be seen from the figure, the two curves show significant differences in absolute stress values and in the general shape of the stress/temperature curve. While the sample shows only minor plastic behavior when heated under N_2-flow, yielding occurs in the heating as well as the cooling branch for the cycle carried out in vacuum. This behavior is reproduced exactly in the second cycle (to 450°C) in vacuum after the nitrogen experiment, while the first cycle shows an intermediate shape.

The difference might be caused by the formation of a thin oxide layer due to air exposure during the sample transfer. This can lead to enhanced strength of a thin film due to the prevention of surface diffusion (8). Apparently the oxide layer was at least partially removed during previous cycling in vacuum and at temperatures up to 450°C. This assumption is supported by the observation that the effect can be reversed by subsequent cycling in vacuum. However, for an unambiguous confirmation chemical analysis has to be carried out in the future.

Influence of catalytic ambient on stress-temperature behavior of copper films

A major purpose to design a wafer curvature system with controlled ambient conditions was the investigation of chemical effects on thin film stress and vice versa. As an example, the industrially important catalytic oxidation of methanol over solid copper (see, e. g. (10-12)) was chosen. In figure 4 a temperature cycle of a 200 nm thick copper film cycled in vacuum is com-

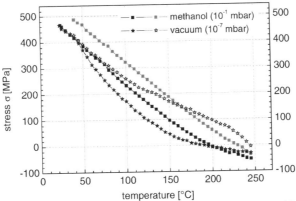

Figure 4: Comparison of temperature cycles performed under different ambient conditions (200 nm polycrystalline Cu); the methanol cycle was performed directly after several vacuum cycles, the last of which is shown in the figure.

pared to results obtained in a methanol ambient.

From the figure it can be inferred that no stress difference occurs in the heating phase for temperatures below 50°C. Above this, however, a significant deviation occurs: compared to the vacuum cycle the methanol-affected curve is less steep and shows less deviation from elastic behavior; hence the stress level reached at 250°C is comparable. Upon cooling, the same phenomenon of a more linear stress-temperature relationship for the methanol cycle can be observed.

The reasons for this significantly different behavior are still unclear. It might be due to adsorbates that form at the film surface. This could lead to an additional passivation layer causing a higher yield stress of the sample. Further measurements are currently under way to test the reliability of these results and to exclude all possible artifacts. If substantiated, these results suggest that the mechanical behavior of thin films may be sensitive to the gas environment. It can be speculated that atmospheric variations can be responsible for so far unexplained deviations from usual stress/temperature curves sometimes found in the literature.

CONCLUSION AND OUTLOOK

In this paper we present a newly developed substrate curvature system capable of measuring thin film stresses in ultra-high vacuum or under controlled atmospheres. First experiments confirmed the reliability and reproducibility of the system. Characteristic differences in stress/temperature curves could be observed in Cu films, depending on whether they were cycled in nitrogen flow, in vacuum or in methanol-rich atmosphere. Since these differences occur mostly at higher temperatures and not for oxidation-resistant materials, the mechanical effects may be due to surface reactions or films. Experiments under UHV-conditions therefore promise to reveal new insights in metal film plasticity.

ACKNOWLEDGEMENTS

We acknowledge the support of the German Research Foundation (DFG) in the frame of the priority program SPP 1091 "Heterogeneous Catalysis" under DE 796/5-4.

REFERENCES

1. M. J. Kobrinsky and C. V. Thompson, *Applied Physics Letters* **73**, 2429 (1998).
2. D. Weiss, H. Gao, and E. Arzt, *Acta Materialia* **49**, 2395 (2001).
3. R.-M. Keller, S. P. Baker, and E. Arzt, *Acta Materialia* **47**, 415 (1999).
4. H. Gao, L. Zhang, W. D. Nix, C. V. Thompson, and E. Arzt, *Acta Materialia* **47**, 2865 (1999).
5. S. Sakong and A. Gross, *Surface Science* **525**, 107 (2003).
6. W. D. Nix, *Metallurgical Transactions A* **20A**, 1989 (1989).
7. P. A. Flinn, *in* "Thin Films: Stresses and Mechanical Properties" (J. C. Bravman, W. D. Nix, D. M. Barnett, and D. A. Smith, eds.), Vol. 130, p. 41–51. Materials Research Society, Pittsburgh, PA, 1989.
8. T. J. Balk, G. Dehm, and E. Arzt, *Acta Materialia* **51**, 4471 (2003).
9. G. Dehm, T. J. Balk, H. Edongue, and E. Arzt, *Microelectronic Engineering* **70**, 412 (2003).
10. C. Ammon, A. Bayer, G. Held, B. Richter, T. Schmidt, and H.-P. Steinruck, *Surface Science* **507-510**, 845 (2002).
11. S. M. Francis, F. M. Leibsle, S. Haq, N. Xiang, M. Bowker, *Surface Science* **315**, 284 (1994).
12. R. Ryberg, *The Journal of Chemical Physics* **82**, 567 (1985).

Mater. Res. Soc. Symp. Proc. Vol. 875 © 2005 Materials Research Society

Laser Lateral Crystallization of Thin Au and Cu Films on SiO$_2$

J.E. Kline and J.P. Leonard, Department of Materials Science and Engineering, University of Pittsburgh, Pittsburgh, PA 15261

Rapid lateral solidification via excimer laser melt processing is demonstrated in 200 nm thick pure Cu and Au films, encapsulated above and below by amorphous SiO$_2$. Mask projection irradiation is used to selectively melt lines 3 to 30 μm wide in the metal films, with lateral solidification proceeding transversely from the edge to the middle of the line. Encapsulation with the SiO$_2$ overlayer and control of the fluence are found to be crucial parameters necessary to prevent dewetting while the films are molten. Transmission electron microscopy reveals large columnar grains with twin structures and other defects typical of rapid solidification.

1 Introduction

Microstructural modification and control in metallic thin films on insulating substrates is a significant materials challenge for a number of emerging device applications. These include: 1) Electrical properties including conductance, scattering losses, and electromigration in sub-100 nm metallic interconnect structures[1]. 2) Compositional and microstructure requirements in electrode attachments for ferroelectric devices[2] and magnetic storage applications. 3) Development of new metal-based actuators, membranes, and other structural components in emerging MEMS/NEMS devices and sensors[3].

Current strategies to effect microstructural control in metal thin films involve parameters associated with the thermodynamics and kinetics of the deposition process, which are typically vapor or electrochemical methods. Heat treatment can also be used to relax stresses, refine crystalline microstructures, redistribute alloy or impurity components, or initiate the formation of new phases. These methods are neither spatially selective, nor direct— they must act through the very thermodynamic and kinetic mechanisms that drive the stochastic development of microstructure—which generally lead to broad grain size distributions over limited ranges.

There is a need for processing techniques that can directly modify the microstructure and properties of metal films after deposition, while avoiding the problems associated with conventional heat treatment. Such a processing technique should allow 1) Microstructural and spatially selective control of metallic sub-micron scale features in two- or three-dimensions, 2) Composition control with engineered and non-equilibrium redistribution of alloying elements, phases, and precipitates, and 3) Compatibility with existing device materials and processes, in particular a low processing temperature, but also should allow control of stress, adhesion, diffusion and interfacial chemistry.

Rapid solidification processing (RSP) is a good example of a potential alternative process, with a demonstrated capability to produce unique properties in metals not otherwise available[4]. A rich array of phenomena have been discovered and investigated using RSP, including grain size reduction and refinement, extending the limits of solid solubility, amorphous and metastable phase formation, as well as eutectic, dendrite and precipitate structures. Conventional RSP has largely centered on bulk processing of pure metals and alloys using techniques such as splat and anvil quenching, melt spinning and twin rolling, levitation, or plasma spray atomization and deposition onto a surface[5]. The resulting material is usually in ribbon, powder, or a thick film aggregate coating. None of these are compatible with high purity micro-device technologies, and all suffer from two major problems. Consolidation of rapidly solidified material into useful structures is difficult without reverting to the equilibrium phase, and secondly, there is limited control over the solidification process and a lack of quantitative in-situ measurements.

Pulsed laser irradiation, like conventional RSP, can rapidly melt metal thin films that will resolidify under a wide range of conditions[6-11]. It has been widely studied in the context of semiconductor materials[12,13], and has provided insights into interfacial kinetics[14] and metastable phase formation[15]. Unlike conventional RSP, pulsed laser melting is a clean, particulate-free, non-contact method compatible with device processing—now commonly used in crystallization of Si thin films on glass[24]. Pulsed laser melting can form rapidly solidified metals of virtually any composition in a thin film geometry, while the films remain integral with the substrate.

Studies of pulsed laser melting in metallic thin films on SiO_2 or other dielectrics are comparatively rare. To our knowledge, only Al[16,17], Au[8,9,18,19], Cu[9,20], Fe[21-23] have been attempted, as well as a few alloys. This can be attributed to the tendency of these films to dewet from the substrate when molten, as well as limited device applications.

In the current work, we demonstrate the laser melting and subsequent formation of rapidly laterally solidified microstructures in pure Cu and Au films on amorphous SiO_2. In these systems, large grains are formed, extending tens of microns laterally across the irradiated region. These results indicate that metallic RSP can be integrated into thin-film micro device technology, potentially leading to many new applications. Furthermore, a quantitative understanding and investigation of the rapid lateral solidification process may open up new areas in fundamental studies of metallic RSP kinetics, composition, and structural development.

2 Experimental

Samples were prepared via sputter deposition of elemental copper and gold films on thermally oxidized (1 μm) Si (001) wafers. Before deposition, the SiO_2 surface was rinsed with acetone, isopropanol and deionized water. Sample films of 200 nm Cu were deposited by magnetron sputtering in a CVC Connexion Sputtering System under a base pressure of approximately 2 x 10^{-8} Torr, working pressure of 5 mTorr Ar, room temperature, 150 W target power. Samples with 200 nm Au films were prepared by magnetron sputtering in a Perkin-Elmer 2400-6J with a base pressure of 8 x 10^{-7} Torr, working pressure of 5 mTorr Ar, room temperature, 50 W target power. Both samples were immediately transferred, in a class 100 clean room, to a Perkin-Elmer 2400-8L RF system, where 580 nm of SiO_2 was deposited in under a base pressure approximately 5 x 10^{-7} Torr, passive heating, 5mTorr Ar working pressure. Voltages were 200W forward (4W reflected), 3V substrate bias, 810V to target, and the deposition rate was 2.57 nm/min.

The SiO_2 capping layer was found to be crucial to prevent dewetting of the metal films during laser melting. Unlike silicon, most metal films adhere poorly to insulators such as SiO_2[25] or Al_2O_3[26], particularly when molten. Hole formation and film agglomeration occur on nanosecond timescales, due to low viscosity of the melt and driven by the higher surface energy of the metals with respect to the underlying substrate[27]. Through a series of experiments, reported elsewhere[28], this encapsulation technique has been developed to fully suppress dewetting in the molten Cu and Au on SiO_2.

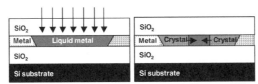

Figure 1.   Schematic of lateral solidification in metal films on SiO_2. a) Excimer laser projection irradiation completely melts the film in a localized region. b) Lateral resolidification from unmelted side regions into the supercooled molten pool.

Films were subjected to pulsed excimer melting using projection irradiation in a single line configuration, as indicated schematically in Figure 1. The optical system consisted of a Lambda Physik EMG-202 KrF excimer laser, a collimating telescope, energy meter, a copper single-slit mask, and a single quartz lens positioned for a 5x demagnification. The resolution was typically 2.0 microns or better, as estimated from the sharpness at the edges of the line. Fluence, as well as transverse and longitudinally and uniformity were carefully controlled[28]. It was found that the best results were obtained in a fluence window situated immediately above the complete melting threshold F_{cm} (430-650 mJ/cm^2 for Au, 400-900 mJ/cm^2 for Cu). This corresponds to condition where all crystallites within the pool are melted, yet avoids significant overheating that can lead to film damage. Irradiation line-widths were varied from 3 to 30 um, while the fluence ranged between 1.0 and 1.2 F_{cm}. Large columnar grains extending laterally from the unmelted polycrystalline edge regions to the center of the line were observed in both materials.

3 Results

TEM samples of the irradiated films were prepared by the liftoff method using undiluted hydrofluoric acid (49%), which also removed the encapsulation while leaving the metal films intact. Films were examined in a JEOL 2000FX 200kV STEM. Examples of the microstructures found in the irradiated line regions are shown in Figures 2 and 3.

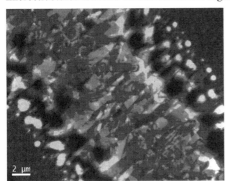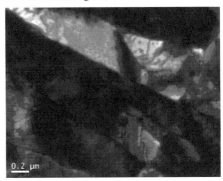

Figure 2.   a) Bright field plan-view TEM micrograph of laterally solidified microstructure in 200 nm Au film after line irradiation (600 mJ/cm^2). b) Magnified view of center line formed by opposing solidification fronts.

Figure 3.   a) Bright field plan-view TEM micrograph of laterally solidified microstructure in 200 nm Cu film after line irradiation (850 mJ/cm^2). b) Magnified view of center line formed by opposing solidification fronts.

In both cases, the films reveal a directionally solidified microstructure in the irradiated line region, flanking the as-sputtered fine-grain polycrystalline regions that were not melted. A boundary running down the center of the line is believed to result from the meeting of opposing lateral solidification fronts. In the case of Au, light and dark spots along the along the edges of the line are due to thickness variations present in the post-solidified metal film. Likewise, small pinholes are visible in the central boundary of the Cu laterally solidified region, possibly due to damage or selective etching during sample preparation.

4 Discussion

Plan view TEM analysis of the Au and Cu films, as shown in Figures 2 and 3, indicates that single elongated grains extend from the edge of the lines to the central boundary. Grains appear to extend vertically through the full 200 nm film thickness. This microstructure is consistent with models for complete melting and lateral solidification in a manner analogous to silicon on SiO$_2$[29,30]. In this model, solidification is initiated by the unmelted side walls, and proceeds into the center of the supercooled liquid pool, eventually meeting the solidification front approaching from the opposite side. In contrast to Si the total solidification distance in Au and Cu is found to be much larger, up to 15 μm in experiments with the widest (30 um) lines studied. This is reasonable, owing to the higher interface velocities possible in rapid solidification of FCC metals[4]. Preliminary numerical investigations using 3DNS have been conducted to estimate the conditions for rapid lateral solidification under conditions similar to the experiments shown in Figures 2 and 3. In both Cu and Au, the lateral solidification velocity is found to range from 30-100 m/sec, believed to be stabilized by the negative slope of the interface response function, the latent heat release at the advancing solidification front, and heat conduction into the substrate. Details will be presented elsewhere.

These results indicate that it is possible to 1) Completely melt metal films that remain continuous on the nanosecond timescales associated with excimer melting. 2) Avoid nucleation at the SiO$_2$-metal interface or in the melt, thereby allowing supercooling and lateral solidification from the unmelted sidewalls.

Additionally it is found that: 1) The laterally solidified grains appear to widen to approximately 1 µm by competitive occlusion of neighboring grains during growth, similar to silicon. This suggests the selection of preferred growth orientations, although additional orientation analysis is needed in order to quantify the degree of texture in these regions. 2) There is no evidence of 'thermal' dendritic growth, which has been observed in rapid solidification of droplets under similar conditions. This is possibly due to stabilization of the interface by the thin-film geometry or the local temperature gradients established at the solidification front. The central seam formed by the opposing growth fronts is generally straight, suggesting a nearly uniform lateral growth velocity for all grains. 3) Grains have significant intra-grain defect structure, bearing further study. Preliminary TEM observations suggest twins or stacking faults present along (111) planes in both Au and Cu, as well as evidence of possible vacancy clusters. 4) Preliminary counts find a density of 3×10^{10} cm/cc for line dislocations, while the density of features that appear to be consistent with the presence of Frank loops are significantly higher, perhaps 10^{12} to 10^{13} cm/cc. Further quantitative study is underway to explore these effects in greater detail.

5 CONCLUSION

The rapid lateral solidification technique has been successfully applied to thin films of elemental Au and Cu. This has been achieved by careful preparation of the metal films and a SiO_2 encapsulation layer that suppresses the liquid-phase dewetting phenomena previously observed in these systems. Pulsed laser melting is a promising new route to fundamental studies of the kinetics of rapid solidification in metals, which has been until now largely limited to droplet supercooling and atomistic simulations. Rapid lateral solidification has numerous advantages, including: 1) Independently adjustable parameters, including the laser fluence, film thickness, substrate thermal conductivity, and substrate temperature. These can allow controlled melting and solidification over a wide range of thermal conditions. 2) Production of large-area, large-grain microstructures in uniformly thick films, amenable to detailed quantitative and statistical analysis of crystallographic orientations, textures, grain boundaries, and defect networks. 3) Access to high purity metal films of virtually any composition (via co-sputtering), ideal for studies of segregation, interface kinetics, and metastable phase formation.

The rapid lateral solidification technique also appears to be a promising route to post-deposition and post-patterned microstructural modification of metal thin films for device applications. In particular, it is believed this process may someday be capable of producing ultra-large single-crystal metal films on insulating substrates, directionally solidified alloys with important anisotropic magnetic and electronic properties, as well as new engineered composition gradients for future devices.

6 ACKNOWLEDGEMENTS

The authors wish to thank the Penn State Nanofabrication Facility and the Carnegie Mellon Center for Magnetic Materials for assistance and equipment used in the preparation of samples.

7 REFERENCES

1. S.P. Murarka, I.V. Verner, R.J. Gutmann, Copper: Fundamental Mechanisms for Microelectronic Applications, Wiley & Sons, (2000).

2. R.E. Jones and S.B. Desu, Mater. Res. Bull. **21**, 55 (1996).
3. T. Fukuda and F. Arai, in <u>Nanotechnology</u>, edited by N. Taniguchi, (Oxford University Press, Oxford UK, 1996). pp. 155-160.
4. D.M. Herlach, Mater. Sci. Eng. R **12**, 177 (1994).
5. H. Jones, Mater. Sci. Eng. A **304-306**, 11 (2001).
6. N.B. Dahotre (ed.), <u>Lasers in Surface Engineering</u>, ASM International, Surface Engineering Series Vol. 1 (1998).
7. E. Schubert, H.W. Bergmann, in <u>Rapidly Solidified Alloys</u>, Ch. 8, edited by H.H. Liebermann (1993).
8. C.A. MacDonald, A.M. Malvezzi and F. Spaepen, J. Appl. Phys. **65**, 129 (1989).
9. J. Boneberg, J. Bischof and P. Leiderer, Opt. Commun. **174**, 145 (2000).
10. J. Bloch and Y. Zeiri, J. Appl. Phys. **61**, 2637 (1987).
11. S. Vitta, A.L. Greer and R.E. Somekh, Mater. Sci. Eng. A **179-180**, 243 (1994).
12. C.W. White, M.J. Aziz, in <u>Surface Alloying by Ion, Electron, and Laser Beams</u>, Ch. 2, (1987).
13. S.J. Moon, M. Lee and C.P. Grigoropoulos, J. Heat Transf. **124**, 253 (2002).
14. J.A. Kittl, P.G. Sanders, M.J. Aziz, D.P. Brunco and M.O. Thompson, Acta Mater. **48**, 4797 (2000).
15. K.K. Desfulian, J.P. Krusius, M.O. Thompson and S. Talwar, Appl. Phys. Lett. **81**, 2238 (2002).
16. J.Y. Tsao, S.T. Picraux, P.S. Peercy and M.O. Thompson, Appl. Phys. Lett. **48**, 278 (1986).
17. S. Williamson, G. Mourou and J.C.M. Li, Phys. Rev. Lett. **52**, 2364 (1984).
18. B.E. Homan, M.T. Connery, D.E. Harrison and C.A. MacDonald, Mat. Res. Soc. Symp. Proc. **279**, 717 (1993).
19. P.B. Comita, P.E. Price and T.T. Kodas, J. Appl. Phys. **71**, 221 (1992).
20. A.J. Pedraza and M.J. Godbole, Metal. Trans. A **23A**, 1095 (1992).
21. S. Vitta, A.L. Greer and R.E. Somekh, Mater. Sci. Eng. **98**, 105 (1988).
22. C.B. Arnold, M.J. Aziz, M. Schwarz and D.M. Herlach, Phys. Rev. B **59**, 334 (1999).
23. H.A. Atwater, J.A. West, P.M. Smith, M.J. Aziz, J.Y. Tsao, P.S. Peercy and M.O. Thompson, Mat. Res. Soc. Symp. Proc. **157**, 369 (1990).
24. A.T. Voutsas, Appl. Surf. Sci. 208-209, **250** (2003).
25. J. Bischof, D. Scherer, S. Herminghaus and P. Leiderer, Phys. Rev. Lett. **77**, 1536 (1996).
26. M.J. Godbole, A.J. Pedraza, D.H. Lowndes and E.A. Kenik, J. Mater. Res. **4**, 1202 (1989).
27. D.J. Srolovitz and M.G. Goldiner, Journal of Metals **3**, 31 (1995).
28. E. Kline, J. Leonard, Mat. Res. Soc. Symp. Proc., **854E**, U11.6.1 (2004)
29. H.J. Kim and J.S. Im, Appl. Phys. Lett. **68**, 1513 (1996).
30. J.S. Im, M.A. Crowder, R.S. Sposili, J.P. Leonard, H.J. Kim, J.H. Yoon, V.V. Gupta, H.J. Song and H.S. Cho, Phys. Status Solidi **166**, 603 (1998).
31. K. Eckler and D.M. Herlach, Mater. Sci. Eng. A **178**, 159 (1994).
32. J.P. Leonard and J.S. Im, Appl. Phys. Lett. **78**, 3454 (2001).
33. S.E. Battersby, R.F. Cochrane and A.M. Mullis, J. Mater. Sci. **35**, 1365 (2000).
34. M. Lee, S. Moon, M. Hatano and C.P. Grigoropoulos, Appl. Phys. A **73**, 317 (2001).
35. L. Wang, H. Liu, K. Chen, Z. Hu, , Physica B, **239**, pp. 267 (1997).
36. D.Y. Sun, M. Asta, J.J. Hoyt, , Phys. Rev. B, **69**, 024108 (2004).

Thin Film Plasticity—
Modeling

Mater. Res. Soc. Symp. Proc. Vol. 875 © 2005 Materials Research Society   O7.4

A New Dislocation-Dynamics Model and Its Application in Thin Film-Substrate Systems

E.H. Tan and L.Z. Sun
Department of Civil and Environmental Engineering and Center for Computer-Aided Design
The University of Iowa, Iowa City, IA 52242-1527, U.S.A.

ABSTRACT

Based on the physical background, a new dislocation dynamics model fully incorporating the interaction among differential dislocation segments is developed to simulate 3D dislocation motion in crystals. As the numerical simulation results demonstrate, this new model completely solves the long-standing problem that simulation results are heavily dependent on dislocation-segment lengths in the classical dislocation dynamics theory. The proposed model is applied to simulate the effect of dislocations on the mechanical performance of thin films. The interactions among the dislocation loops, free surface and interfaces are rigorously computed by a decomposition method. This framework can be used to simulate how a surface loop evolves into two threading dislocations and to determine the critical thickness of thin films. Furthermore, the relationship between the film thickness and yield strength is established and compared with the conventional Hall-Petch relation.

INTRODUCTION

Although dislocation dynamics has been actively researched since 1960, the ability to simulate the evolution of dislocations is still limited by the inherent complexity of this problem. To date, various numerical methods [1-4] have been developed to simulate the evolution and interaction of dislocations. As Gómez-García et al. [1] have indicated, a long-standing problem inherent to these methods is that the selection of both larger and smaller segment sizes in discretization yields unreasonable simulation results; only segment sizes approximating a certain optimal length, which is different for each problem, can lead to reasonable results. Ghoniem et al. [3] also mentioned a peculiar "velocity jump" phenomenon, shown in Fig. 1(a), where the velocity near the pinning points of a straight dislocation line will jump with the increase of segment number. This contradicts numerical-method principles that suggest finer discretization should generally lead to more accurate results. Consequently, simulation results are always significantly affected by the artificial choice of segment sizes, shown in Fig. 1(b), which substantially reduces the reliability of simulation results. The purpose of this paper is to ascertain the reason for this abnormal phenomenon, develop a new dislocation dynamics model to resolve this issue and further apply this new model to thin-film substrate systems.

A NEW DISLOCATION-DYNAMICS MODEL

In order to clarify why the segment-size dependence phenomenon occurs, it is first necessary to review the foundation of these methods. Fundamentally, all the above-mentioned dislocation dynamics methods are based on the governing equation:

$$F_i - B_{ij}V_j = 0 \tag{1}$$

where F_i is the local glide force per unit length acting on a dislocation, B_{ij} is the diagonal drag coefficient matrix that is reduced to a scalar in isotropic cases, and V_j is the local velocity of a dislocation. $B_{ij}V_j$ is also referred to as the drag force. In Equation (1) and subsequent equations herein, the free indices take values from 1 to 3 and the dummy indices are summed from 1 to 3 unless explicit statements are given. As determined from the Peach-Koehler formula, F_i can be written as:

$$F_i = -\left(b_j \sigma_{jk} n_k\right) t_i \tag{2}$$

where b_j is the Burgers vector of the dislocation, σ_{jk} is the local stress tensor, n_k is the unit normal to the slip plane, and t_i is the local in-plane unit normal to the dislocation line. The stress tensor σ_{jk} includes stresses due to lattice resistance, imperfections, applied stresses, and the self-stress due to the dislocation itself. Determination of the first three components of the stress tensor σ_{jk} is relatively simple, although they can occasionally pose some difficult problems in their own rights. The self-stress induced by a dislocation itself is best found using Brown's formula [5] or the Barnett's formula [6] in order to avoid singularity problems.

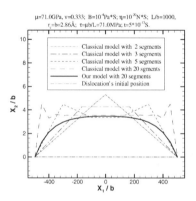

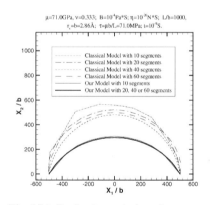

Fig. 1(a) Displacement (velocity) jumps occur near the pinning points of a straight edge dislocation in the first time step with the segment number increase in the classical dislocation dynamics model

Fig. 1(b) During its evolution, the dislocation position depends heavily on how it is discretized in the classical model

Through careful inspection of equations (1) and (2), it can be found that the dislocation motion is entirely determined by the instantaneous local glide force acting on it. Once the material property parameters, the size and position of a dislocation and applied loads are given, the glide force will be completely determined regardless of the velocity variation along the dislocation line, which means equation (1) is essentially an algebraic equation. Mathematically, the velocity can be directly computed from equation (1) without the need for any discretization procedures. If discretization of the dislocation line is employed, as is always the case in dislocation dynamics, nodes near the pinning points will jump due to certain averaging procedures through which the equation is relaxed so that it is satisfied only in the sense of integral over each segment instead of at every point on the dislocation line. The 'seriousness' of the jump depends on the segment size. This is the basis of the abnormal phenomenon.

On the other hand, from a physical perspective, the abnormal segment-size dependence phenomenon occurs because only a minor portion of the interaction between neighboring segments is captured by equation (1). In the elasticity theory, dislocation is perceived as a spatial tube. Although the elastic interaction between neighboring dislocation segments due to the elastic stress field external to the tube is represented by the contribution of self-stress in equation (2), the interaction originating in the non-elastic core region of the spatial tube is not included. In effect, the latter interaction is much more important than the former.

We assume that the non-elastic interaction originating in the dislocation core can be expressed by the Newtonian viscous law (viscous regularization):

$$Q_i(s) = \eta_{ij}\, \partial V_j(s)/\partial s \tag{3}$$

where Q_i is the shear force between neighboring differential dislocation segments, η_{ij} is a diagonal matrix that indicates the second part of the interaction and is reduced to a scalar in isotropic cases, and s is a generalized coordinate, i.e., the distance along the dislocation curve from a point on the dislocation line. Taking the derivative with respect to s on both sides of equation (3) and considering the fact that $\partial Q_i(s)/\partial s = B_{ij}V_j(s) - F_i(s)$, the following dislocation dynamics governing equation can be obtained:

$$\eta_{ij}\, \partial^2 V_j(s)/\partial s^2 + \left[F_i(s) - B_{ij}V_j(s) \right] = 0 \tag{4}$$

This governing equation takes into account both elastic and non-elastic interaction between dislocation segments. Using the Galerkin method and following procedures similar to those of the Finite Element Method, the discretization form of equation (4) can be easily constructed by following procedures similar to those in reference [7].

STRESS CALCULATION IN THIN FILM-SUBSTRATE SYSTEMS

It is difficult to directly obtain stresses induced by dislocation loops in thin film-substrate systems. Alternatively, we decompose this complicated problem into the supposition of two relatively simple sub-problems: a). The occurrence of dislocation loops in an infinite extended bi-material medium, which will create a traction field $T_i(\mathbf{x})$ on the virtual free surface. b). The loading of a traction field $-T_i(\mathbf{x})$ on the free surface of the thin film-substrate system. More details on this method can be found in reference [7].

In our case, the local stress at a point on the dislocation loop is induced by the misfit strain, other dislocations and the dislocation itself; i.e:

$$\sigma_{jk} = \sigma_{jk}^{mis} + \sigma_{jk}^{other} + \sigma_{jk}^{self} \qquad (5)$$

where σ_{jk}^{mis} and σ_{jk}^{other} can be obtained by the Freund's method [8] and the method we have just discussed, respectively. The Brown method is employed to find σ_{jk}^{self}. Then the local force $F_i(s)$ acting on a dislocation line can be easily obtained by equation (2).

APPLICATIONS IN THIN FILM-SUBSTRATE SYSTEMS

Consider a dislocation half loop emerging from the free surface in a thin film-substrate system as shown in Fig. 2. The center point of the initial dislocation half loop is (0, 0, 100nm).

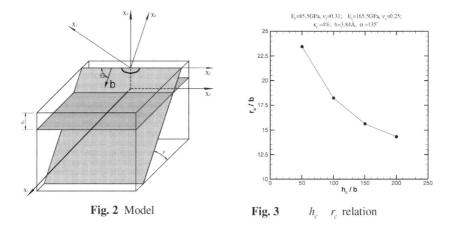

Fig. 2 Model   **Fig. 3** h_c r_c relation

Suppose the thin film is made of GaAs with parameters being Young's modulus $E_f = 85.5GPa$, Poisson ratio $\nu_f = 0.31$ and lattice constant $a_f = 5.653\,\text{Å}$. The substrate is assumed to be made of Si, whose Young's modulus $E_s = 165.5GPa$, Poisson ratio $\nu_s = 0.25$ and lattice constant $a_s = 3.834\,\text{Å}$. Then the misfit between the thin film and the substrate is $\varepsilon_0 = 4\%$. Burgers vector is chosen to be $b = 3.834\,\text{Å}$. The angle between the slip plane of the dislocation loop and the horizontal plane is 60^0. It is well known that, for a given thin film thickness, there exists a critical radius thickness below which the surface loop will shrink and finally leave out of the system and beyond which the surface loop will continue to expand. Although the dislocation velocities and profiles are greatly affected by the values of B and η, the final simulation results of the relationship between the thin film thickness and critical radius are not very sensitive to their specific values provided that they vary in the reasonable range. Fig. 3 shows that the critical radius for a surface dislocation loop to nucleate will decrease when the thin film thickness increases. This means the thin film will be easier to be dislocated with the increase of its thickness in a given external environment. This trend obtained from our simulation results coincide with the experimental results well.

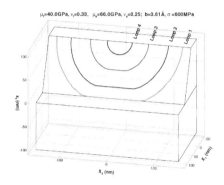

Fig. 4 After 4 loops formed the 1st loop will pass through the interface

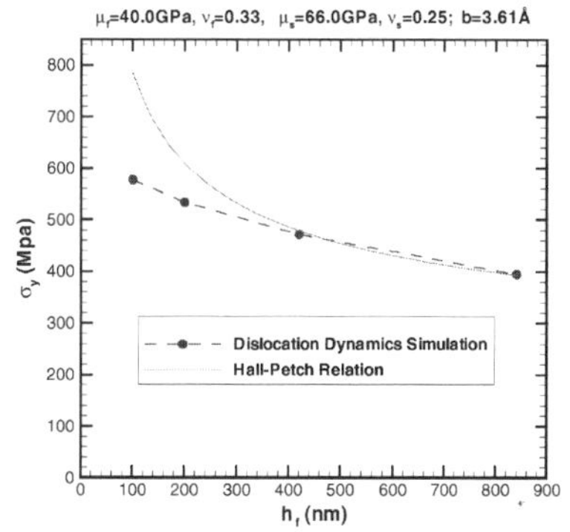

Figure 5 Thin film thickness-yield stress curve

Next, the thin film material is changed to be Copper with Young's modulus $E_f = 106.4GPa$, Poisson ratio $v_f = 0.33$. The misfit between the thin film and substrate is neglected for the study of dislocation pile-ups and thin film work hardening. The interface strength is chosen to be $\sigma^* = 2.28GPa$ as given by Blanckenhagen et al. [9]. Choose the thin film thickness to be $100nm$ and apply an external biaxial stress $\sigma = 600MPa$. As shown in Fig. 4, after the first dislocation loop expands far enough, a second dislocation loop will be nucleated and expand. Similarly, a third loop and a fourth loop are also nucleated and expand. When the first loop reaches the interface, it can not penetrate through the interface because the stress there is less than the interface strength. However, with the consecutive nucleation and expanding of the later loops, the stress at the interface will increase and may finally exceed the interface strength. If this occurs, the first dislocation will get into the substrate and the thin film will yield. The less the thin film thickness, the fewer number of loops for pile-ups will be. The accumulation of stress at the head of the first dislocation on the interface will be less striking. So a higher stress needs to be applied to propel the first loop get through the interface. This is the dislocation pile-up mechanism for work hardening of thin films. We plot our numerical results together with the Hall-Petch relation in Fig. 5. It can be seen that when the thin film thickness is large enough, our results are in good agreement with the Hall-Petch relation. However, when the thin film thickness becomes less and less, the difference between our results and the Hall-Petch relation will become larger and larger. This is because the Hall-Petch relation is not applicable to very small thicknesses where the pile-up contains only several dislocations, while our numerical method based on the new dislocation dynamics model can be applied to both large and small thickness.

CONCLUSIONS

Based on the solid physical background, a new dislocation dynamics model is developed in this paper. The new model fully incorporates not only the interaction through the elastic region outside of the dislocation core but also the interaction through the non-elastic region in the dislocation core among differential dislocation segments. It completely solves the long-standing problem inherent in the classical dislocation model that the simulation results are heavily dependent on dislocation-segment lengths. The simulation based on the new model has clear physical background and is very stable with the change of dislocation segment sizes.

By the applying the new model to thin film-substrate systems, dislocation evolution and pile-ups are successfully simulated and the relationship between yield stress and thin-film thickness are also reasonably predicted. On the other hand, the comparison between our numerical results and the Hall-Petch relation also validates the new model.

ACKNOWLEDGEMENTS

The authors gratefully acknowledge the financial support of NSF under contract no. DMR 0113172.

REFERENCES

1. D. Gómez-García, B. Devincre and L. Kubin, J. Computer-Aided Mater. Design **6**, 157 (1999).
2. H.M. Zbib, M. Rhee and J.P. Hirth, Int. J. Mech. Sci. **40,** 113 (1998)
3. N.M. Ghoniem, S.-H. Tong and L.Z. Sun, Phys. Rev. B **61**, 913 (2000).
4. P.A. Greaney, L.H. Friedman and D.C. Chrzan, Computat. Mater. Sci. **25**, 387 (2002)
5. L.M. Brown, Phil. Mag. **10**, 441 (1964)
6. S.D. Gavazza and D.M. Barnett, J. Mech. Phys. Solids **24**, 171 (1976)
7. E.H. Tan and L.Z. Sun, Proc. of MRS Fall Meeting, December 1-5, 2003, Boston, MA, pp. 47-52.
8. L.B. Freund, J. Mech. Phys. Solids, **5**, 657 (1990)
9. B.V. Blanckenhagen, P. Gumbsch and E Azrt, Modelling Simul. Mater. Sci. Eng. **9**, 157 (2001)

Novel Testing Techniques

Mater. Res. Soc. Symp. Proc. Vol. 875 © 2005 Materials Research Society

Thermal Expansion of Low Dielectric Constant Thin Films by High-Resolution X-Ray Reflectivity

Kazuhiko Omote and Yoshiyasu Ito
X-Ray Research Laboratory, Rigaku Corporation,
Akishima, Tokyo, 196-8666, JAPAN

ABSTRACT

By introducing high precision sample alignment technique, repeatability of incident angle to the sample surface for x-ray reflectivity (XRR) measurement is achieved to be within 0.3 arcsec. As a result, film thickness and density are possible to be measured repeatability within 0.03% and density within 0.26%. This accuracy realized to detect very small change of thermal expansion of thin films. The coefficient of thermal expansions (CTE) for porous low-k films deposited by CVD method were measured up to 400°C. The obtained values are in the range from 40 to 80 $\times 10^{-6}$ K^{-1} and they are very large compare to that of copper (16-20 $\times 10^{-6}$ K^{-1}).

INTRODUCTION

In recent years, new low dielectric constant materials have been considerable interest for reducing dielectric constant of interlayer insulators [1]. For introducing such new materials into microelectronic devices, thermal properties of the film is very important, because of the film will undergo the thermal stresses in the fabricating processes of the devices.

X-ray reflectivity (XRR) technique is well known to measure density and thickness of thin films very accurately [2, 3]. We are trying to measure density and thickness changing with rising to the film temperature by this technique. However, thermal expansion coefficients of materials are typically in the order of 10^{-5} K^{-1} and the changing may be less than one percent, even if the temperature is rising 100°C. Therefore, we need a very accurate XRR measurement for detecting thermal expansion of the film materials. For this purpose, we have developed a high-precision goniometer with high-resolution crystal beam conditioner and analyzer. The repeatability of the measurements has been achieved to be within ±0.3 arcsec.

The samples are deposited by CVD method and including very small pores (about one nanometer in diameter). Experiment was carried out in nitrogen atmosphere up to 400°C after preheating for evaporating adsorbed water. The data were corrected at both heating and cooling processes for confirming the film materials was not deteriorated. The density and thickness of the film was determined by the least square fitting for minimizing the residual error between the observed experimental data and XRR calculation. The estimated error of the obtained density and thickness are less than 0.3% and 0.05%, respectively. From the thickness changing of the film, we could estimate Z-axis (normal direction) coefficient of linear thermal expansion (Z-axis CTE), beside from the density changing, we could estimate coefficient of volumetric thermal expansion, independently.

We will discuss the relation between obtained Z-axis CTE of the film bound to the substrate and that of freestanding state. In addition, we have measured Z-axis and volumetric CTE of copper film on the silicon substrate for confirming reliability of the present technique. The obtained values are consistent with that of reported CTE of copper metal.

ACCURATE SAMPLE ALIGNMENT FOR X-RAY REFRECTIVITY

It is very important for aligning samples to the optimal position for measuring correct XRR data. Incident angle to the sample surface is something ambiguous by usual XRR sample alignment method because sample Z-position may be influence to the observed x-ray incident angle to the surface. In order to eliminate such ambiguity of the incident angle, we have employed an analyzer crystal for observing correct scattering angle. After standard Z-position alignment, we measure scattering angle 2θ of specular reflection. X-ray incident angle to the surface is exactly half of the 2θ angle as shown in figure 1(a). This value may not be affected by the error of Z-position alignment, because analyzer crystal is only sensitive to the angle of the beam and insensitive to the beam position. X-ray reflectivity curve could be measured without analyzer crystal as shown in figure 1(b), once the x-ray incident angle was defined. We could achieve repeatability of the sample alignment within 0.3 arcsec by using the present technique with high precision goniometer system, Rigaku AXIOM-Σ.

We confirmed the effect of Z-position error for the XRR data using the present alignment method. The observed XRR curves are shown in figure 2 and we can see these curves are almost identical even the differences of Z-positions are ± 10 μm. The results of the XRR analysis are listed in Table I. The obtained thickness is varied within 0.16%, if the difference of Z-position is ± 10 μm and the error is to be 0.03% if the Z-position error is within ± 2 μm. In addition, because we defined the x-ray incident angle correctly, critical angle of the film agrees very well in each other. As a result, deviation of the film density is 0.44 %, even the difference of Z-position is ± 10 μm and error is to be 0.26 % when the Z-position error is within ± 2 μm.

The above accuracy of the present XRR analysis enables us to measure very small change of thickness and density in temperature by thermal expansion.

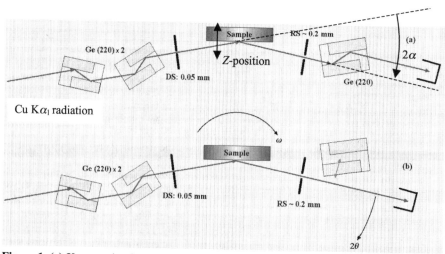

Figure 1. (a) X-ray optics for accurate sample alignment of the present method. (b) That for the XRR measurement. The measurement was done using Cu Kα_1 radiation (8.048 keV).

Table I. XRR analysis results for different Z-position of the sample.

Z-position (mm)	Thickness (nm)	Density g/cm^3
10.003	733.52	1.134
10.011	734.21	1.136
10.013	734.01	1.139
10.015	734.08	1.139
10.023	734.75	1.135

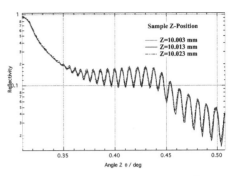

Figure 2. Observed XRR curves for different Z-position of the sample.

THERMAL EXPANSION OF POROUS LOW-*K* FILMS

Three porous low-*k* films were prepared by CVD method with the different flow late of materials gas. The films are composed of Si, O, C, and H atoms. High-temperature XRR measurements were done in the nitrogen flow furnace. Before the measurement, the sample were hold more than two hours at 100°C for evaporating adsorbed water. Then we started to measure XRR curves from lower temperature to higher temperature (heating). We also measured the same XRR curves with lowering the temperature (cooling) for confirming the film was changed reversibly. One of the observed curves is shown in figure 3. Critical angle of the film is clearly

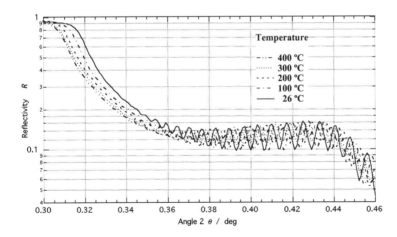

Figure 3. Observed XRR curves with different temperatures. Critical angle is shifted to lower angle with increasing temperature due to thermal expansion.

shifted to lower angle with increasing temperature (lowering the density). It indicates the volumetric thermal expansion of the film. The interference fringe period is also changed by temperature.

The thickness and density are estimated by least square fitting minimizing the error function between experimental and observed XRR curves. The obtained temperature variation of thickness and density are shown in figure 4 and those at room temperature are listed in Table II. It is worth noting that two lower density films (#1 and #2) have much larger thermal expansion rates than that of higher density film (#3). For realizing lower dielectric constant, lower density is necessary. However, lower density film may cause weakness of the film. Thermal expansion of the film may be one of important indicators of film properties.

THERMAL EXPANSION OF THE COPPER FILM

It is interesting to apply the present method to obtain thermal expansion of film for well-known materials. Copper film is very important as wiring material for novel silicon integrated circuits. We have thus measured thermal expansion of copper film for confirming the reliability of the present method. The observed Z-axis and volumetric CTE is 37.6 $\times 10^{-6}$ K^{-1} and 53.1 $\times 10^{-6}$ K^{-1}, respectively. They are measured at the temperature range from 25°C to 400°C.

Table II. Thickness and density of three films at 25°C measured by XRR and average Z-axis and volumetric CTE obtained by the present work.

	Sample #1	Sample #2	Sample #3
Thickness (nm)	755.59	744.63	746.31
Density (g/cm^3)	1.1391	1.1390	1.1897
Z-axis CTE	155 $\times 10^{-6}$ K^{-1}	137 $\times 10^{-6}$ K^{-1}	74.5 $\times 10^{-6}$ K^{-1}
Volumetric CTE	168 $\times 10^{-6}$ K^{-1}	151 $\times 10^{-6}$ K^{-1}	105 $\times 10^{-6}$ K^{-1}

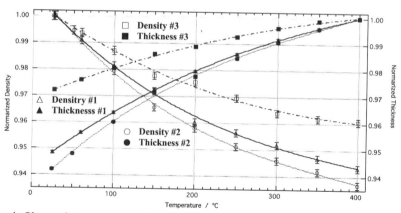

Figure 4. Observed temperature variation of three porous low-k films. Maximum values are normalized at one.

Thermal expansion of copper metal is well known for freestanding state. However, thermal expansion of the film should be restricted by that of the substrate. Let α^f, α_Z, and α_L to be coefficient of linear thermal expansion for freestanding material, Z-direction and lateral direction of film bounded to substrate, respectively.

Thermal expansion of lateral length l_0 is expanded to $l = l_0 (1 + \alpha_L \Delta T)$ with temperature increasing ΔT. It should be different from natural thermal expansion $l^{free} = l_0 (1 + \Delta^f \Delta T)$ and resultant stress is proportional to $(\Delta_L - \Delta^f)\Delta T$. Therefore, Z-axis thermal expansion is calculated by basic elasticity theory as

$$\frac{d}{d_0} = \frac{d^{free}}{d_0} - \frac{2v}{1 v v}\left(\alpha_L \, \alpha \alpha^f\right)\Delta T$$

$$= \left[\frac{1+\Delta}{1\Delta\Delta}\Delta^f \Delta \frac{2\Delta}{1\Delta\Delta}\Delta_L\right] | T \quad (1)$$

$$= / _Z | T$$

where v is the Poisson ratio. The same way, volumetric expansion β is written as

$$\frac{V}{V_0} = \left[\frac{1+v}{1--}\alpha^f + 2\frac{1\alpha2\alpha}{1\alpha\alpha}\alpha_L\right]\Delta T \quad (2)$$

$$= \beta\beta T$$

and we get very simple relation

$$\beta = \alpha_Z + 2\alpha_L \quad (3)$$

From the observed value of Z-axis CTE, 37.6×10^{-6} K^{-1}, we could calculate CTE of copper metal β^f as 20.5×10^{-6} K^{-1} using equation (1) with the reported Poisson ratio of copper $\beta = 0.343$ and assuming that the lateral thermal expansion is the same as that of substrate silicon (4.2×10^{-6} K^{-1}). The obtained value is consistent with reported CTE of Cu metal, $16.5 - 20.3 \times 10^{-6}$ K^{-1} in the temperature range from 20 to 570°C. Therefore, Z-axis CTE may indicate the CTE of the film well.

From the equation (3), lateral expansion dose not depend on the films but depend on the substrate if the film was completely bound to the substrate. However, $\beta - \beta_Z$ is not constant for our observed samples; 15.5×10^{-6} K^{-1} for copper, 13, 14, and 30.5×10^{-6} K^{-1} for low-k samples #1, #2, and #3, respectively. Some of the difference of lateral CTE could be relaxed by the bending of the substrate. However, the difference is too large to relax by wafer bending. Further study is needed for understanding lateral and volmetric thermal expansion of films. For example, it should be useful in combination with crystallographic measurement of lattice constants.

CONCLUSIONS

We introduced a method for accurate sample alignment for high-resolution x-ray reflectivity measurements. The accuracy of the observed thickness is 0.05 % and that of film density is 0.3 %. The coefficients of thermal expansion of porous low-k films were observed directly by this technique and they are very much dependent on CVD deposition conditions of films. In addition, we have confirmed reliability of the present method by means of copper film. The observed Z-axis CTE is consistent with reported value of copper metal, however, we need further study of lateral and volumetric CTE.

REFERENCES

1. For example, *Proceedings of the IEEE 2001 International Interconnect Technology Conference, San Francisco, Calfolnia, USA.*
2. J. A. Nielsen, "Elements of Modern X-Ray Physics," (Wiley, 2001) pp. 61-78.
3. W. Wu, W. E. Wallance, E. K. Lin, and G. W. Lynn, *J. Appl. Phys.* **87**, 1193 (2000).

Mater. Res. Soc. Symp. Proc. Vol. 875 © 2005 Materials Research Society   O8.3

Investigation of local stress fields:
Finite element modelling and High Resolution X-Ray Diffraction

A.Loubens[1,2], C. Rivero[1,3], Ph. Boivin[3], B.Charlet[4], R.Fortunier[2], O.Thomas[1]

1: Laboratoire TECSEN UMR CNRS 6122
Université Paul Cézanne, 13397 Marseille cedex 20, France
2: Ecole Nationale Supérieure des Mines de Saint-Etienne
CMP « Georges Charpak », Avenue des anémones, 13541 Gardanne, France
3 : STMicroelectronics, Zone Industrielle de Rousset 13106 ROUSSET cedex, France
4 : CEA Leti 17 rue des martyrs 38054 grenoble cedex 9 France

Abstract

The influence of local stress fields on the electrical properties of Si-based nanostructures is of increasing concern. The experimental evaluation of stresses at the required scale (few nanometres) remains, however, a very challenging task. We propose a non destructive X-ray diffraction technique for local strain measurements using a laboratory radiation source. This technique provides an alternative route to micro diffraction experiments. High resolution X-ray diffraction is used to analyse the diffraction from the periodic strain field induced in silicon by a periodic array. We analyzed arrays of Si_3N_4 lines (thickness: 149 nm, width: 145 nm, pitch: 169 nm) on silicon, and arrays of single crystal Si lines (period: 0.6 micrometers, width: period/2, thickness: 50 nm) etched in SOI (Silicon On Insulator) and capped with SiO_2 and Si_3N_4. Reciprocal space maps were performed around the Si substrate diffraction lines. A Bartels 4 reflections Ge 220 monochromator was used in combination with a 3 reflections Ge 220 analyser. X ray diffraction rocking curves performed on Si 004 and Si 224 reveal distinct superlattice peaks whose spacing is related to the in-plane periodicity. Reciprocal space maps reveal particular intensity distributions caused by the stress gradient in the transverse and perpendicular directions of the lines. Reciprocal space maps obtained by High Resolution X-Ray Diffraction are compared with maps calculated from displacement fields derived from finite element modeling. Very clear superlattice peaks are, however, observed around the bulk Silicon substrate reciprocal lattice nodes, which indicates a great sensitivity of this method to elastic stresses. The influence of the lines aspect ratio on the reciprocal space maps has been studied both experimentally and through modelling.

Introduction

Mechanical stresses in thin films and nanostructures have very important consequences on the reliability of devices made from such structures. With the general trend of reduced dimensions in microelectronic devices (less than 0.2μm), mechanical stresses are more and more considered a very important issue. Since local stresses need to be known accurately and compared with calculations like Finite Element Modeling (FEM) a lot of research is devoted to the experimental determination of local stress fields. Microdiffraction [1] or microRaman [2] spectroscopy show lateral resolutions of about 0.3 μm in the best cases . Convergent Beam Electron Diffraction (CBED) has the required atomic resolution [3] but needs a painful thinning of the sample down to electron transparency. Moreover, such a procedure is likely to modify the stress state in the object. In this study we evaluate local strain fields by performing high resolution reciprocal space mappings [4,5] on periodic structures(figure 1) (arrays of

Si$_3$N$_4$ lines and Silicon On Insulator lines). The resulting diffraction satellites are very sensitive to the induced periodic strain field. We then compare experimental intensities with those calculated from the displacement field derived from Finite Element Modelling (FEM) calculations. These experiments allow for a direct evaluation of elastic modelling and may provide an alternative route to microdiffraction experiments.

Experimental

All the diffraction measurements have been performed on a 4-circles Philips X'pert MRD diffractometer. The incoming radiation is CuK$_{\alpha 1}$ (λ=0.1540598 nm) with a Ge (220) 4 reflections Bartels monochromator (beam divergence in the scattering plane is 12 arcsec). The diffracted rays are analysed with a 3 reflections Ge (220) analyser. In all the maps which are presented in this article, the scattering vector is $\vec{q} = \vec{k}_d - \vec{k}_i$, where $\vec{k}_i$ and $\vec{k}_d$ are respectively incident and diffracted wave vectors (figure 1a). Reduced components H, K, L are defined as

$$q_x = H * \frac{2\pi}{a} \qquad q_y = K * \frac{2\pi}{a} \qquad q_z = L * \frac{2\pi}{a} \qquad (1)$$

where a is the silicon lattice parameter (a=0.543088 nm).

We analyzed an array of tensile amorphous Si$_3$N$_4$ lines (figure 2a) (with the following dimensions: h=149 nm, w=145 nm, Λ=317 nm) deposited on a Si (001) single crystal. The lines are parallel to the $[1\overline{1}0]_{Si}$ direction. The initial biaxial stress in the nitride thin film before etching is of the order of 1 GPa. Reciprocal space maps were performed, around the 224 Si reflection, i.e. in the plane transverse to the lines.

(001) oriented Silicon-On-Insulator (SOI) blanket films have been prepared with a 45° in-plane rotation (around [001] axis) between the two Si lattices. Si lines are capped with 7 nm SiO$_2$ and 160 nm Si$_3$N$_4$. We have focused on an array with the following dimensions: w = 300 nm, Λ=600 nm, h=50 nm (figure 2b).

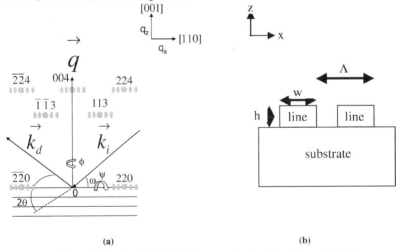

(a) (b)

Figure 1 : (a) Schematic drawing of the diffraction geometry in reciprocal space (b) Real space geometry: h is the thickness, w the width and Λ the period

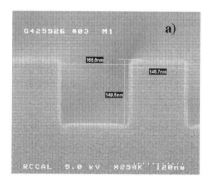
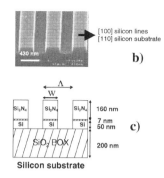

Figure 2 : (a) Scanning Electron Micrograph of an array of nitride lines (thickness: 149 nm, width: 145 nm, period: 317 nm) (b) Scanning Electron Micrograph showing an actual array with a 0.6 μm period (c) Silicon lines capped with 7 nm of SiO₂ and 160 nm of Si₃N₄.

Modelling

Calculations were performed in two dimensions with Femlab®[6], assuming the system to be infinite in the y direction, and thus using x-z plane strain conditions. All the materials are assumed to be linear elastic and isotropic. Calculations were performed on one unit cell with boundary conditions applied to the displacements of symmetric axes: $u_x=0$. A residual stress in Si₃N₄ was used in the calculations and generated with a fictitious thermal loading. From the FEM calculated displacement field, $\vec{u}$, the diffracted intensity I is derived, within kinematical approximation, as the modulus squared of the structure factor F($\vec{q}$)

$$I(\vec{q}) = |F(\vec{q})|^2 = \left| \sum_j e^{i\vec{q}.(\vec{r}_j + \vec{u}_j)} \right|^2 \tag{2}$$

Where $\vec{r}_j$ and $\vec{u}_j$ are respectively the position and displacement of atom j, $\vec{q}$ is the scattering vector. For a system which is periodic in the x direction, the diffracted intensity is given by:

$$I(\vec{q}) = |f(\vec{q})|^2 \left[\frac{\sin^2(\frac{Nq_x\Lambda}{2})}{\sin^2(\frac{q_x\Lambda}{2})} \right] \tag{3}$$

where f($\vec{q}$) is the scattering amplitude from a single period and NΛ the coherence length of the X-ray beam. We used N=5. Calculations were performed in the following way in order to minimize the computer used time while keeping an optimal precision on the diffracted intensity.

- along x, the period Λ is divided into 16 atoms calculation steps,
- along z, the calculation depth is reduced to 500 nm, with 15 atoms steps.

Discussion

Figure 3a shows an experimental Si 224 reciprocal space map measured on the array of Si_3N_4 lines. The periodic satellites elongated in the L direction are a signature of the periodic strain field, which is induced in silicon by the nitride lines. The real space period extracted from the map is 317 nm, in excellent agreement with the period deduced from scanning electron micrographs. A calculated map is shown in figure 3b. In this figure, the displacement field was deduced from FEM calculations. A qualitative agreement between the experimental and calculated maps appears. The strong Bragg peak at H=K=2 and L=4 is absent from the calculated map because only strained silicon is included. An upper-left-to-lower-right oriented diffuse streak can be observed in figure 3a. It is well reproduced in the calculated map (figure 3b). Moreover, one observes the same dissymmetry in the two maps: the upper-left part is more intense than the lower-right part. These features may be understood by analysing the strain state in silicon as calculated by FEM.

Indeed, when σ_0 is tensile ε_{xx} is compressive under the lines and tensile in-between the lines (figure 4a). As a consequence, since $\varepsilon_{yy} = 0$ and because of Poisson effect, we have $\varepsilon_{zz}>0$ under the lines and $\varepsilon_{zz}<0$ in-between. These two strain states correspond to two different locations in figure 3, and thus in the reciprocal space: {upper-left, i.e. low L and high H} and {lower-right, i.e. high L and low H} respectively for {$\varepsilon_{xx}<0$; $\varepsilon_{zz}>0$} and {$\varepsilon_{xx}>0$; $\varepsilon_{zz}<0$}. This explains the orientation of the diffuse streak, which is observed both in experimental and in calculated maps.

The dissymmetry between the two sides of the map is explained by the respective weight of the tensile and compressive regions depicted in figure 4a.

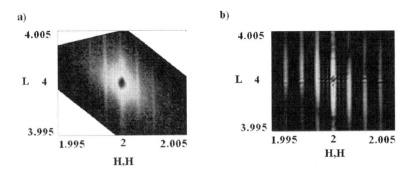

Figure 3 : reciprocal space map around the Si 224 measured (a) and simulated (b) H and L are the scattering vector components in units of $\dfrac{2\pi}{a}$

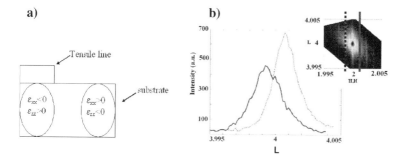

a) [diagram: Tensile line, substrate, $\varepsilon_{xx}<0$ $\varepsilon_{zz}>0$ | $\varepsilon_{xx}>0$ $\varepsilon_{zz}<0$]

b) [graph with Intensity (a.u.) axis from 100 to 700, L axis from 3.995 to 4.005]

Figure 4 : (a) Strain states induced by a tensile line. (b) L scans extracted from Si 224 along the first satellite on the left (dotted line) and on the right (solid line)

This weight is directly linked to the w/Λ ratio. When this ratio is smaller than 0.5, a stronger intensity is expected from areas situated in-between the lines, that is in the upper-left reciprocal space region. At constant w/Λ ratio, the sign of σ_0 produces a higher intensity either on the right side (figure 5a, compressive line) or on the left side (figure 5b, tensile line).

We have also studied experimentally the strain field induced by compressive SOI lines in the silicon substrate. Figure 6 shows a Si 224 map measured from an array of SOI lines (width: 300 nm, period: 600 nm, thickness: 50 nm). The silicon lattices (from the bulk substrate and from the lines) are rotated by 45°, which ensures that there is no superposition between the intensities coming from the two crystals [7]. The periodic satellites show again the signature of the periodic strain field. This outlines the great sensitivity of the technique, since here bulk silicon is screened from the lines by 200 nm of silicon oxide. One observes again in figure 6a an upper-left-to-lower-right oriented diffuse streak. Moreover, in figure 6b, the maximum intensity of the first right and left satellites are the same. This result is in agreement with figure 4a, since in this case w/Λ=0.5.

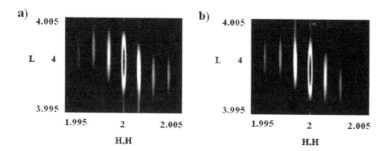

a) [reciprocal space map, L axis 3.995 to 4.005, H,H axis 1.995 to 2.005]

b) [reciprocal space map, L axis 3.995 to 4.005, H,H axis 1.995 to 2.005]

Figure 5 : reciprocal space map around the Si 224 with compressive nitride lines (a) and tensile nitride lines (b) H and L are the scattering vector components in units of $2\pi/a$

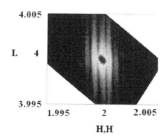
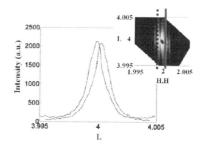

Figure 6 Reciprocal space map around Si 224. L scans extracted from Si 224 along the first satellite on the left (dotted line) and on the right (full line)

Conclusion

High Resolution X Ray Diffraction has been used to investigate the displacement field generated by an array of stressed lines in a silicon substrate. Two different samples have been studied: one array of amorphous tensile nitride lines on silicon substrate, one array of SOI lines capped with nitride. A qualitative agreement is obtained between calculated and experimental reciprocal space maps. The displacement field used in simulated maps is derived from FEM calculations.

Because of the periodic structure of the samples, the maps exhibit periodic satellites along the q_x direction. A tilted diffuse streak is always observed. It is explained by the presence of compressive and tensile regions situated either under the lines or in-between the lines. The relative weight of these regions produces in turn a dissymmetry of the satellites intensity: the tensile region is associated with the small-q_x side of the map, whereas the compressive region corresponds to the large-q_x side. One can thus separate the information about the displacement field by considering different areas in the reciprocal space maps.

In conclusion we have investigated an original method to evaluate local strain fields based on high resolution x-ray diffraction from the deformed crystal. This method applies to single crystals and periodic structures. It yields information on the strain field at the nm scale, thanks to the high reciprocal space resolution. It is thus complementary to micro-beam XRD, which has true real space resolution but at the μm scale.

References

[1]   A.A MacDowell, R.S Celestre, N. Tamura et al., Nuclear Instruments and Methods in Physics Research A **467-468**, 936 (2001).
[2]   I. De Wolf, V. Senez, R. Balboni et al., Microelectronic Engineering **70**, 425 (2003).
[3]   S.Kramer, J. Mayer, C. Witt et al., Ultramicroscopy **81**, 245 (2000).
[4]   Q. Shen and S. Kycia, Phys. Rev. B **55**, 15791 (1997).
[5]   T. Baumbach, D. Lübbert, and M. Gailhanou, J. Appl. Phys. **87** (8) (2000).
[6]   Femlab 2.3 Structural Mechanics Module (Comsol, 2002).
[7]   G. M. Cohen, P. M. Mooney, H. Park et al., J. Appl. Phys. **93**, 245 (2003).

In Situ Characterization Techniques

In-Situ TEM Study of Plastic Stress Relaxation Mechanisms and Interface Effects in Metallic Films

Marc Legros[1], Gerhard Dehm[2], T. John Balk[3]
[1] CEMES-CNRS, 29 rue J. Marvig, 31055 Toulouse - France
[2] Erich Schmid Institute for Materials Science and University of Leoben, Department Materials Physics, Jahnstr. 12, 8700 Leoben - Austria
[3] Dept. of Chemical and Materials Engineering, University of Kentucky, Lexington, KY, USA

ABSTRACT

To investigate the origin of the high strength of thin films, in-situ cross-sectional TEM deformation experiments have been performed on several metallic films attached to rigid substrates. Thermal cycles, comparable to those performed using laser reflectometry, were applied to thin foils inside the TEM and dislocation motion was recorded dynamically on video. These observations can be directly compared to the current models of dislocation hardening in thin films. As expected, the role of interfaces is crucial, but, depending on their nature, they can attract or repel dislocations. When the film/interface holds off dislocations, experimental values of film stress match those predicted by the Nix-Freund model. In contrast, the attracting case leads to higher stresses that are not explained by this model. Two possible hardening scenarios are explored here. The first one assumes that the dislocation/interface attraction reduces dislocation mobility and thus increases the yield stress of the film. The second one focuses on the lack of dislocation nucleation processes in the case of attracting interfaces, even though a few sources have been observed in-situ.

INTRODUCTION

Stresses are known to increase in small-scale systems such as nanocrystalline materials or thin films. This strengthening is qualitatively well understood in term of constrained dislocation motion. For fine-grained materials, the Hall-Petch relation, based on dislocations piling up against boundaries, states that the yield stress varies as one over the square root of the grain size. In the case of thin films on rigid substrates, the so-called Nix-Freund [1, 2] model, itself adapted from the work of Matthews and Blakeslee [3], describes the well-established experimental fact that the strength of thin films varies as one over their thickness [4-6]. In this model, a threading dislocation must deposit an interfacial dislocation as it shears the film. The stress necessary to extend this interfacial dislocation varies as one over the film thickness. For polycrystalline thin films, Thompson combined both models by accounting for the deposition of a dislocation on grain boundaries and also derived a stress dependence that varies as one over the grain size [7]. The need for new concepts was recognized from experimental observations in very fine-grained materials and ultra thin metallic films (less than 100 nm) that the strength started to level off and eventually decreased, at variance from what is expected from the models. This signifies the activation of new deformation mechanisms and their eventual dominance of dislocation-based plasticity.

Another surprising result was that, for a given thickness, Al and Cu films deposited on sapphire demonstrate lower strength than when attached to a SiO_x or SiN_x covered substrate [8]. This discrepancy cannot be attributed to the microstructure only [4], nor to the difference in substrate hardness. Recent transmission electron microscopy (TEM) analysis suggest that the nature of the film/substrate interface –metal on amorphous or metal on crystal- is the relevant factor that governs the strength of the metallic film [9]. The observation of the spontaneous disappearance of interfacial dislocations was made early in Pb films by Kuan and Murakami [10], but the implication of this experimental fact was only recently brought to light by Mullner and Arzt [11] after a similar observation in Al films on oxidized Si substrates. Such observations have since been extensively repeated [9, 12, 13] and interpreted as a dislocation sink effect at the metal/amorphous oxide interface. Of course this phenomenon was not predicted by classic dislocation theory, especially because images forces should prevent dislocations from reaching an interface with a harder material [14]. The sink effect could explain the decrease of dislocation density that is sometimes observed post-mortem [15], but other authors have reported stable dislocation populations before and after thermal cycling [16, 17]. Repulsion of dislocations by the interface is, however, observed in the case of a sapphire substrate [16], even if sources producing dislocation pile-ups can eventually force dislocations into the interfacial network that persists at this crystal/crystal interface [9]. The exchange of dislocations between the interfacial network and sources that are located in the film thickness is thought to maintain a constant dislocation density throughout deformation. Finally, it is interesting to note that when dislocations are really confined by the substrate, as assumed in the Nix-Freund model, the stress values predicted by the model fit the experimental data very well [8]. In the case of an "absorbing" interface, the experimental values are always much higher than predicted.

Because metal/amorphous interfaces cease to act as a strong obstacle to dislocations, one has to find other hardening mechanisms [18]. An emerging hypothesis in small-scale plasticity problems is that dislocation nucleation is the limiting factor [13, 19, 20]. Once nucleated, the dislocation can rapidly shear the small volume, under a fraction of the stress that has been necessary to create it [21]. However, the problem of modeling dislocation sources is very complex and usually requires several assumptions. Friedman and Chrzan [22] have extended the Hall-Petch model to include the effect of dislocation sources and find an inverse grain size dependence of the stress. These authors consider sources at the center of the grains that become blocked by the back stress of emitted dislocations, a hypothesis that is echoed by the observation of sources in a Cu film on sapphire [9]. In the case of films deposited on an amorphous substrate, no active dislocation source located midway through the thickness of the film has been observed so far. Grain boundaries could act as nests for sources, but the rare examples do not mention sufficient activity [23, 24]. The dislocation behavior described by Balk et al in Cu films is rather different because, in this case, emitted dislocations do not mediate strain since their glide is parallel to the interface. In fact, this parallel dislocation glide is thought to be the signature of a diffusion process through grain boundaries as recently modeled by Gao et al. [25]. In this hypothesis, the diffusion of material from the film surface into crack-like grain boundary wedges manages to completely relax the thermal stress in very thin, uncapped films.

In the present work, in-situ TEM is used as the main tool to investigate the different hardening processes that control the plasticity of Al and Cu films subjected to thermal stresses. The dislocation-interface interaction will be analyzed in the case of passivated Al films and epitaxial Cu films. Special attention will be given to dislocation nucleation processes.

EXPERIMENTAL

The origin of stress variation in the following experiments lies in the difference between the coefficients of thermal expansion of the metallic films and their underlying semiconductor or ceramic-type substrates.

Pure Al films. Passivated pure Al films were sputter-deposited on 620 μm thick and 10 cm diameter (100) oxidized Si wafers at 100°C at the Lincoln Laboratory (Cambridge, MA). The Al films studied here are 1 μm thick. The amorphous oxidation layer (18.5 nm SiOx) on the wafer served as a diffusion barrier between Al and Si. A thicker 1 μm SiO$_x$ passivation layer was deposited on the Al film by low-pressure chemical vapor deposition (LPCVD) at a temperature of 300°C. The sample studied here has been thermally cycled several times to 450°C in a Tencor laser profilometer [26].

Cu films. 800 nm thick epitaxial Cu films on sapphire were prepared at the Max Planck Institute for Metals Research (Stuttgart, Germany) using ultra-high vacuum molecular beam epitaxy. Cu was deposited at room temperature on previously cleaned (0001) α-Al$_2$O$_3$ substrates at a rate of 0.5 nm/min. The epitaxial relationship between Cu and sapphire is (111)Cu//(0001)Al$_2$O$_3$ and <211>Cu//<$\bar{1}$010>Al$_2$O$_3$. The film is composed of two variants in twin orientation [27].

Both configurations are summarized in figure 1: the Cu film presents one crystal/crystal interface with the sapphire substrate and one free surface; the Al film is bound by two crystal/amorphous interfaces with SiOx.

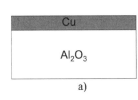 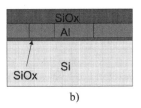

a)   b)

Figure 1. Sketch of both systems studied here: a) Epitaxial 1μm Cu on sapphire, b) large-grained polycrystalline 1 μm Al sandwiched between two amorphous SiO$_x$ layers. The passivation is also 1 μm thick.

In-situ TEM samples were prepared using tripod polishing. Final polishing was obtained with 5 to 10 minutes ion milling on a Gatan PIPS operating at 5keV. In-situ TEM observations were performed at 200kV on a JEOL 2010 microscope equipped with a SIS CCD camera. Thermal cycles from room temperature to 450°C for Al or to 490°C for Cu were completed at a rate varying between 20 and 30°C/min.

RESULTS

General microstructural evolution

Figure 2 presents four TEM cross-sectional images of both types of samples at the beginning of an in-situ test and after several in-situ thermal cycles. The as-processed Cu film (see figure 2a) typically exhibited a higher dislocation density than the passivated Al film (see figure 2b), which had already been thermally cycled. The high initial dislocation densities in figures 2a and 2b can be partly attributed to the TEM foil preparation. Also, because the Al film has previously been cycled, it exhibits very large grains. Most of the grain boundaries are perpendicular to the interface (as shown on the right hand side of figures 2b and 2d), and the aspect ratio of the grains is very large, making comparison with the single crystal relevant. Strictly speaking, the Cu film is not perfectly single crystalline since it contains a few misaligned grains and two differently orientated crystalline domains (variants) that are separated by Σ3 (twin) boundaries. Also the fact that a passivation layer was added on top of the Al film maintains the initial flatness (< 50 nm) of the Al film upper surface. After cycling, most of the Al grains are almost dislocation free as seen in figure 2d, which corresponds to a reduction of dislocation density by 2 to 3 orders of magnitude. The reduction seen in the case of Cu is much lower (figure 2c).

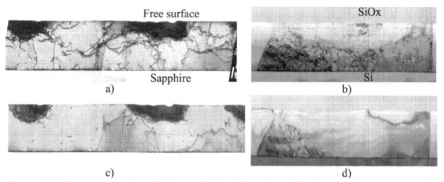

a)   b)

c)   d)

Figure 2. Microstructural evolution in Cu/sapphire after 5 in-situ cycles and in passivated Al/oxidized Si after 3 in-situ cycles. a,c) Cu/sapphire, b,d) passivated Al on Si. a,b) Initial microstructure; c,d) microstructure after in-situ thermal cycling. Note that the sapphire substrate is very transparent at the bottom of the Cu film and is only partly visible in a). Both films are 1 μm thick.

After several cycles, Cu films exhibit a low but somewhat uniform dislocation density. Al grains present very different deformation microstructures. In grains limited by straight grain boundaries (GB) (such as on the right of the Al grain in figures 2b, 2d), the dislocation density is usually very low, as in figure 2d.

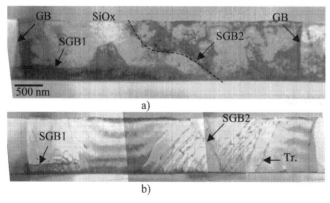

a)

b)

Figure 3. An Al grain containing sub-grain boundaries (SGBs) a) before and b) after in-situ thermal cycling. Note the many slip traces (Tr.) originating from SGB2.

In other grains, where some sub-grain boundaries (SGBs) are present (figure 3a), the dislocation activity is intense and many slip traces are visible after a cycle (figure 3b). In these grains, the initial dislocation density can be considered to be restored at the end of the thermal cycle, even if the dislocations are concentrated around these sub-grain boundaries. Note that SGBs, similar to GBs, tend to reduce their length and progressively become perpendicular to the interfaces (see for instance SGB2 in figure 3).

Hardening mechanisms and nucleation processes

Moving dislocations are good probes for local stress measurement [28]. We have measured their radii during in-situ cycles and studied their configuration with respect to the interfaces. Figure 4 gathers still pictures taken during a test or extracted from video sequences.

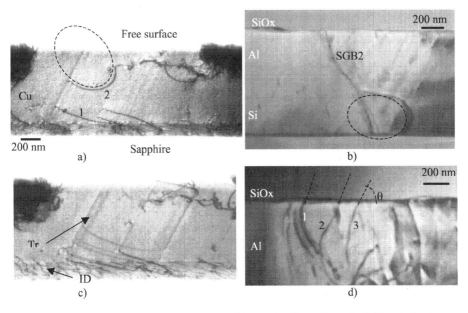

Figure 4. Gliding dislocations in a,c) Cu/sapphire and b,d) passivated Al film during in-situ cycles. a,c) T=25°C, b) T= 200°C, d) T=400°C. Dashed ellipses mark the radius of curvature of mobile dislocations projected in their (111) slip plane. The great radius of the ellipse corresponds to the actual curvature of the dislocation. Figure a, c) show dislocations emitted by a source Dashed lines in d) indicate the angle θ formed between dislocation lines 1 to 3 and the interface.

Figures 4a and 4b illustrate different nucleation processes in the Cu and Al film, respectively. In the case of Cu, the source is located close to the surface and consists of a spiraling dislocation segment (labeled 2 in figure 4a), probably moving around a second dislocation that is sessile. It provides a very good configuration for stress measurement because, in this position, dislocation 2 is distant from most other dislocations and from both the surface and interface. However it probably feels the back stress of the previously emitted dislocation that lies close to the interface (labeled 1). This image was taken at room temperature, after a thermal cycle, while the stress was relaxing after having reached its maximum. The measured radius of curvature corresponds to half the thickness of the film. The stress needed to curve a dislocation with a Burgers vector b to a radius R is given by:

$$\sigma = \mu b / R \qquad [1]$$

where μ is the shear modulus of the material. Here, the resolved stress needed to bow the dislocation in figure 4a is about 25 MPa. The radii of curvature of dislocations in figure 4c are much larger than the film thickness. This is because a pile-up is formed against the interface that is here not easily penetrable. The interface is inclined and visible in figures 4a and 4c through the network of interfacial dislocations (labeled ID). The back stress partially compensates the thermal stress, causing the radius of the dislocation to increase locally. Also, this sequence has been taken after some dislocations were emitted, which means that the stress was locally and partially relaxed.

Compared to the stress measured using the wafer curvature method and the Stoney formula, those deduced from TEM images seem much lower. First, the stresses measured from the dislocations' radii of curvature are resolved shear stresses. Secondly, the dislocations are not completely isolated: dislocation 2 in figure 4a is preceded by dislocation 1, which has not yet interacted with the interfacial network. This network may also create long-range stresses, as does the sub-grain boundary in Figure 4b. Finally, given that the stress in Al was measured from dislocations at 350°C and not at room temperature, the calculated stress is comparatively high.

a)

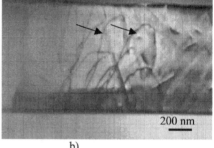

b)

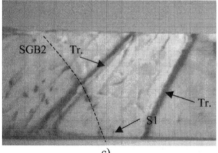

c)

Figure 5. Dislocation nucleation processes in a passivated Al film. a,b) T=360°C, two dislocations (pointed by arrows) progress towards the passivation layer. Both images are separated by Δt=1min.
c) T=25°C, the source in S1 nucleates very fast dislocation loops that shear the entire Al film thickness and produce thick slip traces (Tr.)

In the case of the Al film, none of these spiral sources have been observed so far. The nucleation process illustrated in figure 4b is characteristic of this system: a dislocation bulges from the middle of a sub-grain boundary and usually expands instantly to span the entire thickness of the film (see figure 5c for instance). In figure 4b, an intermediate position has been captured on video, during cooling off. The radius of curvature is about 250 nm, leading to a resolved stress of 30 Mpa. When compared to the average stress measured with the wafer curvature method [29], this local stress is larger, probably because it corresponds to a nucleating stress. We can also consider threading dislocations that have been nucleated several seconds earlier and interact with the interface while shearing the Al film. Their radius of curvature is generally large compared to the film thickness. When intersecting the SiO_x interface, gliding dislocations form on average an angle of about 60° (figure 4d). This value is occasionally lower, but seldom below 30°, a direct indication that this interface is attractive. This attraction embodies a potential hardening mechanism, as discussed later.

In the passivated Al film, most of the threading dislocations originate from sub-grain boundaries (see figure 3). In contrast to normal grain boundaries, sub-grain boundaries separate two regions of the same grain that have very close orientations (<1-5°) and are formed by a loose network of dislocations. When oriented parallel to the interface, as in figures 3 and 5a, the constituent dislocations untangle from SGB1and shear both the upper (figure 5b) and lower parts of the Al grain. This untangling happens progressively, and it seems that this process can adapt itself to the thermal-strain gradients. Figure 5c exemplifies the other case when a SGB is essentially perpendicular to the interface. There, the emission of dislocations requires more stress and usually happens in bursts. A source labeled S1 is located at the intersection between SGB2 (not in contrast here but marked with a dashed line) and the interface. Note that this is slightly different from figure 4b where the loop was emitted from SGB2 at about the middle of the film. The loop in S1 grows first laterally along the interface. Once it has reached a width of about 140 nm, a dislocation is emitted and shears the whole grain, producing the slip traces at the top and bottom of the foil. If we assume that the radius of the observed small loop in figure 5c corresponds to an equilibrium between the applied stress and the line tension (equation 1), the local resolved shear stress is about 55 Mpa, which is very high. Because the emission happens in less than 1/25s, which is the time resolution of the video camera, not all the dislocation intermediate positions have been recorded, but the thickness of the slip traces indicates that the process repeated itself several times. Note that no trace of a dislocation pile-up is visible at the SiO_x interfaces. Eventually, the emitted dislocation hit SGB2 or points of the interface nearby and produced an avalanche of new emissions. This avalanche mechanism has been observed several times and is at the origin of the slip trace pattern seen in the vicinity of SGB2 in figure 3. The straightforward conclusion of this observation is that stress relaxation depends primarily on the nucleation of dislocations. Once emitted, dislocations do not encounter obstacles that require more energy than the one necessary to create them.

DISCUSSION

The recent comparison of the thermal stresses generated in Al and Cu films deposited either epitaxially on sapphire or as polycrystalline films on an amorphous barrier was at first surprising: for a given thickness, the metallic film attached on the harder substrate exhibited a RT stress lower than its counterpart attached on nitride- or oxide-coated silicon [8]. This reduction is much larger than expected from grain size effects only [4]. Explanations based on the Nix-Freund model face the dilemma that dislocations reaching the interface should be repelled [14]. This repulsion force is higher in the case of sapphire, which is the stiffer substrate, and this system should thus exhibit a higher strength. The present observations indicate that this scenario actually works only in the case of epitaxial films: the density of dislocations remains constant throughout cycling, with the activation of efficient sources. It has also been observed that the interfacial dislocation network can serve as a dislocation reservoir [9, 16]. Finally the interface is repulsive, as demonstrated by the piling up necessary to force dislocations into it (figure 4c). Isolated dislocations have radii of curvature that are of the order of the film thickness. Note that a resistant interface also bolsters those models of dislocation sources that presume the pile-up back stress would block further nucleation. A question remains concerning the axis (if any) around which these sources operate and what makes them absent in Al films.

When the film/substrate interface is a metal/amorphous interface, the radii of curvature of the dislocations are not controlled by their interaction with the interface, which acts as a sink. As a result, dislocations traveling perpendicularly to the interface have to drag the equivalent of a pinning point that would have a varying strength, depending on the local roughness of the interface (figure 4d). This strength is characterized by the angle between the dislocation and the interface, which is, at 400°C, 60° on average. Figure 6 is a sketch of this configuration: if we assume a threading dislocation gliding in a film bounded by two attractive interfaces that would act as a constant pinning point, the angle of the dislocation relative to the interface will be constant and equal to θ, with:

$$2\cos \theta = h/R \qquad [2]$$

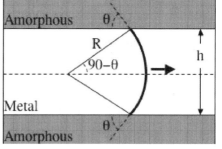

Figure 6. Dislocation moving between two sink interfaces. The strength of the interface, which acts as a pinning point and exerts a drag force, determines θ for a given film thickness h and radius of curvature R.

Of course, this configuration where θ is always greater than 0° (θ=0° corresponds to the repelling interface or to the Nix-Freund model) leads to a radius R that is larger than half the film thickness. The associated stress needed to move a dislocation between two "dragging" interfaces will therefore always be lower than in the Nix-Freund model, itself giving values that underestimate the experimental strength of metal films with amorphous bounds. In the case of figure 4d, for instance, an angle of 60° leads to an apparent radius of curvature of 1 μm, and thus a stress of about 5 Mpa. This "interface dragging" mechanism therefore cannot account for the high strength of thick Al films bounded by amorphous interfaces. A potential effect of this mechanism could arise in ultra thin films (h ≈ b), provided that threading dislocations are still active.

The only hardening factor remaining in these systems is therefore the lack of dislocations, i.e. the difficulty of nucleating dislocations. Multiplication mechanisms are very hard to model since they are not well known, hardly controllable and depend on many parameters that are themselves inaccurate. The criterion put forward by Friedman and Chrzan [22] is for instance inadequate here: there is never any back-stress induced by dislocations piling-up against amorphous interfaces since these interfaces absorb dislocations. Moreover, no efficient sources were found, except at the intersection of a SGB and the interface. At this very location, it is probable that a stress concentration exists, as evidenced by the avalanche of nucleation events initiated by a single source. The question remains as to why this interface/SGB configuration operates while a perpendicular GB connected to the same interface does not? We have also seen here that sub-grain boundaries could unleash some of their dislocations, but this process may be easier in the case of a cross-sectional thin foil where the SGB intercepts free surfaces. Besides, dislocations are absorbed by the interfaces and low-energy grain boundaries (see [13]), and since SGBs contain only a limited number of dislocations, this process dies off after a few thermal cycles. When a SGB is connected at both ends to more stable GBs or to interfaces such as SGB2, emission of dislocations appears to be more difficult, even at the interface. Finally, as the proportion of high-energy boundaries diminishes with thermal cycling [30], the number of potential locations for dislocation sources is also lowered.

Gao et al. [25] have considered that grain boundaries perpendicular to the interfaces could act as diffusion wedges that would have a crack-like configuration. In this constrained diffusional creep model, the stress concentration would be maximized at the bottom of the GB, where it connects to the interface, and therefore nucleate dislocations close to the interface. Balk et al. [31] have observed dislocations, gliding parallel to the interface, that could be the signature of this diffusion mechanism. In this particular case, the film has to be very thin (less than 400 nm) and unpassivated, which is not the situation in the present work. There are however indications that diffusion processes are very active even in bare Al films (although self-passivated) [15], and diffusion is also invoked to explain the repeated observation of the sink effect at amorphous interfaces [11-13]. If diffusion is active at the interface, it would also partially relax the local stresses and therefore limit the potential creation of dislocation sources, and explain why dislocation nucleation at the interface is rare. Following this prospect, the unusually high strain rates (as compared to wafer curvature experiments) employed in this study may also have helped triggering such interfacial sources.

CONCLUSION

In-situ thermal cycles performed inside the TEM allowed direct and real-time observations of dislocation behavior in thin metal films under stress. Al films bordered by two amorphous barriers and Cu films with an epitaxial interface with sapphire were studied. We concentrated our observations on dislocation hardening and nucleation mechanisms. The epitaxial interface of Cu films is repulsive and this repulsion represents the main hardening mechanism in this system. The Al films have amorphous interfaces that act as a dislocation sink. Threading dislocations have to 'drag' this interface as a continuum of pinning points, but this "interface dragging" mechanism cannot explain the high stresses reached by Al films that have a thickness of several hundreds of nm. The strength of these films is therefore probably due to the lack of efficient dislocation nucleation processes. Two of them have been reported here: the untangling of sub-grain boundary dislocation networks and the nucleation of dislocation loops at the intersection between a low angle boundary and the interface. This latter mechanism is obviously a controlling one since, once emitted, dislocations instantly shear the entire film thickness and eventually induce avalanches of new emissions. One reason for the insufficient number of dislocation sources at the metal/amorphous interface and at the grain boundaries could be the existence of a substantial amount of self-diffusion mechanisms. These could partially relax the high local stress needed to trigger dislocation sources, and also explain the sink effect of amorphous interfaces and grain boundaries.

ACKNOWLEDGMENTS

M.L. would like to acknowledge financial support from the Stressnet network. The authors thank Prof. E. Arzt from the Max Planck Institute for Metals research, where parts of this research project were carried out.

REFERENCES

1. L.B. Freund, J. Appl. Mech., 54 553 (1987).
2. W.D. Nix, Metall. Trans. A, 20A 2217-2245 (1989).
3. J.W. Matthews and A.E. Blakeslee, Journal of Crystal Growth, 29 273-280 (1975).
4. R.-M. Keller, S.P. Baker, and E. Arzt, J. Mater. Res., 13 1307-1317 (1998).
5. R. Venkatraman and J.C. Bravman, J. Mater. Res., 7 2040 (1992).
6. R. Venkatraman, J.C. Bravman, W.D. Nix, P.W. Davies, P.A. Flinn, and D.B. Fraser, J. Electron. Mater., 19 1231-1237 (1990).
7. C.V. Thompson, J. Mater. Res., 8 237-238 (1993).
8. G. Dehm, T.J. Balk, H. Edongue, and E. Arzt, Microelectronic Engineering, 70 412 (2003).
9. M. Legros, G. Dehm, T.J. Balk, E. Arzt, O. Bostrom, P. Gergaud, O. Thomas, and B. Kaouache. Plasticity-related phenomena in metallic films on substrates. in Multiscale Phenomena in Materials Experiments and Modeling Related to Mechanical Behavior. (2003). San Francisco, Vol. 779, pp. 63-74.
10. T.S. Kuan and M. Murakami, Metall. Trans. A, 13 383-391 (1982).
11. P. Müllner and E. Arzt. Observation of dislocation disappearance in aluminum thin films and consequences for thin film properties. in Thin films - Stresse and mechanical properties. (1998). Boston, MA: Mat. Res. Soc. Symp. Proc. 505, Warrendale, PA, Vol. 505, pp. 149-54.
12. G. Dehm, D. Weiss, and E. Arzt, Materials Science and Engineering A, 309-310 468 (2001).
13. M. Legros, K.J. Hemker, A. Gouldstone, S. Suresh, R.M. Keller-Flaig, and E. Arzt, Acta Materialia, 50 3435 (2002).
14. M.L. Ovecoglu, M.F. Doerner, and W.D. Nix, Acta Metallurgica, 35 2947 (1987).
15. B. Kaouache, P. Gergaud, O. Thomas, O. Bostrom, and M. Legros, Microelectronic Engineering, 70 447 (2003).
16. G. Dehm, B.J. Inkson, T.J. Balk, T. Wagner, and E. Arzt. Influence of film/substrate interface structure on plasticity in metal films. in Dislocations and deformation mechanisms in thin films and small structures. (2001). San Francisco: Mat. Res. Soc. Symp. Proc., Vol. 673, pp. 1-12.
17. R.-M. Keller, W. Sigle, S.P. Baker, O. Kraft, and E. Arzt. In situ TEM investigation during thermal cycling of thin copper films. in Mat. Res. Soc. Symp. Proc. (1997). Boston, MA, Vol. 436, pp. 221-226.

18. H. Gao, L. Zhang, and S.P. Baker, Journal of the Mechanics and Physics of Solids, 50 2169-2202 (2002).
19. B. von Blanckenhagen, P. Gumbsch, and E. Arzt, Modelling and Simulation in Materials Science and Engineering, 9 157-169 (2001).
20. J.R. Greer, W.C. Oliver, and W.D. Nix, Acta Materialia, 53 1821 (2005).
21. H. Van Swygenhoven, Science, 296 66-67 (2002).
22. L.H. Friedman and D.C. Chrzan, Philosophical Magazine A, 77 1185-1204 (1998).
23. K. Owusu-Boahen and A.H. King, Acta Materialia, 49 237-247 (2001).
24. G. Lucadamo and D.L. Medlin, Acta Materialia, 50 3045-3055 (2002).
25. H. Gao, L. Zhang, W.D. Nix, C.V. Thompson, and E. Arzt, Acta Materialia, 47 2865 (1999).
26. Y.-L. Shen and S. Suresh, Acta Met. Mat., 43 3915-3926 (1995).
27. G. Dehm, C. Scheu, M. Ruhle, and R. Raj, Acta Materialia, 46 759-772 (1998).
28. A. Couret, J. Crestou, S. Farenc, G. Molenat, N. Clement, A. Coujou, and D. Caillard, Microscopy Microanalysis Microstructures, 4 153-170 (1993).
29. M. Legros, K.J. Hemker, A. Gouldstone, S. Suresh, R.-M. Keller-Flaig, and E. Arzt, Acta Materialia, 50 3435-3452 (2002).
30. B. Kaouache, Propriétés mécaniques et microstructures de films minces métalliques, in Laboratoire de Physique des Matériaux. 2002, Université Nancy I: Nancy.
31. T.J. Balk, G. Dehm, and E. Arzt, Acta Materialia, 51 4471 (2003).

247

Mater. Res. Soc. Symp. Proc. Vol. 875 © 2005 Materials Research Society

Fiber-optics Low-coherence Integrated Metrology for In-Situ Non-contact Characterization of Wafer Curvature for Wafers Having Non-uniform Substrate and Thin Film Thickness

Wojciech J. Walecki, Alexander Pravdivtsev, Kevin Lai, Manuel Santos II, Georgy Mikhaylov, Mihail Mihaylov, and Ann Koo

Frontier Semiconductor, 1631 North 1st Street, San Jose CA 95112

Abstract. We propose novel stress metrology technique for measurement of local values stress tensor components in the coated wafers. New metrology is based on fiber-optic low coherence interferometry and can be applied to study stress not only in semicondiuctor wafers but in wide variety applications spanning from semiconductor to construction industry where measurements of plates covered by thin film encountered in flat panel displays, solar cells, modern windows.

Keywords: Thin-Film Metrology; Diagnostics; In-Situ, Real-Time Control and Monitoring, Thin Film Stress Metrology, Interferometry.
PACS: 68.60.-p, 68.90.+g, 83.85.St, 95.75.Kk

INTRODUCTION

Low coherence optical interferometry [1] has been proven to be an effective tool for characterization of thin and ultra-thin semiconductor silicon [2], [3], [4], compound semiconductor wafers [5], and Micro-Electro- Mechanical Systems (MEMS) structures [6] - [10]. In this paper we describe extension of this method to characterization of strained silicon wafers, SOI and other novel structures.

Traditional Stress Metrologies

The metrology of stress and topography of silicon wafers is a mature field. Available solutions are based on large aperture optical interferometers, capacitance, and laser scanning tools.

Free space optical interferometers for 300 mm wafer metrology are expensive due to high cost of large aperture optics and high cost of precision mechanical mounts. Due to large dimensions, and large mass of optical components they are rather difficult to integrate for in-situ applications, and their

integration with other complementary metrologies is usually quite cumbersome.

Recently local stress metrology tool based on curvature-sensitive measurement was proposed by Rossakis et al [11]. This very interesting method allows measuring accurately curvature of the wafer when achieving very high throughput. It suffers however from the same limitations as traditional phase shifting optical interferometry – it requires use of large optics and is not capable to neither provide information on neither wafer nor film thickness.

Some of the most successful tools for wafer topography measurements are based on capacitance measurements. This method however is not suitable for measuring thickness of semi-insulating and insulating materials, and does not provide insight in the internal structure of layered systems such as SOI. Laser scanning tools are offering several advantages over two earlier mentioned technologies.

Laser scanning tools are easier to adapt in integrated environment, and have great potential for in-situ application. This technology does suffer however from limitations of accuracy inherent to scanning technology due to finite spacing between photo-elements used for measuring angle of deflected beam. Typical accuracy of surface height measurement of the curved sample is of the order of 1-2 μm.

Advantages of Low Coherence Interferometry as Probe for Local Stress Tensor

Proposed in this paper metrology does not suffer from limitations inherent to existing techniques. Furthermore it allows to measure substrate thickness, wafer topography and film thickness in the same tool. It is readily scalable and the same sensor technology can be used for measurements of small and large plates.

THEORY OF MEASUREMENT

In this section we discuss our analytical model allowing us to determine local stress tensor components in film residing on wafer. We closely follow analysis presented by K. Roll [12], which we generalize for the case of wafers and films having non-uniform thickness. We consider approximately plane substrate residing in (x, y) plane, and having of position dependent thickness $H_s(x, y)$.

Substrate is covered by thin film having thickness $H_f(x, y)$. The structure is assumed in the original stress state (prior deformation) in which z components of the stress tensor $\sigma_{i,k}^0(x, y, z)$, where $i, k = x, y$ vanish. Following [12] and [13] we assume that deformation of the film can be described by the displacement vector $\vec{V}(x, y, z)$ which is linear in z. (this assumption follows from Kirchhoffs "hypothesis of linear elements"). We assume that stress does not depend on z in substrate and film. We introduce components of displacement tensor at the surface of the structure: $u_x(x, y)$, $u_y(x, y)$, and $W(x, y)$. Shape of the surface is given by:

$$z = W(x, y) \qquad (1)$$

We introduce local coordinate ξ measuring distance to surface of the wafer, which is more convenient to use than z due to bending of the wafer.

The components of the strain tensor are given by [12], [13]:

$$\varepsilon_{i,k}(x, y, \xi) = \alpha_{i,k}(x, y, \xi) + \Delta\alpha_{i,k}(x, y, \xi) + \ldots$$
$$\ldots + \xi \cdot \beta_{i,k}(x, y, \xi) \qquad (2)$$

where: $\alpha_{i,k} = \frac{1}{2}\left(\partial u_i / \partial x_k + \partial u_k / \partial x_i\right)$,

$\Delta\alpha_{i,k} = \frac{1}{2}\left(\partial W / \partial x_i\right)\left(\partial W / \partial x_k\right)$,

$\beta_{i,k} = -\partial^2 W / \partial x_i \partial x_k$.

As a result of Kirchhoffs hypothesis shear components $\varepsilon_{x,z}$ and $\varepsilon_{y,z}$ vanish. It can be shown that if deflection of the plate is small with comparison to thickness of the substrate second term in the Equation (2) can be neglected [12]. Following analysis presented in [12], and using Hooke's law we obtain stress in the substrate ($\rho = s$) and in the film ($\rho = f$):

$$\sigma_{i,k}^{(\rho)}(x, y, \xi) = (\rho - 1)\sigma_{i,k}^0 + \ldots$$
$$\ldots + \frac{E_\rho}{1 + m_\rho}\left(\varepsilon_{i,k} + \frac{m_\rho}{1 - m_\rho}\left(\varepsilon_{x,x} + \varepsilon_{y,y}\right)\delta_{i,k}\right) \qquad (3)$$

where E_ρ is Young's modulus, m_ρ is Poisson's ratio and ρ is index numerating substrate, and $\delta_{i,k} = 1$ when both indices are same or $= 0$ otherwise.

Both stresses in film and in substrate canbe obtained using Equation (3). The conditions of static equilibrium allow us to relate stresses in substrate to stresses in film. By applying conditions of equilibrium of forces and torques at every point (x, y), from equations (1) and (2) in the linear regime K. Roll [12] has obtained following expression for the local stress in the film:

$$\left\langle \sigma_{i,k}^{(2)} \right\rangle = \frac{E_s H_s^2}{6(1 + m_s)H_2} \cdot \ldots$$
$$\ldots \cdot \left[\frac{\partial^2 W}{\partial x_i \partial x_k} + \frac{m_s}{1 - m_s}\delta_{ik}\left(\frac{\partial^2 W}{\partial x^2} + \frac{\partial^2 W}{\partial y^2}\right)\right] \qquad (4)$$

where $\langle ... \rangle$ denotes averaging over coordinate ξ. In his paper K. Roll implicitly assumed that thicknesses of the wafer H_1 and film H_2 are constant and do not depend on (x, y).

We have noticed that this assumption was actually never used in derivation of Equation (4). Equation (4) can be generalized to following form in which H_s and H_f are explicit functions of (x, y), provided that $H_s = H_s(x, y)$, and $H_f = H_f(x, y)$ are slowly varying functions of (x, y):

$$\langle \sigma_{i,k}^{(2)} \rangle = \frac{E_s \left[H_s(x, y) \right]^2}{6(1 + m_s) H_f(x, y)} \cdots$$

$$\cdots \left[\frac{\partial^2 W}{\partial x_i \partial x_k} + \frac{m_s}{1 - m_s} \delta_{ik} \left(\frac{\partial^2 W}{\partial x^2} + \frac{\partial^2 W}{\partial y^2} \right) \right]$$

(5)

It is important to notice that for parabolic distortion $W = \frac{1}{2r}\left(x^2 + y^2 \right)$, and H_s, H_f constant the Equation (5) becomes equivalent to well known Stoney equation [14]:

$$\langle \sigma^{(2)} \rangle = \frac{E_s H_s^2}{6(1 - m_s) H_f} \frac{1}{r} \qquad (6)$$

In practical application we map shape of the top surface of the wafer $W(x, y)$, and thickness of the substrate and film $H_s(x, y)$, $H_f(x, y)$ respectively, and calculate stress tensor components $\langle \sigma_{i,k}^{(2)} \rangle$ using Equation (5).

EXPERIMENTAL

All measurements were performed using FSM 413–300 semiautomatic wafer thickness and topography mapping tool equipped with Echoprobe low coherence high precision interferometer [2].

The wafer thickness and topography were measured using standard Echoprobe sensor.

The reproducibility of the measurement of wafer thickness $H_s(x, y)$ is ± 0.2 µm, the absolute accuracy of the measurement of wafer thickness was approximately ± 0.8 µm. The reproducibility and accuracy of the measurement of wafer shape $W(x, y)$ was approximately ± 0.7 µm.

To demonstrate our metrology we have measured stress in thin film comprising of single layer of 2000 Å (nominal thickness) Cu film deposited on standard 300 mm diameter Si wafer oriented having surface perpendicular to [100] direction. Substrate thickness was measured $H_s(x, y)$ and shown that TTV was below 2 µm.

RESULTS

The substrate thickness and substrate topography and thickness were measured by the same 413 tool by two separate optical probes [7] in approximately 200 different points of the wafer. Since measurements are performed only at finite number of points we have approximated shape of the wafer by Since Equation (5) involves second derivative, and since $W(x, y)$ is slowly varying function of it's argument we approximated shape of the measured wafer surface by expanding it basis of Zernike's [16], [17] polynomials in cylindrical coordinates (r, ϕ) where r is normalized so $r \leq 1$.

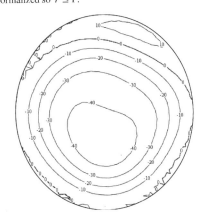

FIGURE 1. Measurement of wafer topography and thickness using Echoprobe Technology. Values on iso-lines are expressed in µm. Wafer is concave in the middle.

We keep all expansion terms for up to some arbitrary number n_{MAX} :

$$W(r,\phi) =$$

$$= \sum_{n=0}^{n_{MAX}} \sum_{m=0}^{n} \left[A_n^m \, {}^oU_n^m(r,\phi) + B_n^m \, {}^eU_n^m(r,\phi) \right] \quad (7)$$

where

$$\begin{Bmatrix} {}^oU_n^m(r,\phi) \\ {}^eU_n^m(r,\phi) \end{Bmatrix} = R_n^m(r) \cdot \begin{Bmatrix} \sin(m\phi) \\ \cos(m\phi) \end{Bmatrix} \quad (8)$$

where: the radial function $R_n^m(r)$ is defined for n and m integers $n \geq m \geq 0$ by following expression for $n - m$ even :

$$R_n^m(r) =$$

$$= \sum_{l=0}^{(n-m)/2} \frac{(-1)^l (n-l)!}{l! \left[\frac{1}{2}(n+m) - l \right]! \left[\frac{1}{2}(n-m) - l \right]} r^{n-2l} \quad (9)$$

and $R_n^m(r) = 0$ otherwise.

Coefficients A_n^m and B_n^m in Equation (7) are calculated using fact that Zernike's functions are orthogonal and complete basis set as shown in Equation (8). In practical calculation double integral has been replaced by discrete summation over measured surface area.

$$\begin{Bmatrix} A_n^m(r,\phi) \\ B_n^m(r,\phi) \end{Bmatrix} = \frac{(n+1)}{\varepsilon_{m,n}^2 \pi} \int_0^1 \int_0^{2\pi} W(r,\phi) \cdots$$

$$\cdots \begin{Bmatrix} {}^oU_n^m \\ {}^eU_n^m \end{Bmatrix} d\phi r dr \quad (10)$$

where

$$\varepsilon_{m,n} = \begin{cases} \frac{1}{\sqrt{2}} & \text{for } m = 0, n \neq 0 \\ 1 & \text{otherwise} \end{cases}$$

The shape of the wafer surface is then expressed in

Cartesian coordinates $W(x,y)$, and stress tensor components are calculated using Equation (5). This allows us to differentiate series of Zernike functions rather than approximate second order derivatives in Equation 5 by finite differences.

The above outlined procedure allows us to find values of stress tensor components $\sigma_{X,X}$, $\sigma_{X,Y}$, and $\sigma_{Y,Y}$ everywhere on the wafer. For particular dataset presented in Figure 1 we have found that xx component of stress tensor at the center of the wafer was $\sigma_{X,X}(0,0) = 1.5 \cdot 10^9$ dyn/cm^2, while $\sigma_{Y,Y}(0,0) = 2.0 \cdot 10^9$ dyn/cm^2, the off-diagonal component is about two orders of magnitude smaller ($| \sigma_{X,Y}(0,0) | < 2.0 \cdot 10^7$ dyn/cm^2). We defer detailed error analysis, and discussion of mapping related issues to separate paper, at this point we estimate that for better for 300 mm wafers having bow in range of 40 - 100 μm and thickness of substrate of 0.780 mm.

CONCLUSIONS

We have proposed new non-contact method of measuring local stress of wafers with deposited thin films. The proposed method takes into account local thickness variation of substrates and measured films. It allows to extract from measured data $\sigma_{X,X}$, $\sigma_{X,Y}$, and $\sigma_{Y,Y}$ components of the stress tensor in film.

REFERENCES

1. D. Huang, E. A. Swanson, C. P. Lin, J. S. Schuman, W. G. Stinson, W. Chang, M. R. Hee, T. Flotte, K. Gregory, C. A. Puliafito, J. G. Fujimoto, "Optical coherence tomography," Science 254, 1178-1181 (1991).

2. W. Walecki, "Non-Contact Fast Wafer Metrology for Ultra-Thin Patterned Wafers Mounted on Grinding Dicing Tapes", SEMI® Technology Symposium: International Electronics Manufacturing Technology IEMT), San Jose, July 16, 2004.

3. W.J. Walecki, R. Lu, J. Lee, M. Watman, S.H. Lau, and A. Koo, "Novel Non-contact Wafer Mapping Metrologies for Thin and Ultrathin Chip Manufacturing Applications", 3rd International Workshop on

Semiconductor Devices – Manufacturing and Applications, Munich, Germany, November 25, 2002.

4. W. J. Walecki, V. Souchkov, K. Lai, P. Van, M. Santos, A. Pravdivtsev, S. H. Lau, and A. Koo, "Novel Noncontact Thickness Metrology for Partially Transparent and Nontransparent Wafers for Backend Semiconductor Manufacturing" in *Progress in Compound Semiconductor Materials IV—Electronic and Optoelectronic Applications,* edited by G.J. Brown, M.O. Manasreh, C. Gmachl, R.M. Biefeld, K. Unterrainer (Mater. Res. Soc. Symp. Proc. 829, Warrendale, PA , 2004), B9.31.

5. W. J. Walecki, K. Lai, V. Souchkov, P. Van, SH Lau, and A. Koo, "Novel noncontact thickness metrology for backend manufacturing of wide bandgap light emitting devices", phys. stat. sol. (c) 2, No. 3, 984– 989 (2005).

6. W. Walecki, F. Wei, P. Van, K. Lai, T. Lee, SH Lau, and A. Koo, "Novel Low Coherence Metrology for Nondestructive Characterization of High Aspect Ratio Micro-fabricated and Micro-machined Structures "Reliability, Testing and Characterization MEMS/MOEMS III, edited by D. M. Tanner and R. Ramesham, Proc. SPIE 5343, 55 (2003).

7. Walecki, Low Coherence Interferometry Based Metrologies for MEMS Manufacturing, SEMI Technology Symposium: Innovations in Semiconductor Manufacturing (STS: ISM), July 13, San Francisco, 2004.

8. W. Walecki, F. Wei, P. Van, K. Lai, T. Lee, V. Souchkov, S.H. Lau, and A. Koo "Low Coherence Interferometric Metrology for Ultra-Thin MEMs Structures", in *Nanoengineered Assemblies and Advanced Micro/Nanosystems,* edited by David P. Taylor, Jun Liu, David McIlroy, Lhadi Merhari, J.B. Pendry, Jeffrey T. Borenstein, Piotr Grodzinski, Luke P. Lee, and Zhong Lin Wang (Mater. Res. Soc. Symp. Proc. 820, Warrendale, PA , 2004), 08.8.

9. Wojciech J. Walecki, Frank Wei, Phuc Van, Kevin Lai, Tim Lee, SH Lau, and Ann Koo , "Novel Low Coherence Metrology for Nondestructive Characterization of High Aspect Ratio Micro-fabricated and Micro-machined Structures", Reliability, Testing, and Characterisation of MEMS/MOEMS III, edited by Danelle M. Tanner, Rejeshuni Ramesham, Proceedings of SPIE Vol. 5343 p. 55-62 (SPIE, Bellingham, WA, 2004).

10. W. J. Walecki, V. Souchkov, K. Lai, T. Wong, T. Azfar, Y. T. Tan, P. Van, S. H. Lau, and A. Koo, "Low-coherence interferometric absolute distance gauge for study MEMs structures" in *Reliability, Testing and Characterization MEMS/MOEMS IV,* edited by D. M. Tanner and R. Ramesham, Proc. SPIE 5716, 23 (2005).

11. A. J. Rosakis, R. Singh, E. Kolawa and N. Moore, Jr "Coherent Gradient Sensing Method and System for Measuring Surface Curvatures", Issued 2000, US Patent # 6,031,611 2.

12. K. Roll, Journal of Applied Physics, vol. 47, No. 7, p. 3224 (1976).

13. L. D. Landau and E. M. Lifschitz, "Theory of Elasticity" (Pergamon, New York 1970).

14. G. G. Stoney, Proc. Roy. Soc, London, Ser A 82, 172, (1909).

15. P.T.B. Shaffer, *Appl. Opt.* 10 (1971), 1034-1036.

16. Born, M. and Wolf, E. "The Diffraction Theory of Aberrations." Ch. 9 in *Principles of Optics: Electromagnetic Theory of Propagation, Interference, and Diffraction of Light, 6th ed.* New York: Pergamon Press, pp. 459-490, 1989

17. Eric W. Weisstein. "Zernike Polynomial." From *MathWorld*--A Wolfram Web Resource. http://mathworld.wolfram.com/ZernikePolynomial.html

Adhesion and Fracture of
Thin Films

Mater. Res. Soc. Symp. Proc. Vol. 875 © 2005 Materials Research Society

In-situ Observations on Crack Propagation along Polymer/glass Interfaces.

W.P.Vellinga, R.Timmerman, R.van Tijum, J.Th.M. De Hosson
Materials Science Centre, University of Groningen
Nijenborgh 4, 9747 AG, Groningen, The Netherlands

ABSTRACT

The propagation of crack fronts along a PET-glass interface is illustrated. The experimental set-up consists of an Asymmetric Double Cantilever Beam in an optical microscope. Image processing techniques used to isolate the crack fronts are discussed in some detail. The fronts are found to propagate inhomogeneously in space and time, in forward bursts that spread laterally along the front for some distance. In some cases the forward movement of a crack can be almost entirely due to the lateral movement of forward steps (analogous to "kinks") along the crack front.

INTRODUCTION

Interest in the mechanical properties of polymer-metal laminates is rising since they appear in an ever increasing number of applications, as diverse as car panels and high-tech displays. Of course the mechanical integrity of the interfaces in such applications is important. As the failure of such an interface usually implies the propagation of a crack along it, this is a clear target for investigation. In the field of crack propagation there is growing interest in bringing together theory used in continuum mechanics with ideas from statistical physics. This seems especially helpful in the area of lifetime prediction where quantities accessible from continuum mechanics (stresses, stress intensity factors, energy release rates) may be related to the interplay between disorder and stress aided thermally activated processes [1]. Crack propagation in itself is of interest as recently intriguing results on crack front instabilities and scaling behavior have been encountered in experiment and theory [2,3]. It has been observed in experiment that the response of stressed disordered media may show bursts of widely distributed magnitude, pointing to an internal avalanche dynamics.

The work presented here is related to the works mentioned above but is original in a number of aspects, ranging from the system studied to the data treatment involved and the actual dynamics observed. We present observations on the propagation of cracks along polymer-glass and polymer-metal interfaces.

EXPERIMENT

The experimental method is an Asymmetric Double Cantilever Beam test. The actual set-up is based on a miniature tensile stage (Kammrath&Weiss) which fits in a reflection optical microscope. Driving speeds for the cracks are typically 10 µm/s. Samples consist of Poly-Ethylene Terephthalate (PETG) spin-coated on steel, thickness, dried in a convection oven at 80°C for a few hours, and subsequently pressure bonded to a glass support (for 240s, at 140°C and 1.5 MPa). For a schematic picture of the experiment see figure 1. The crack front is observed through the glass with a CCD camera (1376*1032 pixels, 3*8 bit) at a rate of 1 Hz

ADCB

Excellent papers dealing with Asymmetric Double Cantilever Beam (ADCB) experiments and issues related to it can be found in the literature [e.g. 4]. Here we only repeat the main results necessary to understand our experimental results. In ADCB experiments one deals with the geometry shown in figure 1. In practice a >> h_1, h_2. Also at all times, for the remaining adhered portion of the beam L, L >>a should hold.

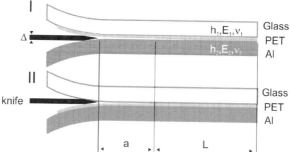

Figure 1. Schematic drawing of an ADCB experiment. Δ crack opening, a crack length, h_i height, E_i Young's modulus, v_i Poisson's ratio, subscript i=1,2 refer to glass, Al respectively. In I the crack propagates along the Al-PET interface, in II along the PET-glass interface. Both situations have been encountered in practice and showed quite similar propagation behavior. Here we show only results of situation II.

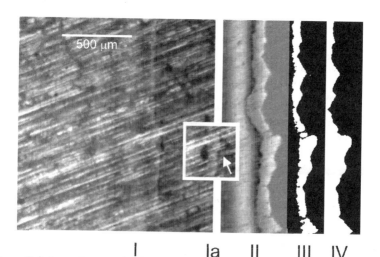

Figure 2. I: Image I_0 at t_0, showing metal roughness, wedge fringes where laminate has separated (left), brightness change at crack front. Ia: part of image I_{30} at t_0+30 (s). Arrow indicates necessary displacement vector to bring background into register (exaggerated). II: Pixelwise subtraction I_{30}–I_0 after image registration. III: Thresholding. IV: Front filling (see text).

In ADCB we measure the energy release rate for a certain interface structure, sample geometry and phase angle, simply by measuring a, as shown in figure 1, and using the approximate formulas shown below:

$$G = \frac{\left(3\Delta^2 E_1 E_2 h_1^3 h_2^3\right)\left(C_2^2 E_1 h_1^3 + C_1^2 E_2 h_2^3\right)}{8a^4 \Lambda^2} \tag{eq.1}$$

with $\Lambda = C_1^3 E_2 h_2^3 + C_2^3 E_1 h_1^3$ and $C_i = 1 + 0.64 \dfrac{h_i}{a}$ [4] .

The usefulness of the approximation given above, depends on the elastic mismatch, which should not be too large. A typical unprocessed image of a crack front is shown in figure 2-I.

Image processing

Various image correlation and analysis techniques may be necessary to extract the crack front shape as a function of time and position. We use a phase correlation algorithm for the registry of the images. In the experiment the position of the knife is fixed, and the sample is clamped in one of the clamps of the tensile stage. The movement of the clamps may not be entirely smooth, and may also not be entirely aligned with the coordinate system of the image. Image correlation allows us to determine the movement of the sample between two images, and subsequently the movement of the front with respect to the sample during the same time can be determined. The translation vector between two images I_i and I_j is indicated by the position of the maximum of the phase correlation function

$$\vec{x}\left(I_i, I_j\right) = \vec{x}(\max\left(\frac{F\left(I_i\right)F\left(I_j\right)^*}{\left|F\left(I_i\right)F\left(I_j\right)^*\right|}\right)) \tag{eq. 2}$$

where F denotes the discrete Fourier transform. For these experiments I_i and I_j were sub-images of 512*512 pixels ahead of the moving front. The image series is divided into sets of n images, the first of which are "base" images. The displacement vectors between these base images are first calculated. Subsequently, the displacement vectors between images in a certain set and the base image of the set are calculated. Finally the displacement of a certain image N with respect to the first follows from

$$\vec{x}\left(I_1, I_N\right) = \sum_{i=1}^{N\backslash n} \vec{x}\left(I_{in}, I_{(i-1)n}\right) + \vec{x}\left(I_{(N\backslash n)*n}, I_N\right) , \tag{eq.3}$$

where \ denotes integer division (see fig 2-Ia). In order to isolate the fronts we subtract two images that are n frames apart (fig 2-II). If the correlation is good this gets rid of the background, dominated by the surface texture of the metal effectively. If not, strongly correlated noise remains that complicates the subsequent determination of the front position. The main stages in the determination of the front position are shown in fig.2-III and fig 2-IV. The subtraction image is noise filtered and thresholded, which leaves a solid area behind the front. However, the front usually shows holes (fig 2-III). There are two main reasons for this to occur: firstly the front may not or hardly have moved in the time between the acquisition of the two images (*intrinsic* hole), secondly the correlation procedure may come up with a displacement vector that is in error, which combined with the background texture leads to missing lines (*extrinsic* holes). We now fill up the front using the following reasoning. Intrinsic holes signify that the front has not moved, so for the positions inside the hole we should take the last front position that was successfully calculated. Considering that the extrinsic holes are due to random errors in the correlation procedure, and are not expected to occur on the same spot twice, front positions for coordinates in the extrinsic holes are also filled with the last available front position. Denoting thresholded images by T and front files by C the following is performed in a pixel-wise fashion:

$$C_{i+1} = OR(T_{i+1}, C_i) \qquad \text{(eq.4)}$$

Subsequently the image behind the front is filled (Fig.2-IV), and the front itself is isolated using a very simple edge detection scheme, subtracting an eroded version of the image from itself.

RESULTS AND DISCUSSION

Front propagation mode

In figures 3 to 5 we present results of experiments carried out on 2 different laminates. Figure 3 shows results on 1 laminate, figures 4 and 5 on another laminate. In both cases the crack was found to propagate along or very close to the PET-glass interface, The front dynamics are displayed in two different ways. Figures 3 and 5 show local front position with respect to the mean front position and local forward acceleration, as a function of time. Figure 4 shows the successive bursts occurring along the crack front. The front shown in figure 3 was moving at a constant mean speed of 5 μm/s, the front shown in figure 4 and 5 was moving spontaneously (and slowing down) after the knife had been stopped some time before the measurements shown.
The main observation in figures 3 to 5 is that in general the crack propagation is inhomogeneous in time as well as in space. Whereas the crack movement is smooth on a macroscopic scale, on a microscopic scale crack movement occurs because parts of the front become unstable and move ahead of the mean crack position. Subsequently this disturbance spreads laterally along the front for some distance.
Behavior similar to that shown in figure 3 has been found by us for cracks propagating along PET-metal interfaces (to be published), and in the literature for polymer-polymer interfaces [2,3]. This seems to indicate that such crack front movement is in fact quite common. Figure 4 shows that in extreme cases the lateral movement associated with a disturbance may span the entire crack front (= sample width). In such cases the forward movement of a crack front in fact consists almost entirely of lateral movement of forward steps (analogous to "kinks" in dislocation movement) moving along the front. Figure 5 illustrates this in a way analogous to figure 3. From figure 5-II it is clear that apart from the lateral movement forward bursts do also occur, and that there exists a strong spatial correlation between the positions of forward bursts.

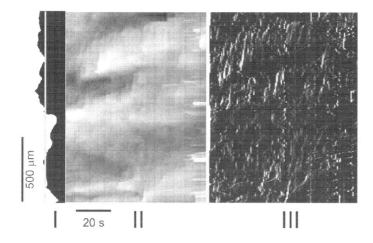

Figure 3. I: Example front. Grey line indicates the mean front position. II: Position with respect to mean front position as a function of time. Light parts: front locally ahead of mean front position, dark parts: front locally behind mean front position. III: Forward acceleration as a function of time and position along front. $G = 2$ J/m^2.

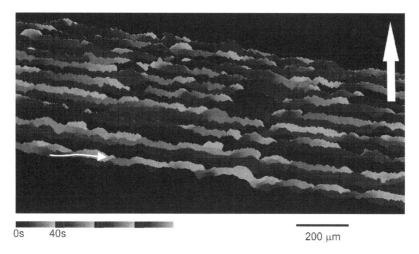

Figure 4: Front progression. Bursts of movement occurring between two subsequent images have been color coded. Colors recur after 40 seconds. Clearly, most of the forward movement in this case derives from movement of forward steps in the front (somewhat analogous to "kinks" in dislocation movement) parallel to the front. This is further clarified by the small arrow. The large arrow indicates the macroscopic propagation direction. $G = 4$ J/m^2

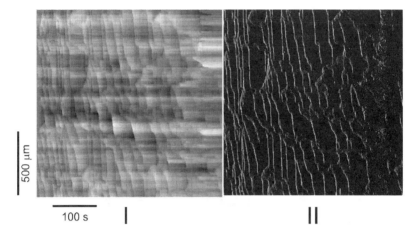

Figure 5: The measurements shown figure 4, illustrated as in figure 3. All displacements with respect to front position at time zero. I: Position with respect to mean front position as a function of time. II: Forward acceleration as a function of time and position along front. $G = 4 \text{ J/m}^2$.

Another remarkable feature of the movement shown in figure 4 and 5 is the occurrence of a characteristic forward step size. The reason for this is not clear. However, we note that in the literature arrays of regularly spaced shear bands have been observed in front of cracks propagating along interfaces between glassy polymers [e.g. 4]. The occurrence of such a structure ahead of the crack might possibly lead to the observed regularity in crack movement. More detailed investigation of this aspect, of the scaling properties of crack front, possible relation of propagation mode and values for G, and G as a function of driving speed are outside the scope of this paper, and will be discussed elsewhere.

CONCLUSIONS

Crack fronts along a PET-glass interface were found to propagate inhomogeneously in space and time. The fronts move in forward bursts that spread laterally along the front for some distance. The mode of propagation implies that the opening always has mode III components. In some cases the forward movement of a crack can be almost entirely due to the lateral movement of forward steps (analogous to "kinks") along the crack front.

ACKNOWLEDGEMENTS

This work is supported by IOP (project IOT 01001) and STW (project GTF.4901).

REFERENCES

[1] S.Santucci, L.Vanel, and S.Ciliberto, Phys. Rev. Lett. **93**, 095505 (2004)
[2] J.Schmittbuhl, K.J.Måløy, Phys. Rev. Lett. **78**, 3888–3891 (1997)
[3] K.J.Måløy, J. Schmittbuhl, A. Hansen and G. G. Batrouni *International Journal of Fracture*, **121**, 9 (2003)
[4] B.Bernard, H.R.Brown,C. J. Hawker, A.J. Kellock, T. P. Russell, Macromolecules 6254-6260, 32 (1999)

Mater. Res. Soc. Symp. Proc. Vol. 875 © 2005 Materials Research Society   O10.4

Strain Mapping on Gold Thin Film Buckling and Silicon Blistering

P. Goudeau[1], N. Tamura[2], G. Parry[1], J. Colin[1], C. Coupeau[1], F. Cleymand[3], H. Padmore[2]
[1]Laboratoire de Métallurgie Physique, UMR 6630 CNRS, Université de Poitiers, SP2MI, Téléport2, Bd M. et P. Curie, BP 30179, F-86962 Futuroscope Chasseneuil cedex
[2]Advanced Light Source, Lawrence Berkeley National Laboratory, 1 Cyclotron Road, MS 2-400, Berkeley, CA, 94270, USA
[3]Laboratoire de Science et Génie des Surface, Ecole des Mines de Nancy, UMR CNRS 7570, Parc Saurupt, F-54042, Nancy

ABSTRACT

Stress/Strain fields associated with thin film buckling induced by compressive stresses or blistering due to the presence of gas bubbles underneath single crystal surfaces are difficult to measure owing to the microscale dimensions of these structures. In this work, we show that micro Scanning X-ray diffraction is a well suited technique for mapping the strain/stress tensor of these damaged structures.

Keywords: Thin film, residual stresses, delamination, micro X-ray diffraction, strain mapping

INTRODUCTION

Biaxial compressive residual stresses are usually present in thin films deposited at room temperature by direct ion beam sputtering. The stress magnitude is often very high (larger than the elastic limit of the same material in the bulk state) and thus spontaneous delamination phenomenon such as 1D – wrinkling or 2D - buckling may appear for a critical film thickness (relaxation of the stored elastic energy) when extracting the sample from the deposition chamber [1]. In the same way, thin film adherent to a substrate and placed in an axial compression may also induce straight line wrinkles over a critical applied stress. Combined with Atomic Force Microscopy, this kind of experiment allows for determining the adhesion energy from the evolution under compression of the geometrical parameters which define buckling (width, deflection and film thickness) [2].

Although a lot of theoretical works have been done to develop mechanical models and calculations (elasticity of thin plates, fracture mechanics, finite element and analytical calculations) with the aim to get a better understanding of driving mechanisms giving rise to this phenomenon, only a few experimental works have been done on this subject to support these theoretical results and almost nothing concerning local stress/strain measurement mainly because of the small dimension of the buckling (few tens of micrometer).

An equivalent and interesting problem is related to blistering and splitting of hydrogen implanted silicon [3]. Models have been suggested to understand the growth of micro-cracks and then to evaluate the critical stress at the edge above which breakage occurs. The size of the blisters is typically around 10 μm in width and thus strain/stress measurements at such small scales are also difficult.

In this work, we use micro beam X-ray diffraction (micro-XRD) available on synchrotron radiation sources as a local probe (spatial) for analysing stress/strain fields. Following our previous studies on spontaneous buckling [4], we investigated delamination of gold films and also the local surface curvatures in Silicon (100) wafers due to the presence of implanted H_2 bubbles. The X-ray diffraction results are correlated with finite element simulations.

EXPERIMENTAL DETAILS

Two types of samples have been studied. For delamination studies, gold films were deposited by ion beam sputtering on 630 μm thick silicon (100) and LiF single crystal substrates. The gold films thicknesses are 630 and 150 nm respectively. The metallic film is perfectly adherent to the LiF susbtrate. For blistering, (100) silicon wafers were implanted with H_2^+ molecules at 160 KeV (80 KeV for H^+ ions) with a dose of $2.5\ 10^{16}$ ions / cm^2 ($5\ 10^{16}$ for H^+). This implantation has been followed by an annealing treatment at 250°C for 1H to obtain bubble formation and thus very localised silicon bending or blistering [3]. The corresponding implanted depth calculated with the TRIM code (Projected range + longitudinal straggling) gives the silicon thickness which is bent by the H_2 bubbles. The value according to our experimental conditions is 800 nm. The damage structures have been characterized prior to XRD measurements by optical microscopy to identify precisely markers which would be easily found with the X-ray beam for x-y stage calibration and also by AFM for extracting geometrical parameters of wrinkles and blisters. A Au/LiF sample set has been deformed during a compression test and studied in situ by AFM measurements [2]. Wrinkles are clearly visible on the post-mortem sample surface as well as dislocation step on the LiF substrate where gold film has been removed.

Micro Scanning X-ray diffraction experiments have been done at the ALS on beam line 7.3.3. [5]. Monochromatic X-ray beam with a photon energy of 6 keV has been chosen for investigating gold thin film samples while polychromatic x-ray beam (5-14 keV) has been used for experiment calibration (sample to detector distance and CCD detector tilt angles) and strain measurement on silicon blistering. Back reflection mode is operating for the diffraction measurements as well as fluorescence scans which are used for localising the markers on the sample surface. The sample to detector distance is generally rather small (about 35 mm) which allows a wide diffraction data integration range in the configuration 45°/90° respectively for the incident x-ray beam angle and the angular detector position. The diffraction patterns are analysed using specific software developed for laue pattern (white beam).

RESULTS AND DISCUSSION

Gold films: We performed at first a scan with a small step of 2 micron on a circular blister resulting from the presence of residual compressive stresses of about -400 MPa in the gold film after deposition on the silicon wafer. The procedure for doing this measurement and the results are given in figure 1. The optical image (fig 1.a) allows us to know precisely the location of the buckles to be scanned. The coordinates are defined using a reference point with is given by a step corner realized during the thin film deposition. The gold fluorescence scan in fig. 1 (b) gives the x-y stage coordinates of this reference point. The gold film is textured with diffracting rings showing the maximum intensities at the pole directions (2Θ axis in the vertical direction and ± X axis along the ring on each side of the 2Θ axis). We extracted from the (220) ring the d-spacing and plotted its evolution during the x-y scan on fig. 1 (d). We clearly see a decrease of this value at the top of the buckle which is slightly correlated with a strain/stress relaxation of the compressive stress present in the adherent region. Let us note that the stress-free (220) d-spacing calculated from bulk gold lattice parameter (0.4078 nm) is 0.1442 nm and the angle between (111) planes and (220) is X=35.3° (pole direction).

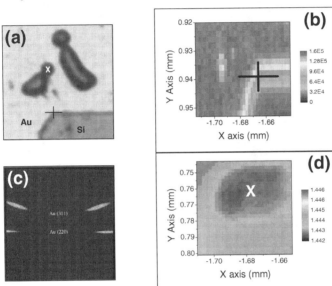

Figure 1: Strain mapping on a gold blister 40 μm width. (a) optical image (400 x 400 μm) of the buckle region and the reference corner marked with the sign +, (b) gold fluorescence scan allowing to obtain the coordinates of the reference point (+), (c) 2D diffraction pattern of the <111> textured gold film and (c) (220) d-spacing variation on the buckle marked with the white cross.

The second example concerns induced delamination after performing an axial compressive test on a LiF single crystal substrate covered with a thin gold film. This work was principally done for studying the effect of metallic film on the emergence process of dislocations nucleated in LiF substrates [2]. The gold film has been partially removed in order to observe by AFM the slip line structures. Straight sized wrinkles appear above a plastic strain of 1.71 % with a deflection and a width estimated to about 0.9 and 9.4 μm respectively. The optical image of the sample surface is shown on fig. 2 (a). The reference point for x-y stage calibration corresponds to the white symbol + drawn on the optical image and also the gold fluorescence scan fig. 2 (b). The diffracted intensities in fig 2 (c) are similar but weaker than the one in fig. 1 (c) due to the lower thickness of the film deposited on LiF (ratio ¼). The variation of the (220) d-spacing during the x-y scan is shown fig. 2 (d). The d value is decreasing from the bottom to the top of the wrinkle, in the same way than what is observed for spontaneous buckling fig 1 (d). However, the absolute value of (220) d-spacing is weaker as well as the amplitude of the relaxation. This film was almost stress free after deposition; buckling is created after plastic deformation of 1.7% which certainly modifies drastically the gold microstructure (elastic domain < 0.3 %). This effect may then explain the lattice parameter values obtained.

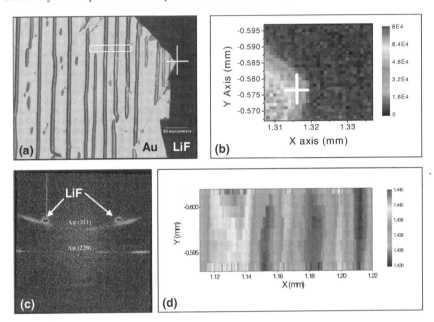

Figure 2: Induced delamination of gold film on LiF substrate: (a) optical image of the gold film surface after 1.7 % plastic deformation; the white rectangle images the zone where XRD scan will be done (b) Gold fluorescence scan of the reference point (c) Diffraction diagram showing the presence of a <111> fibber texture (d) x-y scan performed in the white rectangle of figure (a) with a 1 μm step.

Silicon blisters: The main difficulty here is that the entire material is single crystal silicon and thus it appears difficult to differentiate with XRD the bulk wafer signal from the signal due to Silicon blisters. However, XRD is sensitive to crystallographic orientation and then to surface bending. In that preliminary study, we used a 1 μm^2 white beam with a 1 μm step to scan the strain field associated with Si (100) blistering. The geometry of blisters is shown in fig. 3 (a) as well as the breakage effect in (b).

Following an equivalent procedure to the one described above for gold films buckling, XRD measurements have been done over a single blister 9.8 μm in width and 135 nm in height. The Laue pattern obtained when the x-ray beam is at the top of the blister is shown in fig. 4 (a). We clearly observed a ring around the (004) Laue spot of Si bulk structure. In order to get better precision, we increased the sample to detector distance from 35 to 235 mm in order to zoom in on the region close to the (004) Bragg spot. A puzzle or mosaic image of the x-y scan is given on fig 4 (b). The strain analysis of this ring over the Bragg spot is still in progress since specific development software needs to be implemented. However, we can get an idea of the strain variation in this Si blister by using Finite Element simulation. The theory of thin plates can be used since the thickness of the silicon (0.8 µm) is large compared to the blister width (~ 10 µm). The results obtained with ABAQUS are given in fig. 5. The mean stress is a maximum at the blister edges (blisters break at the edge – see fig. 3 (b)) and the stress gradients are important (a).

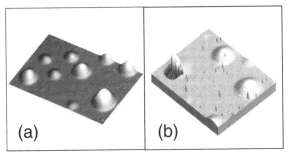

(a)   (b)

Figure 3: AFM images of the silicon blisters and the effect of breakage in (b) which leads to the presence of small silicon pieces at the sample surface.

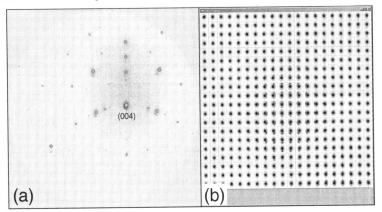

(a)   (b)

Figure 4: White beam diffraction measurement on the Si blister: (a) diffraction diagram taken at the top of the blister (b) Mozaic image showing the x-y scan performed over the Si blister.

A monochromatic energy scan of (004) Si planes has been done in the energy range of 6318-6733 eV with 5eV steps, a detector position at 60°, a sample to detector distance at 235 mm, an X-ray beam size larger than the blister width (~ 10 µm), and an exposure time of 60s. The x-y stage is fixed at the position at the top of the Si blister. A horizontal curved line is then observed moving from the top to the bottom of the detector with a change in curvature. The complete analysis of diffraction patterns needs specific software data treatment for monochromatic wavelength which is in progress.

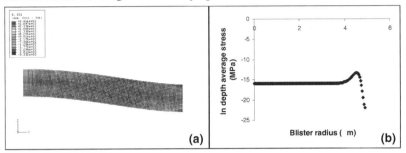

(a)   (b)

Figure 5: Finite Element simulation of stresses in Silicon blister: (a) cross-section along the radius showing the in depth stress gradient in MPa (b) Mean stress variation along the radius.

CONCLUDING REMARKS

Micro Scanning X-ray diffraction with either a white or a monochromatic x-ray beam is a powerful technique for exploring microscale damaged structures. Improvements in data treatment are essential for extracting strain/stress tensors from measurements done on Si blisters; software developments are in progress. Furthermore, 3D measurements should be done for extracting in depth stress gradient.

In the case of thin film buckling, the next experimental step will consist in combining compression tests to micro XRD scans for in situ studying the early stage of delamination.

ACKNOWLEDGMENTS

The Advanced Light Source is supported by the Director, Office of Science, Office of Basic Energy Sciences, Materials Sciences Division, of the U.S. Department of Energy under Contract No. DE-AC03-76SF00098 at Lawrence Berkeley National Laboratory. The authors would like to thank Philippe Guerin for thin film deposition and hydrogen implantations in silicon wafers.

REFERENCES

1. C.Coupeau, P. Goudeau, L. Belliard, M. George, N. Tamura, F. Cleymand, J. Colin, B. Perrin and J. Grilhé, Thin Solid Films **469-470**, 221 (2004).
2. C. Coupeau, F. Cleymand and J. Grilhé, Scripta mater. **43**, 187 (2000).
3. L.-J. Huang, Q.-Y. Tong, T.-H. Lee, Y.-L. Chao and U. M. Gosele, Electrochemical Society proceedings **19**, 1373 (1998).
4. P. Goudeau, P. Villain, N. Tamura, H. Padmore, Applied Physics Letters **83**, 51 (2003).
5. N. Tamura, A. A. MacDowell, R. Spolenak, B. C. Valek, J. C. Bravman, W. L. Brown, R. S. Celestre, H. A. Padmore, B.W. Batterman and J. R. Patel, J. Synchrotron Rad. **10**, 137 (2003).

Mater. Res. Soc. Symp. Proc. Vol. 875 © 2005 Materials Research Society   O10.5

Interfacial Adhesion of Pure-Silica-Zeolite Low-k Thin Films

Lili Hu[1], Junlan Wang[1], Zijian Li[2], Shuang Li[2], Yushan Yan[2]

[1]Department of Mechanical Engineering
[2]Department of Chemical and Environmental Engineering
University of California, Riverside, CA 92521

ABSTRACT

Nanoporous zeolite thin films are promising candidates as future low dielectric constant (low-k) materials. During the integration process with other semiconductor materials, the residual stresses resulting from the synthesis processes may cause fracture or delamination of the thin films. In order to achieve high quality low-k zeolite thin films, the evaluation of the adhesion performance is important. In this paper, laser spallation technique is utilized to investigate the interfacial adhesion of zeolite thin film-Si substrate interfaces prepared using three different processes. The experimental results demonstrate that the nature of the deposition method has a great effect on the resulted interfacial adhesion of the film-substrate interfaces. This is the first time that the interfacial strength of zeolite thin films-Si substrates is quantitatively evaluated. The results have great significance in the future applications of low-k zeolite thin film materials.

INTRODUCTION

Low dielectric constant (low-k) materials have been identified as one of the most important challenges for interconnects of the future generation integrated circuits (IC). Among the several recently developed low-k materials, zeolite thin films have unique advantages over polymer and amorphous porous silica based low-k materials [1]. As a class of inorganic nanoporous crystalline materials [2, 3], zeolites are frameworks of aluminosilicates consisting of corner-sharing tetrahedrons of SiO_4 and AlO_4, where the ratio (Si +Al)/O must equal to 1/2. Nanostructured pure-silica-zeolite (PSZ) thin films have been recently developed and demonstrated to be a promising low-k material [4]. While the high porosity of zeolite thin films significantly reduces the dielectric constant, the strong crystalline framework structure maintains a comparatively higher mechanical strength than other amorphous porous silicas. The molecular sized pore is much smaller than IC feature size thus electric breakdown is much less of a concern. PSZ is chemically compatible with dense silica thus is potentially integratable with existing IC technology.

Previous studies demonstrated a viable low-k material must have mechanical reliability in order to withstand chemical mechanical polishing (CMP) and wire bonding process without fracturing or delaminating and the CMP survivability is a combined factor of modulus, hardness, adhesion or toughness [5, 6]. Thus for any potential low-k candidate, their mechanical properties must be thoroughly investigated. This paper focuses on the investigation of the interfacial adhesion of PSZ thin film-Si substrates using laser induced thin film spallation technique.

EXPERIMENTAL PROCEDURES

Sample Preparation

The sample preparation involves three steps. In the first step, zeolite films of different thickness are deposited onto Si substrates. In the second step, an Al film around 0.4 μm thick is thermally evaporated onto the back surface of the Si substrate. The third step involves spin coating of a thin waterglass confining layer on top of the Al layer. For the laser spallation experiment described below, the confined Al film serves as an energy-absorbing layer to generate laser induced thin film delamination.

Three different methods for zeolite thin film deposition are involved in this work, namely, spin-on, in-situ growth and seeded growth [7]. Details of the deposition process have been reported elsewhere [4]. A brief description of each method is provided here to help understand the influence of deposition method on the measured adhesion result. For all three processes, a very important and critical step is the synthesis of the zeolite nanoparticle suspension.

Spin-on

In order to prepare the zeolite nanoparticle suspension, a synthesis solution with a molar composition of $1TPAOH/2.8SiO_2/22.4EtOH/40H_2O$ (TPAOH and EtOH stand for tetrapropylammonium hydroxide and ethanol, respectively) was prepared. The clear solution thus obtained was aged in a capped plastic vessel for 3 days at room temperature followed by increasing temperature to 80°C for 5 days. During the whole process constant stirring was provided. The nanoparticle suspension thus obtained was centrifuged at 5000rpm for 20min to remove big particles, and then passed through 0.2 μm PTFE filters. The zeolite nanoparticle suspension is then spun on to Si wafers at 3300 rpm for 20 seconds at room temperature using a Laurell spin coater (model WS-400A-6NPP/LITE). The resultant film is a mixture of crystalline pure-silica-zeolites and amorphous silica.

In-situ Growth

In this method, the synthesis mixture was prepared by slowly adding tetraethylorthosilicate (TEOS) to a solution of tetrapropylammonium hydroxide (TPAOH) and distilled water under stirring. After aging for 4 hours under stirring at room temperature, the synthesized solution has a final molar composition of $0.32TPAOH:1TEOS:165H_2O$. Zeolite crystallization on Si wafer is carried out at 165°C for 2 hours. The samples are recovered and thoroughly washed with de-ionized water and blown-dry in nitrogen stream. Typical zeolite film thickness obtained from this process is around 0.4 μm. The final product is a fully crystalline pure-silica-zeolite MFI [4].

Seeded Growth

Following a reported procedure [8], a seeding layer is first introduced onto the Si substrate surface. An aqueous suspension containing silicalite seeds crystals of particle size around 80 nm. The substrate is immersed in the suspension for 2 minutes and then dried in ambient air. After repeating this procedure three times, the synthesis solution with final molar composition of $40SiO_2:9TPAOH:9500H_2O:160EtOH$ is loaded to a 45ml Teflon-lined Parr autoclave where the seeded silicon wafer was placed at the bottom. Crystallization of zeolite on seeded-Si wafer is carried out at 175°C for 5-15 hours. The samples are recovered and thoroughly washed with de-ionized water and blown-dry with nitrogen. Typical zeolite film thickness obtained using this process is around 5 μm. This method also produces a fully crystallized pure-silica-zeolite film [4].

Adhesion Measurement using Laser Spallation

A schematic of the tensile laser spallation for zeolite thin film interfacial testing is shown in **Figure 1**. The experimental setup consists a Q-switched Nd:YAG laser as a energy source, a focal lens for adjusting the laser fluence and the multilayer thin film sample prepared in the previous step. An infrared laser pulse (wavelength of 1064 nm) from the Nd:YAG laser is focused onto a 1-2 mm spot on the Al absorbing layer confined between the transparent waterglass layer and the backside of the Si substrate. The laser pulse is uniform in space and Gaussian in time with a maximum pulse energy of 300 mJ and a pulse duration around 5 ns. After the Al layer absorbs the laser energy, its sudden expansion causes a compressive longitudinal stress wave of rise time comparable to that of the laser pulse to be emitted. The wave propagates towards the test film, then reflects back from the film free surface and finally loads the test interface in tension.

The out of plane displacement of the test film is monitored by a Michelson interferometer. The stress at the interface is then derived from the film displacement using standard wave mechanics. Details of the interferometric displacement measurement can be found in reference [10]. In all of the adhesion tests performed in this work, the loading area is much bigger than the total thickness of the test sample. Therefore, the compressive stress propagating from the absorbing layer towards the test film can be calculated using simple one-dimensional wave mechanics:

$$\sigma_{sub} = -\frac{1}{2}(\rho c_L)_{sub}\frac{\partial u}{\partial t} \tag{1}$$

where u is the measured out-of-plane displacement of the film surface, t is time, ρ and c_L are the density and longitudinal wave speed of the substrate, respectively. The actual stress at the thin film-substrate interface can be calculated by

$$\sigma_{int} = \sigma_{sub}(t-\frac{h}{c}) - \sigma(t+\frac{h}{c}) \tag{2}$$

$$\approx -(\rho h)_{film}\frac{\partial^2 u}{\partial t^2} \quad \text{(for small film thickness } h\text{)}$$

where ρ and c are the density and longitudinal wave speed of film, respectively, and h is the film thickness.

A typical experimental data on a 2.2 μm thick seeded growth zeolite film including the out-of-plane displacement, substrate and interface stresses are shown in Figure 2. A maximum displacement of 960 nm is reached within 10 ns. The substrate stress has a similar profile to that of the YAG laser pulse, and the maximum compressive stress of 2.51 GPa was achieved. The corresponding maximum interface tensile stress at the zeolite-substrate interface was around 587 MPa.

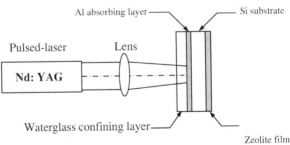

Figure 1. Schematic of tensile laser spallation for zeolite thin film adhesion testing.

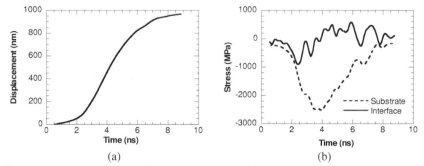

(a)   (b)

Figure 2. Experimental data obtained on a 2.2 μm thick seeded growth zeolite film on Si substrate: (a) out-of-plane displacement; (b) substrate and interface stresses.

Material properties of Si substrate used in the calculation are 2.39 g/cm^3 in density and 7910m/s in longitudinal wave speed, respectively. Zeolite thin films from both in-situ and seeded growth are fully crystallized and have a similar density of 1.84 g/cm^3. The spin-on zeolite film, which is a composite of crystalline and amorphous silica, has a slightly different density estimated around 1.1 g/cm^3.

ADHESION MEASUREMENT RESULTS

Spin-on and Seeded-growth Films

The interfacial adhesion of zeolite thin film-Si substrate deposited using spin-on and seeded-growth methods were successfully measured using the setup described in the previous section. Figure 3(a) shows a typical damage pattern observed on a 0.4 μm thick spin-on zeolite film at a stress level well above the critical stress. The localized damage is well controlled by the YAG laser beam diameter. By varying the energy fluence from the pulse laser, different damage level was achieved, including no delamination, spallation initiation and complete removal. The critical interface stress at the delamination initiation was defined as the interfacial strength. From ten measurements on a 0.4 μm thick spin-on zeolite film, the interfacial strength is determined to be 111 ± 9 MPa.

A 5 μm thick as-grown zeolite film was originally adopted for adhesion study. Thicker films

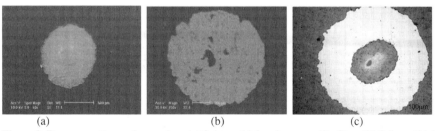

(a)   (b)   (c)

Figure 3. Damage patterns observed on (a) 0.4 μm thick spin-on zeolite film; (b) 2.8 μm thick seeded-growth zeolite film; (c) 0.35 μm thick in-situ growth zeolite film.

are easier to be spalled as expected from Eq. (3) (thicker film results in higher interface stress). The 5 μm thick zeolite film was completely removed even at modest laser energy level. The interferometric measurement was not able to obtain the critical stress for failure initiation for this film even at very low laser energy. In order to observe the failure initiation at slightly higher energy level, the thick zeolite film was mechanically polished with a 0.05 μm alumina slurry to a smaller thickness between 2-3 μm, with which a range of delamination status from perfect bonding, slight delamination to full removal were obtained. Nine measurements on a 2.8 μm thick seeded-growth film yielded an interfacial strength of 714 ± 68 MPa, which is much higher than that of the spin-on film. Figure 3(b) shows the damage pattern observed on the 2.8 μm thick film at a stress level above the critical stress.

In-situ Film

The interfacial strength measurement for the in-situ growth zeolite film is somewhat more complicated than that of the spin-on and seeded-growth films. Due to the nature of the in-situ growth process, typical film thickness obtained from this method is between 0.3~0.4 μm and the interface between in-situ film and Si substrate is much stronger than that of the spin-on film. Because the interface stress at the film-substrate interface is proportional to the film thickness (Eq. 3), with a smaller film thickness, the stress obtained at the interface of the in-situ film is significantly lower than the critical stress needed to delaminate the interface even at the highest YAG laser fluence. Motivated by a previous shock wave study [10], a 500 μm thick pulse-shaping material - fused silica - is inserted between the Al energy absorbing layer and the Si substrate to evolve the original compressive Gaussian stress pulse into a shock wave to enhance the tensile loading at the thin film interface. The double side polished fused silica plate was attached to the back surface of Si substrate using a thin layer of super glue after the deposition of zeolite film followed by the thermal deposition of Al absorbing layer and the spin coating of the waterglass confining layer.

Figure 4 contains the experimental data obtained on a 0.35 μm thick in-situ film after using fused silica as pulse shaping material. With the pulse shaping material, a completely different substrate stress profile was obtained. A linear acceleration ramp followed by a decompression shock is formed after the wave travels through the fused silica substrate and the shock profile is maintained after traveling through the Si substrate.

Although the peak substrate stress is low due to the shock evolution and attenuation [9], the interface stress is around 400 MPa. By carefully adjusting the level of the deposited laser energy fluence, successful delamination of in-situ film was obtained. From 6 measurements, the interfacial strength for in-situ film was found to be 326 ± 17 MPa. Figure 3(c) shows a typical damage pattern observed on in-situ zeolite film with the aid of fused silica as pulse shaping material.

In Figure 5, the interfacial strength for all three types of zeolite film-Si substrate interfaces are plotted together. Overall, the seeded-growth film demonstrated the highest interfacial strength, the spin-on film demonstrated the lowest strength while the in-situ film in-between.

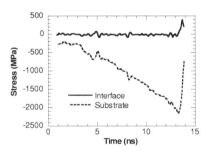

Figure 4. Substrate and interface stress of zeolite film-Si substrate sample with fused silica as pulse-shaping material.

DISCUSSION AND CONCLUSIONS

Laser spallation technique was applied to characterize the interfacial adhesion of zeolite thin film-Si substrate interfaces synthesized from three different processes. Experimental results showed that the adhesion of the in-situ and seeded growth films on Si substrates is higher than that of the partially crystallized spin-on films. The bonding between in situ and seeded growth films and the silicon substrate are established during the hydrothermal synthesis process while the bonding for spin-on films is achieved by a post-synthesis annealing at 400 °C. While the former two methods give out fully crystalline films, the spin-on method gives partial crystalline films. Between the two

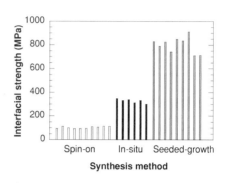

Figure 5. Interfacial adhesion of zeolite film-Si substrate interfaces prepared using the three different synthesis methods.

crystalline films, seeded-growth films demonstrated a much higher interfacial adhesion than the in-situ growth films. Therefore a direct conclusion can be drawn that the synthesis method is the major factor determining the adhesion between zeolite thin films and the Si substrates, and the films by direct synthesis have stronger adhesion with silicon substrate.

For in-situ zeolite film, due to the small film thickness, the stress generated at the interface is not sufficient to introduce interface delamination. The use of fused silica as a pulse shaping material is beneficial in obtaining the critical interfacial strength of in-situ zeolite films of this thickness range. It is worth mentioning that this is the first time a pulse-shaping material is introduced into the traditional laser spallation technique and the results are very promising.

REFERENCES

1. Y. Yan, S. Li, and Z. Li, *Zeolite News Letters* **20** (2003) 111.
2. D.W. Breck, *Zeolite molecular sieves: Structure, chemistry, and use*. New York, Wiley. New York, 1974.
3. W.M.M. Ch. Baerlocher, D.H. Olson, *Atlas of zeolite framework types*. Amsterdam; Elsevier. New York, 2001.
4. Z. Wang, A. Mitra, H. Wang, L. Huang, and Y. Yan, *Adv. Mater.* **13** (2001) 1463.
5. J.B. Vella, A.A. Volinsky, I.S. Adhihetty, N.V. Edwards, and W.W. Gerberich, *Mat. Res. Soc. Symp. Proc.* **716** (2002) B12.13.1.
6. A. Volinsky, J. Vella, and W. Gerberich, *Thin Solid Films* **429** (2003) 201.
7. Y. Yan, S. Li, and Z. Li, *Zeolite News Letters* **20**(3) (2003).
8. M.C. Lovallo and M. Tsapatsis, *AIChE Journal* **42**(11) (1996) 3020.
9. J. Wang, R.L. Weaver, and N.R. Sottos, *Experimental Mechanics* **42**(1) (2002) 74.
10. J. Wang, R.L. Weaver, and N.R. Sottos, *Journal of Applied Physics* **93**(12) (2003) 9529.

Mater. Res. Soc. Symp. Proc. Vol. 875 © 2005 Materials Research Society   O10.6

Environmental Effects on Crack Characteristics for OSG Materials

Jeannette M. Jacques, Ting Y. Tsui, Andrew J. McKerrow, and Robert Kraft
Silicon Technology Development, Texas Instruments Inc., Dallas, TX 75243

ABSTRACT

To improve capacitance delay performance of the advanced back-end-of-line (BEOL) structures, low dielectric constant organosilicate glass (OSG) has emerged as the predominant choice for intermetal insulator. The material has a characteristic tensile residual stress and low fracture toughness. A potential failure mechanism for this class of low-k dielectric films is catastrophic fracture due to channel cracking. During fabrication, channel cracks can also form in a time-dependent manner due to exposure to a particular environmental condition, commonly known as stress-corrosion cracking. Within this work, the environmental impacts of pressure, ambient, temperature, solution pH, and solvents upon the channel cracking of OSG thin films are characterized. Storage under high vacuum conditions and exposure to flowing dry nitrogen gas can significantly lower crack propagation rates. Cracking rates experience little fluctuation as a function of solution pH; however, exposure to aqueous solutions can increase the growth rate by three orders of magnitude.

INTRODUCTION

The integration of low-k dielectric films is required to maintain and improve device performance in future technologies. For intermetal dielectrics, the class of materials known as organosilicate glass (OSG) has emerged as the principal candidate. During the manufacturing process, catastrophic fracture due to channel cracking is a potential failure mechanism. [1] The driving force for channel cracking is dependent on several material properties, with the film modulus, density, and residual tensile stress serving as key factors. [2-3] Channel cracks can also form in a time-dependent fashion due to exposure to specific environmental conditions. [2] These mechanisms are commonly known as environmentally-assisted or stress-corrosion cracking.

A typical cohesive failure mode for thin films under stress is channel cracking. Channel cracks are defined as extending through the film thickness and propagating perpendicular to the substrate-film interface. [2] Studies by Cook et al. [3, 4] showed that the crack propagation rate V in the reaction-limited regime can be expressed by Equation 1. where σ represents the residual film stress, h the film thickness, and E the plane-strain modulus. The variable k is Boltzmann's constant, T is temperature, and n denotes the area density of bonds fractured at the crack tip during crack propagation. Z is a constant defined by crack geometry and V_o is a fitting constant. The density of bonds fractured during crack separation n is analogous to the average film density ρ. [5]

$$V = V_o \, e^{\frac{\pi Z}{4nkT}\left(\frac{\sigma^2 h}{E}\right)} \tag{1}$$

The majority of studies [6-13] regarding the cracking and failure of silicon-based structures have employed double cantilever beam geometries and cleavage techniques. Limited studies [2-3] have utilized channel crack propagation as a means of assessing mechanical stability. Many authors stress the importance of molecular structure in determining solution corrosivity. [2, 7-13] A siloxane

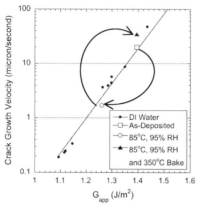

Figure 1: Crack Growth Velocity versus Applied Strain Energy Release Rate for samples under different processing conditions.

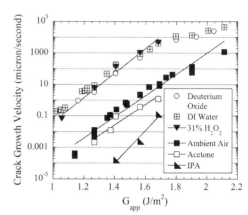

Figure 2: Crack Growth Velocity versus Applied Strain Energy Release Rate for OSG samples.

backbone (Si-O-Si) renders silicon-based materials susceptible to stress-corrosion cracking. [2] Solution molecules must fit between Si-O bonds at the crack tip in order to effectively break silonal bonds and promote crack propagation. [13] It has been suggested [14-16] that water may have a greater impact on crack growth than temperature, demonstrating the importance of understanding the behaviors of low-k dielectrics during exposure to common manufacturing conditions. In this study, the responses of OSG films were characterized for a range of environmental conditions. The impacts of ambient, solution composition, solution pH, temperature, and pressure on channel crack propagation rates in blanket, low-k dielectric films are discussed at length. The results demonstrate that the well-established fracture theories and relationships pertaining to structured silicate glasses can be extended to channel cracks formed in novel dielectric materials.

EXPERIMENTAL CONDITIONS

OSG films were deposited onto 200 mm silicon wafers via plasma enhanced chemical vapor deposition (PECVD), using a tetramethylcyclotetrasiloxane (TMCTS) precursor. Temperature and % relative humidity (RH) measurements were recorded by a Traceable Hygrometer/Thermometer Unit. A diamond scribe was used to initiate channel cracks in blanket OSG films. Scratches exceeded 2 cm in length and completely penetrated the films. Crack growth was monitored and measured using electronic digital calipers and optical microscopy. All recorded crack lengths exceeded a minimum length/thickness ratio of 30. Channel crack length was determined to be independent of crack proximity for the materials studied. Certified Fisher Chemicals buffer solutions were used to investigate pH effects on channel crack growth rates. Environmental pressure controlled experiments were conducted in an ultra low pressure XPS chamber with base pressure capabilities of less than 10^{-8} T.

RESULTS AND DISCUSSION

Throughout these studies, we make direct comparisons between channel crack growth behaviors for a variety of environmental conditions. The intrinsic cohesive strength of our materials was not impacted during testing by these diverse treatments. In Figure 1, the cracking

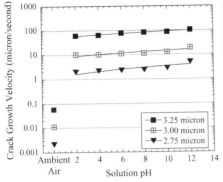

Thickness (microns)	Pressure (Torr)	Crack Growth Velocity (μm/sec)
3.25	760	0.060
3.25	635	0.0016
3.25	1×10^{-9}	0.0010
3.25	1×10^{-6}	0.0003
3.0	760	0.0100
3.0	635	0.0004
3.0	1×10^{-9}	$<1\times10^{-6}$

Figure 3: Crack Growth Velocity as a function of Solution pH for OSG films of Variable Thickness.

Table I: Crack Growth Velocity of OSG films as a function of exposure pressure and film thickness.

behavior of OSG films in DI water are illustrated for three different types of processing conditions: (1) as-deposited, (2) 85°C/95% RH, and (3) 85°C/95% RH followed by 350°C N₂ bake. The as-deposited film represents material that was tested immediately after deposition. Additional processing occurred directly after film deposition, whereby films were exposed to a temperature of 85°C and 95% RH for two hours. The baking process followed for some samples, proceeding at 350°C for 15 minutes under an inert dry nitrogen ambient. Samples depicted in Figure 1 received some or all of these processing steps prior to testing in DI water. Moisture absorption at elevated temperatures relaxes the films to lower stress states and water undergoes complete desorption at high exposure temperatures. A complete cycle is apparent as we move from an as-deposited state, to 85°C/95% RH, and then 85°C/95% RH and 350°C Bake, where we return to the as-deposited conditions. All three of these processing conditions follow the same trend line as OSG samples of variable thickness tested after a limited period of post-deposition storage, suggesting comparable fracture strengths. The empirical stress and plane-strain modulus values used throughout our G_{app} calculations for as-deposited films are 65 MPa and 10 GPa, respectively. G_{app} is the applied strain energy expressed as ($\sigma^2 h/E$). The data in Figure 1 illustrate that the OSG intrinsic fracture toughness is not altered by moisture absorption or exposure to elevated temperatures, as defined by the post-deposition processes.

Channel crack propagation characteristics depend strongly on the environmental testing conditions. Ambient testing was conducted at a temperature of 22°C, 44% RH, and pressure of 760 T. Figure 2 illustrates the crack growth velocities of OSG films verses the applied strain energy release rate. The data follow exponential relationships, in agreement with past works focusing on the reaction rate controlled regime. [2, 7, 9, 15, 16] Exposure to aqueous solutions raised the crack rates by approximately 1000X, as compared to ambient testing conditions. Crack growth velocities were lowest for samples exposed to organic solvents, such as acetone and isopropyl alcohol (IPA). In previous experiments, Guyer et al. [10-12] observed delamination cracking rates to be up to three orders of magnitude higher in lower concentration hydrogen peroxide (H₂O₂) solutions (< 3 wt%), as compared to water. However, we did not observe a measurable difference between channel cracking rates for samples submerged in DI water and 31 wt% H₂O₂.

Cracking rates appear to be primarily a function of environmental water content, as water was the main constituent in our DI water and H₂O₂ solutions. However, the acetone used in these

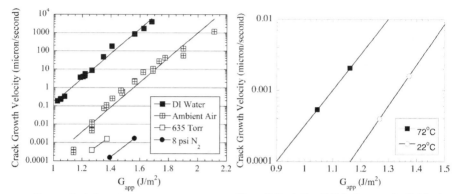

Figure 4: Crack Growth Velocity versus Applied Strain Energy Release Rate for OSG samples.

Figure 5: Crack Growth Velocity verses Applied Strain Energy Release Rate as a function of Temperature.

studies had a moisture content of 6260 ppm and IPA had only 4480 ppm water, as verified through titration techniques. Once exposure conditions become dominated by the presence of water, no measurable changes in the cracking rates are observed. It is interesting to note that the diffusivity of water molecules in solution is double that of H_2O_2 molecules, inferring a greater reactivity. [10] Our results support the early work of Weiderhorn et al. [9] which states that crack motion is governed by the rate of reaction between water and the silicon-based glass and that the reaction mechanism at the channel crack tip is not altered by changing the moisture content.

The channel cracking rates for OSG films during exposure to deuterium oxide and DI water are also included in Figure 2. Deuterium oxide is an isotopic form of water, commonly referred to as heavy or deuterated water. It is more viscous than water and has greater molecular weight. The self diffusivity of deuterium oxide molecules is 12% lower than that of water. [17] Throughout the G_{app} range studied, OSG samples immersed in D_2O and DI water behave similarly, with cracks exposed to D_2O propagating at a slightly slower velocity than in DI water. This is in agreement with prior studies [13], whereby double cantilever beam bulk glass specimens produced rates consistently lower within D_2O than in water. The separation of two distinct regimes can be observed at a crack growth velocity of roughly 1×10^4 micron/second – a transition between reaction rate controlled and diffusion rate controlled regimes.

Previous adhesion studies by Guyer at al. [10-12] demonstrated a distinct correlation between solution pH and crack propagation characteristics. Cracking rates were generally higher for basic solutions, decreasing with increasing solution acidity. Rates were reported to drop by as much as 100X over the full range of pH values. [10-12] During the course of our experiments, however, only a minimal dependence on pH was demonstrated. As seen in Figure 3, the crack propagation velocity for OSG films increases by a factor of less than 3X as the solution pH is increased from 2 to 12. A distinct linear relationship exists between the crack rate and solution pH. Samples exhibit roughly one order of magnitude increase in the growth velocity for each 0.25 μm increase in film thickness. The pH buffer solutions did not impact the films, as verified through FTIR and XPS analysis. Collectively, our results demonstrate that crack propagation characteristics may depend strongly on the testing geometry, as channel cracks do not always react similarly to cracks monitored in cantilever beam structures under identical environments stimuli.

Testing Environment	ppH_2O (Torr)
DI Water	19.46
Ambient Air, 760 Torr	8.56
Pressure of 635 Torr	7.15
Flowing Dry N_2	< 7.15

Table II: Partial Pressure of Water (ppH_2O) at $22^\circ C$

The effects of environmental pressure on OSG channel cracking were also investigated. For an exposure temperature of $22^\circ C$, the cracking rates of 3.25 and 3.00 micron OSG samples are shown in Table I. It is important to note that as the environmental pressure decreases, the partial pressure of water (ppH_2O) is also effectively lowered. By decreasing the pressure from 760 to 635 T, cracking rates drop by more than 95%. Under high vacuum conditions of 1×10^{-6} and 1×10^{-9} T, crack growth is greatly retarded; measurable rates are lowered by more than 99% compared to atmospheric pressures. Figure 4 shows the effects of dry N_2 gas flow rate upon the crack growth velocity of OSG films at $22^\circ C$. For reference, data for DI water, ambient air, and pressure conditions of 635 T are also included. By raising the N_2 flow rate from 0 psi (i.e. Ambient Air) to 8 psi, cracking rates are lowered by more than 100X. The increase in N_2 flow pressure also serves to further reduce the environmental partial pressure of water. The ppH_2O for each of the environmental conditions shown in Figure 4 are given in Table II. These results are in agreement with prior studies that have shown moisture content to be a governing factor for crack growth. [9, 14-16] Under low pressure conditions, the entire energy of the bridging structural bond must be overcome for crack propagation to occur. [13] On the other hand, when water is present a charge transfer mechanism offers a lower energy reaction path for growth [13] and lowers the stress required for crack extension. [18] All things being equal, as the environmental water content increases, so does the propensity for channel cracking.

Temperature is another factor governing crack growth in silicon-based dielectric glasses. The temperature response of OSG channel cracks was examined at 635 T for temperatures of $22^\circ C$ and $72^\circ C$. The crack growth velocity V can be described as follows [7] where K_{IC} is the stress intensity factor, R is the universal gas constant, and T is the exposure temperature. The variables V_o, E_a, and b are experimental constants obtained by fitting data via the least squares method.

$$V = V_o \, e^{\frac{-Ea + bK_{IC}}{RT}}$$ (2)

Figure 5 shows the impact of exposure temperature on channel cracking in OSG. As temperature increases for a set G_{app}, the crack rate also increases. The activation energy for OSG channel crack propagation was determined to be 116.8 kJ/mol. A $50^\circ C$ shift in temperature can alter the measured crack growth velocity by 1.5 orders of magnitude at 635 T. Weiderhorn et al. observed a similar shift for various glasses. [7, 8]

CONCLUSIONS

The results presented throughout our studies emphasize the key environmental parameters governing channel crack propagation in OSG materials. Many of the well-established sub-critical fracture theories and empirical relationships formulated for bulk, structured silicate glasses have been shown to apply directly to thin OSG films. In general, channel cracks behave similarly to cracks formed and monitored under more conventional testing techniques. [6-8, 10-13, 15, 16] Thus illustrating that channel crack analysis is a viable method for evaluating and characterizing the sub-critical fracture properties of silica-based thin film glasses. Environmental water content appears to be the most important factor in determining sub-critical channel cracking rates. Moisture levels can be influenced by additional exposure parameters such as solution type, pressure, and dry ambient flow rate. Immersion within aqueous solutions raises crack growth velocities by three orders of magnitude; while solution pH imparts only a minimal influence upon channel crack behavior, adjusting rates by less than 3X.

ACKNOWLEDGMENTS

One of the authors, Jeannette Jacques, would like to thank Judy Shaw and Dr. Ben McKee of Texas Instruments Inc. for providing the opportunity to participate in novel research projects as a graduate intern. These works would not have been possible without the materials analysis support of Richard Kuan and Lucyna Carrasco, as well as the processing assistance of Chris Brainard, Jim Burris, Rhida Dalmaico, and Darrell Ingram.

REFERENCES

1. X.H. Liu, T.M. Shaw, M.W. Lane, R.R. Rosenberg, S.L. Lane, J.P. Doyle, D. Restaino, S.F. Vogt, and D.C. Edelstein. IEEE Proceedings of the International Interconnect Technology Conference 2004 (Pg.93-95), San Francisco, CA.
2. R.F. Cook and E.G. Liniger, J. Electrochem. Soc. **146** (12), 4439 (1999).
3. R.F. Cook and Z. Zuo, Materials Research Society (MRS) Bulletin, January 2002, 45.
4. R.F. Cook, Mat. Sci. and Eng., A, **260**, 29 (1999).
5. J.M. Jacques, T.Y. Tsui, A.J. McKerrow, and R. Kraft, Mater. Res. Soc. Symp. B3.8.1-6 (2005). (Submitted for Publication)
6. S.M. Wiederhorn and H. Johnson, J. Appl. Phys. **42** (2), 681 (1971).
7. S.M. Wiederhorn, H. Johnson, A.M. Diness, and A.H. Heuer, J. Amer. Ceram. Soc. **57** (8), 336 (1974).
8. S.M. Wiederhorn and L.H. Bolz, J. Amer. Ceram. Soc. **53** (10), 543 (1970).
9. S.M. Wiederhorn, J. Amer. Ceram. Soc. **50** (8) 407 (1967).
10. E.P. Guyer and R.H. Dauskardt, Nature Materials **3**, 53 (2004).
11. E.P. Guyer and R.H. Dauskardt. IEEE Proceedings of the International Interconnect Technology Conference 2003 (Pg. 89-91), San Francisco, CA.
12. E.P. Guyer and R.H. Dauskardt. IEEE Proceedings of the International Interconnect Technology Conference 2004 (Pg. 236-238), San Francisco, CA.
13. T.A. Michalske and S.W. Freiman, Nature **295**, 511 (1982).
14. M.W. Lane, J.M. Snodgrass, and R.H. Dauskardt, Microelectronics Reli. **41**, 1615 (2001).
15. T.I. Suratwala, R.A. Steele, G.D. Wilke, J.H. Campbell, and K. Takeuchi, J. Non-Crystalline Solids **263&264**, 213 (2000).
16. T.I. Suratwala and R.A. Steele, J. Non-Crystalline Solids **316** (1), 174 (2003).
17. J.F Thomson, *Biological Effects of Deuterium* (The Macmillan Company, New York, 1963).
18. T.A. Michalske and B.C. Bunker, J. Am. Ceram. Soc. **76** (10), 2613 (1993).

Fatigue and Stress in
Interconnect Metallization

Mater. Res. Soc. Symp. Proc. Vol. 875 © 2005 Materials Research Society   O11.2/B7.2

TEM-Based Analysis of Defects Induced by AC Thermomechanical *versus* Microtensile Deformation in Aluminum Thin Films

R.H. Geiss, R.R. Keller, D.T. Read and Y.-W. Cheng
Materials Reliability Division, National Institute of Standards and Technology
325 Broadway
Boulder, CO 80305-3328, USA

ABSTRACT

Thin films of sputtered aluminum were deformed by two different experimental techniques. One experiment comprised passing high electrical AC current density through patterned Al interconnect lines deposited on SiO_2/Si substrates. The other consisted of uniaxial mechanical tensile deformation of a 1 μm thick by 5 μm wide free standing Al line. In the electrical tests approximately 2×10^7 W/cm^2 was dissipated at 200 Hz resulting in cyclic Joule heating, which developed a total thermomechanical strain of about 0.3 % per cycle. The tension test showed a gauge length fracture strain of only 0.5 % but did display ductile chisel point fracture. In both experiments, certain grains exhibited large, > 30°, rotation away from an initial <111> normal orientation toward <001>, based on electron backscatter diffraction (EBSD) measurements in the scanning electron microscope (SEM). Transmission electron microscopy (TEM) analysis of specimens from both experiments showed an unusually high density of prismatic dislocation loops. In the mechanically-tested samples, a high density of loops was seen in the chisel point fracture zone. In cross sections of highly deformed regions of the electrical test specimens, very high densities, > 10^{15}/cm^3, of small, < 10 nm diameter, prismatic loops were observed. In both cases the presence of a high density of prismatic loops shows that a very high density of vacancies was created in the deformation. On the other hand, in both cases the density of dislocations in the deformed areas was relatively low. These results suggest very high incidence of intersecting dislocations creating jogs and subsequently vacancies before exiting the sample.

INTRODUCTION

As we transition into a world of smaller dimensions – the nanoworld – it becomes increasingly more difficult to reliably measure the mechanical properties of materials with nano-dimensions (< 100 nm). New measurement tools are needed in the rapidly growing field of nanomaterials. In particular, information about mechanical properties such as elastic modulus, ultimate tensile strength, fatigue life, maximum strain, adhesion and the relation of defect structures to these properties is critical to successful development of new materials. Such information is also needed to assess integrity or reliability in many applications; for example, multilayer electronic interconnects and solder joints. The difficulty of fabricating complex systems requires the use of predictive modeling in order to achieve cost and time savings. However, modeling can correctly predict system performance only if the property data used as input are accurate at the relevant length scales. Furthermore, in heterogeneous systems it is often the localized variation in properties that causes failure (void formation, fracture, *etc*.). Thus it is increasingly important to assess not only the "average" sample properties, but also to obtain data relating to the spatial distribution in properties.

Many existing methods for mechanical-property measurements have drawbacks: they are often destructive, not quantitative, limited to specialized geometries, require samples which are difficult to fabricate, and so on. Currently, one of the most commonly used tools for this purpose is nanoindentation [1]. However, existing nanoindentation techniques face measurement challenges as dimensions continue to shrink. In such systems, the volume sampled by nanoindentation may be too large for adequate analysis due to the large radius of the indenter tip. The lateral resolution of a typical Berkovich diamond indenter used is a few hundred nanometers and the relatively large loads applied may also be an issue.

We are pursuing nanomechanical measurements from a different point-of-view by using electrical measurements to determine the mechanical properties of thin films on a substrate. As part of this program we compared the defect structures in samples exposed to two very different types of tests. In one test thermomechanical fatigue was introduced by electrical means and in the other a uniaxial tension test was performed. The microscopic deformation behavior was evaluated in the SEM and with EBSD and is reported elsewhere in these proceedings [2]. The defect structures which remained after failure were studied by TEM.

The literature contains an abundance of TEM studies discussing defects in thin films and in thin foils prepared from bulk samples, but none from alternating current thermomechanically fatigue (ACTMF) tested Al-Si interconnect films on a substrate, or from 1 μm thick Al films fractured in tension. There have been numerous high voltage SEM studies of void formation and migration in direct current electromigration experiments [3], an entirely different type of test.

EXPERIMENT

AC tests were carried out on non-passivated, single-level structures composed of patterned and etched Al-1Si lines sputtered onto thermally oxidized silicon, using conventional processing parameters. We used a NIST test pattern originally designed for electromigration and thermal conductivity measurements. Testing was conducted on a 4-point probe station using 100 Hz sinusoidal alternating currents with zero DC offset. Tests were conducted continuously until the lines became electrically open. Current densities (rms) applied to individual lines ranged from 11 to 16 MA/cm^2. For the sample whose microstructure is presented here, testing was done with an AC current density of 12.2 MA/cm^2 at 100 Hz, The line was 800 μm long by 3.3 μm wide and 0.5 μm thick. In this test approximately 2 x 10^7 W/cm^2 was dissipated at 200 Hz resulting in cyclic Joule heating. Based on prior work [4], when a current density of 12.2 MA/cm^2 was applied, the base specimen temperature, as monitored using a thermocouple attached directly to the die, indicated a rise of < 10 K during testing. However, the low frequency AC signals led to temperature cycling superimposed onto the base die temperature, with an amplitude of approximately 100 K, at a frequency of 200 Hz, corresponding to a power cycling input into the line which developed a total thermomechanical strain of about 0.3 % per cycle. Based on the TEM observations of loop diameters, discussed later, an upper bound on the temperature rise of ~170 °C [5] can be established. For more details of the AC test see [2]. The uniaxial tension tests were conducted on sputtered Al test sections 200 μm long, 1 μm thick and 5 μm wide at room temperature (RT). The elongation rate was approximately 10^{-4}/s. The tension test showed a gauge length fracture strain of only 0.5 % but did display ductile chisel point fracture.

Specimens for TEM were prepared from severely deformed regions after the ACTMF electrical test by focused ion beam (FIB) preparation of sections along the test line. In the tensile tests, the free fractured end from the samples was mounted on a Si_3N_4 window by simply pressing the non-tested end down firmly along the outer frame.

The TEM experiments were carried out in a microscope with a LaB_6 gun at 200 kV. Images were recorded on a 1K x 1K CCD camera mounted below the viewing chamber. Sample manipulation was done using side entry holders with single-tilt, double-tilt and tilt-rotate capabilities. The fracture tip samples pressed onto the Si_3N_4 window grids were very fragile and usually studied using the single tilt holder, which is better designed for loading more delicate samples.

EBSD studies of samples from both experiments were done quasi *in-situ*. A summary of the ACTMF experiment is presented elsewhere in these proceedings [2]. Grain orientation maps from a few samples in the tensile experiments were obtained prior to pulling and also during the deformation by stopping a few times prior to fracture and acquiring maps from the complete test section with EBSD. After fracture we could only get data from the part of the sample that was still fixed to the deformation jig.

RESULTS

TEM examination of failed specimens from both experiments showed a very high density of prismatic dislocation loops in areas relatively devoid of forest dislocations, see Figures 1 and 2. Figure 1 is typical of a severely deformed grain after failure in the AC test. The average diameter of the loops is 8 ± 4 nm. Each loop contains about 10^3 vacancies assuming a loop to be a collapsed disc of vacancies. The loop density is about 3.3 x 10^{15}/cm^3, which compares with ~ 10^{15}/cm^3 reported for Al sheets quenched from 600 °C into iced brine [6]. The vacancy concentration is about 10^{-4}, which also compares favorably with the prior TEM results from quenched Al sheets [6]. The loop diameter of 8 nm also suggests the maximum temperature reached during the thermal cycle is less than 170 °C [5]. Figure 2, from the tip at a site of chisel point fracture in a tensile sample, shows somewhat larger loops, about 14 ± 5 nm in diameter. The loop density here is ~ 7 x 10^{14}/cm^3. This is an order of magnitude increase over that in undeformed samples. There is also a decrease in average loop diameter after fracture. Initially the average loop diameter was 23 ± 10 nm, but after fracture it was ~ 14 ± 4 nm. Calculations using these numbers yield an initial vacancy concentration of 5 x 10^{-5} while after fracture at the chisel point tip it increased to about 10^{-4}. This latter result is similar to that obtained from the ACTMF tested samples. As expected, the dislocation density in the Al tensile specimens increased from the initial state of ~ 2 x 10^{13}/m^2, (see Figure 3), to ~10^{15}/m^2 after fracture, (see Figure 4).

EBSD measurements showed a rotation of about 30°, from near the <011> to near the <112>, of the tensile axis in grains at the fracture tip This is as would be expected from classical single crystal tensile deformation of face centered cubic metals.

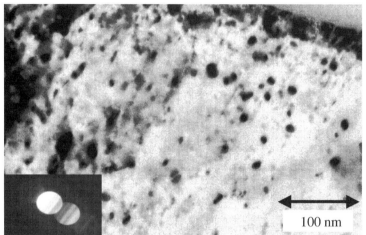

Figure 1.~10 nm prismatic dislocation loops in a failed site of an Al-1Si ACTMF sample.

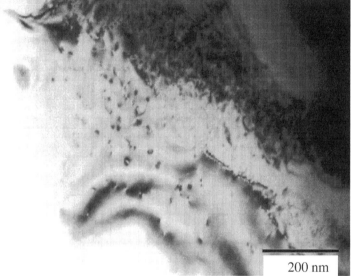

Figure 2. ~15 nm prismatic dislocation loops at the chisel point fracture tip in a tensile sample.

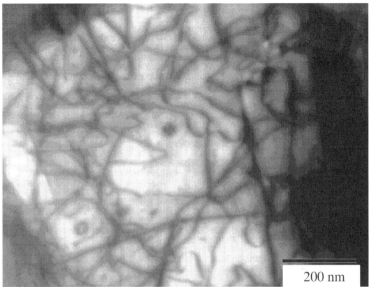

Figure 3. Typical dislocation and loop structure in a tensile sample prior to testing.

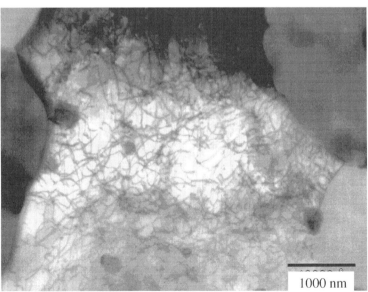

Figure 4. Dislocation loops and tangles in the bulk of a tensile sample after deformation.

DISCUSSION

TEM studies of defects in sputtered aluminum films, deformed using two very different experimental techniques, show striking similarities. In both ACTMF and tension tests very high concentrations of prismatic dislocation loops were observed in locations generally void of dislocations. In both instances we propose that a large number of dislocations on multiple glide systems intersected and created vacancies which subsequently coalesced to form the loops. The loops are prismatic on (111) planes with a ½[110] Burgers vector normal to the plane of the loops and were not very mobile at the temperatures of the experiments, approximately 170 °C for the ACTMF experiment and RT for the tension experiment. Thus the loops remained while the dislocations continued to slip to either to free surfaces or to grain boundaries resulting in the formation of dislocation free zones with a high density of loops. EBSD analysis of both experiments showed large grain rotation in the deformation zones most likely caused by a large number of dislocations intersecting grain boundaries and increasing the degree of misorientation across a boundary. The vacancy concentrations in the deformed samples are consistent with those in prior reports on quenched samples and suggest that a concentration of 10^{-4} may be an upper limit to that which can remain in aluminum at room temperature. The unique observations that we have made in samples from both our experiments are the presence of a very high density of prismatic dislocation loops and the absence of dislocations. In the AC tested samples this observation is general throughout the whole of the deformed regions. In the fractured sample this observation is restricted to fracture tip areas.

The similarities in the defect microstructures after testing, namely the high density of prismatic dislocation loops, suggest a similar deformation mechanism proceeded in both experiments and that AC electrical thermomechanical fatigue testing might be used to measure the mechanical properties of constrained dimension metal structures.

ACKNOWLEDGMENTS

We thank the NIST Office of Microelectronics Programs for support. The work is a contribution of the U.S. Department of Commerce and is not subject to copyright in the U.S.A. We also want to acknowledge Dr. Dudley Finch for providing the FIB sample.

REFERENCES
1. W. C. Oliver and G. M. Pharr, *J. Mater. Res.* **7**, 1564 (1992)
2. R. R. Keller, R.H. Geiss, Y.-W. Cheng and D.T. Read, (these proceedings)
3. S-H. Lee, J.C. Bravman, J.C. Doan, S. Lee and P.A. Flinn, *J. Appl. Phys.* **91**, 3653 (2002)
4. R. R. Keller, R. Mönig, C. A. Volkert, E. Arzt, R. Schwaiger, and O. Kraft, in *6th International Workshop on Stress-Induced Phenomena in Metallization*, edited by S. P. Baker, M. A. Korhonen, E. Arzt, and P. Ho, (AIP Conference Proceedings 612, American Institute of Physics, New York, pp.119-132, 2002)
5. P.B. Hirsch, J. Silcox, R.E. Smallman and K.H. Westmacott, *Phil. Mag.* **3**, 897 (1958)
6. J. Silcox and M.J. Whelan, *Phil. Mag.* **5**, 1 (1959)

Mater. Res. Soc. Symp. Proc. Vol. 875 © 2005 Materials Research Society   O11.3/B7.3

Employing Thin Film Failure Mechanisms to Form Templates for Nano-electronics

Rainer Adelung, Mady Elbahri, Shiva Kumar Rudra, Abhijit Biswas*, Seid Jebril, Rainer Kunz, Sebastian Wille and Michael Scharnberg

Faculty of Engineering, Christian-Albrechts-University of Kiel,
Kaiserstr. 2, D-24143 Kiel, Germany
electronic mail: ra@tf.uni-kiel.de

ABSTRACT

Recently, we showed that thin film stresses can be used to form well aligned and complex nanowire structures [1]. Within this approach we used stress to introduce cracks in a thin film. Subsequent vacuum deposition of metal leads to the formation of a metal layer on the thin film and of metal nanowires in the cracks of the film. Removal of the thin film together with the excess metal cover finishes the nanowire fabrication on the substrate. As stress can be intentionally introduced by choosing an appropriate thin film geometry that leads to a stress concentration, the cracks and consequently the nanowires can be well aligned. Meanwhile, we have demonstrated how to form thousands of parallel aligned nanowires, x-and y-junctions or nanowires with macroscopic contacts for sensor applications, simply by applying fracture mechanics in thin films. Christiansen and Gösele called this approach "constructive destruction" in a comment in Nature Materials [2]. This gives a hint how to overcome some problems of the approach, arising from the limits of thin film fracture. A generalization of the fracture approach by being "more destructive" can overcome this limitations. For example, it is difficult to form pairs of parallel wires with a nanometer distance of the pair, but a micrometer separation between the individual pairs. Structures like this are useful for many contact applications including sensor arrays or field effect transistors. As well as thin film fracture, thin film delamination can be well controlled by fracture mechanics. Our latest experiments show that the combination of both, fracture and delamination, forms an ideal shadow mask for vacuum deposition. Cracks with delaminated sides were used as templates for the deposition of pairs of parallel wires consisting out of different materials with only a few 10 nm separation. First, a metal was sputter deposited under an angle of approx. 45° through the delaminated crack, which was used as a shadow mask. Afterwards, a second deposition metal is deposited under the opposite 45° angle with respect to the sample normal, having the crack located in the middle between both deposition sources. The angle, the delamination height and the crack width determine the separation of the nanowire contacts. We present several examples which show how these mechanisms of mechanical failure of thin films can be turned into useful templates for various nanostructures. We will focus here on two thin film systems, that can be easily deposited in every lab. These are wet chemically deposited photo-resist and flash evaporated amorphous carbon. These examples are compared with finite element simulations of the thin film stress with the ANSYS program. Moreover, we show how the delamination cracks can be also used as masks for the removal of material. Channals with a width down to 20 nm produced by ion beam sputtering are shown.

*Present Address: Institute for Shock Physics, Applied Sciences Laboratory, Washington State University, Spokane, WA 99210, U.S.A.

INTRODUCTION

Several self-organized mechanisms have been shown based on thin film fracture mechanics and the interplay between diffusion and nucleation. In [3] we showed fracture of the three atom-wide layers of layered crystals and that it is possible to form aligned nanostructures by using the crystal symmetry. The further development of the approach was demonstrated with the fabrication of nanowires on different materials by the thin film fracture [1] and vacuum deposition. Nearly at the same time, it was demonstrated by another group that also wet chemical generation of aligned nanowires is possible, in this case by electroless nickel [4] deposition. In both cases the main principle is the same. A thin film is structured by predetermined breaking points and strain fields are used to create aligned nanostructures by fracture of the thin films. In this way it is possible to form nanowire arrays over large samples on the wafer scale. In principle, there is no upper limit of the sample size on which the nanowires can be deposited.

EXPERIMENTAL

Fabrication steps

Figure 1 illustrates the principle how to create aligned nanowires by thin film fracture. First, a thin film will be deposited on a substrate. The thin film can be microstructured in order to create pre-determined breaking points, see fig. 1a. Fig. 1b shows the sample after applying tensile stress to the thin film. The stress should be high enough to lead to mechanical failure, so that the thin film cracks. By using a geometry as shown in fig. 1, even ill defined stress will lead to a fracture along the "weakest point" of the sample. In order to fill the created mold, material can be deposited by vacuum sputter deposition or other thin film deposition techniques. This will result in a filling of the crack with material, but also in excess material on top of the thin film, see fig. 1c. This excess material can be removed by removing the thin film in a "lift off" process. Two examples for removal of the thin film are chemical removal by etching or mechanical removal in an ultrasonic bath. The lift off with ultrasonic sound is possible if the thin film is brittle and the nanowire material is well attached to the substrate and ductile, e.g., it is a metal. For this example, the whole procedure will lead to a structure as shown in figure 1d. Here a nanowire is created between two larger "contact-pads", which could be used if a metal was chosen as nanowire material (Compared with experiment in [1])

a)   b)   c)   d)

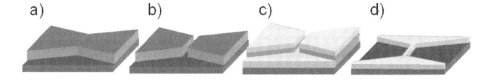

Figure 1. Different steps of the nanowire fabrication a) microstructured thin film on a substrate b) fracture of the thin film. c) deposition of nanowire material. d) structure after "lift off" process.

Sample preparation

The thin films used here are deposited by spin coating (PMMA) or spraying (positiv 20, Contact Chemie, Iffezheim, Germany) in the case of photoresist. The carbon deposition was performed in a Balzers Bal-tec SCD 050 carbon coater under pre-vacuum conditions by flash evaporation (sublimation of a carbon thread by ~200 W, 20 A).

The nanowire material used here is a metal, either Au or Cu. Both are deposited in a vacuum chamber by sputter deposition.

Microscopy

Optical microscopy was carried out with a Reichert-Jung Polylite SC optical microscope with up to 1,000× magnification. A Phillips XL 30 Scanning electron microscope (SEM) was used to analyze the surface geometry in detail. Fig. 2 a was taken by M. Schossig from the GKSS research center with a backscatter detector of field emission SEM. Images were taken at 30 Pa chamber pressure to avoid charging.

Sputter etching

Sputter etching was performed in a Krypton atmosphere at $2.5*10^{-4}$ mbar with a 13mA current from an ion tech sputter gun (ION TECH, INC., Fort Collins, Colorado, USA) for 30 minutes.

RESULTS AND DISCUSSION

Nanowires on large scales

Figure 2a shows an image of a Au nanowire. The nanowire is produced on a flexible commercial polymer substrate (ink jet transparency foil). In this case spray on photoresist was used. After the photoresist film dried, cracks are generated as a template for the nanowires. This is done by bending the film under a defined angle, resulting in an array of parallel cracks. Those cracks are turned into nanowires as described above. Figure 2b shows a lower magnification SEM image of the sample shown in Figure 2a. The uniaxial stress as a result of the bending creates quite regular gaps between individual nanowires. On this sample, the wire array is limited to an area of approximately 1cm x 0.5cm. Note that the procedure is in principal not limited to certain sample dimensions. Upscaling to larger sample dimensions could be easily achieved by using production methods like roll to roll processes and corona discharge for polymer metallization. In this way, it should be possible to generate rolls of polymer foil, continuously coated with parallel nanowires. These samples show asymmetric conductivity, and due to the nanoscale of the metal wires, a high fracture toughness. First experiments show a high robustness of the wires against deformation. Those results will be published in detail elsewhere.

Also, more complex forms can be produced by the sequential application of uniaxial stress. Figure 2c and d show the result of the repeated application of uniaxial stress in two different directions. Figure 2c shows an image where the propagation of a second crack was blocked by a previously formed one, resulting in a y-shaped connection. The wires here are of more complex shape than those in Figure 1a, as Au was deposited first and on top of it Cu. In principle it should be possible to produce more complex layered wires by piling several metals on top of each other. In contrast, the figure 2d shows an x-junction where a Cu wire crosses the underlaying Au wire. Connections like that are desired for contacts with a very small contact area. By an evaporation step between Cu and Au deposition, functional materials like switchable molecules could be brought into the junction and used as e.g. electrical switches, like in [5].

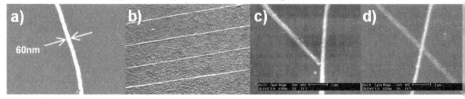

Figure 2. SEM images of metal nanowires on a polymer substrate. a) Part of a Au nanowire with the diameter of ~60 nm in a high resolution SEM taken by M. Schossig (GKSS research center). b) lower magnification of a different part of the sample from (a), showing the parallel alignment of the wires (sample coated with carbon). c) y-structure as a result from the application of different stress. d) x-junction of wires made out of different material (Au wire under a Cu wire).

Ways to form parallel nanowires

But even though nanowires can be used to cover large sample areas and to build x- and y – junctions, not every individual structure can be formed that easily. The whole approach is based on thin film fracture, and this is also what limits the approach. The driving force for thin film cracks is the energy gain due to the relaxation of the strained film, the gain of elastic energy. Therefore, in a homogeneous film, there is no formation of "double cracks", i.e. the formation of two parallel cracks very close together without any other cracks around them. A cracked film, attached to a substrate, relaxes an area beside the crack of a typical width proportional to the crack depth, the substrate constrains the elastic deformation [6]. Two parallel wires in a distance smaller than the relaxation width are not stable, they would either unite or separate, depending on the strain field.

This holds true for a homogeneous thin film. In inhomogeneous thin films, the cracks can be guided by the inhomogeneous strain field within certain limits. By simulating the strain fields with a finite element program (ANSYS), predictions can be made about the shape of the expected fracture. Figure 3a shows a cross section of the strain field simulation in a triangular film. A baseline expansion generates the film stress. The darker the image, the higher is the strain and the higher is the probability of fracture around that point. This suggests that a thin film like the one in Fig. 3b will fracture in the depicted way. This can be verified experimentally.

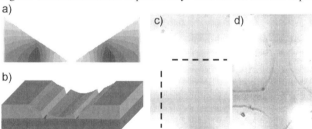

Figure 3. Inhomogeneous carbon thin films. a) finite element simulation of a cross section of a strained film. b) Sketch of the expected fracture behaviour. c) Optical microscopy image (dashed lines ~5μm). d) Optical microscopy image of the fractured sample after a dip in liquid nitrogen.

The structure in fig. 3c is the optical micrograph of a patterned carbon thin film. The bright structures are elevated, while the structures with less contrast are substantially thinner. The dashed lines indicate where to find a geometry similar to the one in figure 3a and b. The result of stress by quenching using liquid nitrogen is shown in figure 3d. The sample fractures in the middle between the highest and the lowest point.

But the control gained in this way is relatively poor and not yet suited to form useful structures like two electrodes contacting each other. [In addition, in order to bring the cracks closer by such an approach tends to be inefficient. Because the mask, that is providing the thin film structure must have elements on the nanoscale, deleting the strongest advantage of the fracture technique, which is the miniaturization effect of the thin film fracture].

Delaminated cracks as shadow masks

Other techniques for fabrication on the nanoscale do not have such disadvantages. Shadow masks can be used to form well defined nanostructures. By deposition under different angles through such a mask different nanostructures can be created with a separation of only a few nanometers [7]. The general concept for nanostructure generation by vacuum deposition with such a shadow mask is shown in figure 4.

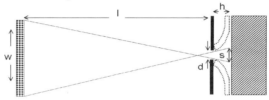

Figure 4. Deposition of nanostructures through a shadow mask. The nanostructure size s is given by the opening d in the mask, the distance h of the mask to the sample, the distance l from the source to the mask (~ to the sample) and the evaporator width w by the following : $s \approx d + ^w/_l * h$

The most important feature is that the mask placed close to the substrate surface. The size s of the nanostructure is approximately given by ~s ≈ $d + ^w/_l * h$, see fig. 4. The structure generated by the shadow mask is always bigger than the opening d of the mask. The broadening of the structure depends on the distance of the mask from the surface times the fraction of the width of the evaporation source divided by its distance to the mask (approximately the distance from the evaporation source to the sample). Typically the fraction is limited by the design of the vacuum chamber and gives values in the order of 1/100 (e.g. evaporation width 1 mm, sample distance 10 cm). This means that a shadow mask placed 1μm above the sample creates a broadening of ≈10 nm. This can be seen as disadvantage of the technique as the shadow mask has to be thin enough or of special conical shape to be placed properly. Moreover, the mask structure is already a nanostructure, so the structure creation procedure is very complex and expensive.

Here, we want to combine the advantages of the shadow mask technique with the advantages of the thin film fracture template approach in order to demonstrate how complex structures can be made by the help of structures created by mechanical failure. If the thin film not only fractures by the procedure but also the sides of the crack delaminate, a structure like the dotted lines in figure 4 will be formed. Delamination occurs typically if there is some shear stress in the thin film and the adhesion between substrate and thin film is weak.

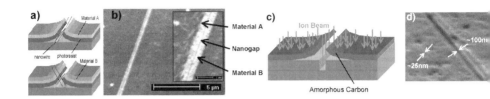

Figure 5. Generation of nanostructures with the help of delaminated cracks. a) Principle of angular deposition into the delaminated cracks. b) Wires formed in that manner (Material A=Cu, Material B=Au). c) principle of nanochannel formation. d) SEM image of nanochannels in Au.

The feasibility of the approach might be demonstrated in fig 5. Here, Au and Cu were evaporated under different angles. Two parallel wires with a small nanosized gap between them were found. Further experiments will show the reproducibility of the approach.

Instead of being used for deposition, the delaminated cracks could be also used to serve as a mask for the removal of material (see Fig. 5c). For such experiments, we chose glass with a 50 nm Au film as substrate. Amorphous carbon was chosen to serve as mask, because it has a relatively low sputter yield. Compared to Au, the carbon sputters 4 times slower in our experiment. Creating stress by dipping the samples into liquid nitrogen results again in a crack network. After a subsequent sputtering of approximately 80 nm, nanochannels were found. A first image is shown in figure 5d. The channel dimensions are in the order of a few 10 nm. Further experiments are necessary in order to characterize the limits for the channel width and a possible use as contacts for nano-sized objects.

CONCLUSION

In this paper, several mechanisms were shown to create nanostructures. All are based on structures of mechanical failure. The systematic examination of the properties of typically unwanted effects like fracture or delamination is necessary in order to turn them into tools for the generation of nanostructures.

Acknowledgements
The authors thank Prof. Franz Faupel for stimulating discussions and the possibility to work in his laboratory. We gratefully acknowledge financial support by the German Science Foundation (DFG) in the framework of the Priority Program 1165 for contract AD 183/4-1 .

REFERENCES

1 R. Adelung, O.C. Aktas, J. Franc, A. Biswas, R. Kunz, M. Elbahri, J. Kanzow, U. Schürmann, and F. Faupel, Nature Materials, **3**, 375 (2004).

2 S. Christansen and U. Gösele, Nature Materials, **3**, 357 (2004).

3 R. Adelung, L. Kipp, J. Brandt, L. Tarcak, M. Traving, C. Kreis, and M. Skibowski, Appl. Phys. Lett **74**, 3053 (1999).

4 B. E. Alaca, H. Sehitoglu, and T. Saif, Appl. Phys. Lett. **84**, 4669 (2004).

5 Y. Chen, D. A. A. Ohlberg, X. Li, D. R. Stewart, R. S. Williams, J. O. Jeppesen, K. A. Nielsen, J. F. Stoddart, D. L. Olynick, and E. Anderson, Appl. Phys. Lett. **82**, 1610 (2003).

6 R.F. Cook and Z. Suo, MRS Bulletin, **27**, 45 (2002).

7 M. M. Deshmukh, D. C. Ralph, M. Thomas, and J. Silcox, Appl. Phys. Lett. **75**, 1631 (1999).

Mater. Res. Soc. Symp. Proc. Vol. 875 © 2005 Materials Research Society   O11.4/B7.4

Degradation of Fracture and Fatigue Properties of MEMS Structures under Cyclic Loading

Jong-jin Kim and Dongil Kwon
School of Materials Science and Engineering, Seoul National University, Seoul 151-742, Korea

ABSTRACT

The fracture and fatigue properties of LIGA nickel MEMS structures were evaluated by microtensile and fatigue test methods. A microtensile/fatigue device was developed and specimens with feature size ten micrometers were used. The fatigue property was derived from displacement amplitude – the number of cycles to failure curve by applying a dynamic load with a piezoelectric actuator. The tensile/fracture properties after various cyclic loading were also measured. Both fatigue and tensile test results showed a cyclic softening phenomenon.

INTRODUCTION

It is well known that mechanical properties of thin films differ from those of the bulk materials [1]. Information on the mechanical properties of thin films has become indispensable in the design of microelectromechanical systems (MEMS), where thin films are structural as well as electrical materials. In particular, MEMS such as gyroscopes, optical switches or micro-mirrors are subjected to cyclic stresses and often function under constant displacement conditions. Therefore, an understanding of the behavior of cyclically loaded films is essential in designing a new product and assessing its reliability. Much attention has been given to tensile properties of microsamples, but limited research has been performed on the fatigue resistance of micrometer-scale materials [2-5], despite its importance in evaluating the lifetime and reliability of microdevices.

In recent years, interest in LIGA (Lithographie, Galvanformung, Abformung) nickel MEMS structures has proliferated rapidly because of their high damage tolerance. The LIGA process was developed to overcome large sustained forces or torques that cannot be sustained in thin, two-dimensional components. The mechanical properties of LIGA nickel MEMS structures have been studied [6-12], but few reports on their fatigue behavior are available in the literature. Although stress-life behavior has been measured [13,14] and the underlying fatigue mechanisms have been investigated [15,16], no attempts have apparently been made to investigate low cycle fatigue properties and degradation of mechanical properties under cyclic loading.

In this study, equipment with piezoelectric actuators and laser-speckle interferometry was used to evaluate the effect of cyclic loading on the mechanical properties of LIGA nickel MEMS structures.

EXPERIMENTAL DETAILS

Ni specimens 20 μm thick, 10 μm wide and 600 μm long were fabricated by a surface micromachining process including nickel electroplating with a LIGA-like or UV-LIGA technique. A 50 nm Cu seed layer was deposited on a Corning glass 7740 wafer; then thick photoresist (PMER-LA900) was spun to a thickness of 25 μm and prebaked at 110 for 360 seconds. To form a mold for nickel electroplating, the thick photoresist was exposed to UV light at 18 mW for 180 seconds, and nickel was then electroplated at 50 with stirring. The nickel-sulphamate electroplating bath was used, consisting of 110 g/L $NiSO_47H_2O$, 6 g/L $NiCl_26H_2O$, and 30 g/L boric acid at a current density of 3 mA/cm^2. After the residual seed layer was etched, the glass substrate was isotropic-etched in liquid hydrofluoric acid to obtain the free-standing nickel structure. Figure 1 shows a schematic diagram of the specimen and the piezoelectric-driven uniaxial stress-strain measurement system. The force resolution is 0.2 mN and the displacement range of the closed-loop piezoelectric actuator is 0-90 μm, with a displacement measurement resolution of 1.8 nm.

To measure the exact strain in the gauge section, one-dimensional electron speckle interferometry was used, as described in [17]. We used UV adhesive to grip the device to the microtensile system; the grip was transparent glass to pass UV light and cure the underlying UV adhesive layer effectively and quickly [18]. Capillary forces kept the devices in contact with the glass grip until the glue was cured, and a three-axis micropositioning system and CCD with remote microscope were used to locate the glass grip appropriately on the devices.

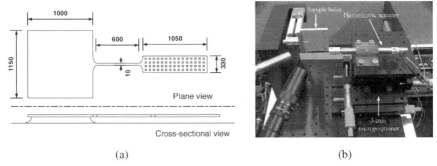

(a)   (b)

Figure 1. (a) Schematic drawing of Ni specimen (unit: μm) and (b) photograph of piezoelectric stress-strain measurement system.

RESULTS AND DISCUSSION

Microtensile test

First of all, microtensile tests were performed to determine the fatigue test condition. The strain rate was 5×10^{-3} s^{-1} and five specimens were tested. The yield strength was determined as 223.4±10.8 MPa at 0.2% offset using an experimentally measured modulus value of 88.7±2.2 GPa. The tensile strength and fracture strength were 312.1±17.6 and 296.4±14.8 MPa, respectively.

Fatigue test

We conducted displacement-controlled fatigue tests at a frequency of 10 Hz. Since the specimen cannot support a compressive load, tension-tension-type fatigue tests were utilized. The applied waveform was sawtooth shape. The test condition for low cycle fatigue was determined on the basis of the microtensile test results. The mean displacement was 8 μm and displacement amplitudes were 1, 2, 3, and 4 μm. Load, displacement and time were recorded during the test. Typical data from fatigue testing under displacement control (figure 2) exhibits cyclic softening, i.e. a decrease in load with increasing number of cycles. Cyclic softening develops after a number of load cycles have occurred. The earlier load cycles exhibit elastic strain behavior, however after a number of cycles, depending on the type of material and the stress range, material will exhibit local microplastic strain or deformation. Cyclic softening is related to dislocation behavior which is mainly dependent on stacking fault energy [19]. Therefore, it seems that this phenomenon in our study is attributed to high stacking fault energy of Ni; as stacking fault energy is higher, cross slip becomes easier and then dislocation movement becomes more active. The fatigue test results plotted in figure 3 show that number of cycles to failure decreased with increasing displacement amplitude. Future study will include further investigation of the mechanism of cyclic softening and the effect of displacement amplitude on fatigue lifetime.

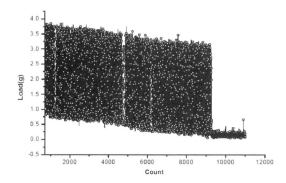

Figure 2. Typical fatigue test data under displacement control.

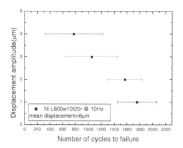

Figure 3. Variations in the number of cycles to failure with displacement amplitude.

Microtensile test after cyclic loading

To investigate the effect of cyclic loading on mechanical properties, we conducted microtensile testing after cyclic loading. Three cycles (25, 200 and 500) were chosen and cyclic loading was applied. Mean displacement was 1.2 μm and displacement amplitude was 1.0 μm. After cyclic loading, specimens underwent microtensile testing as described above. The Young's modulus and yield strength results did not vary significantly with the number of cycles. However, about 10% degradation in tensile and fracture strength occurred as the number of cycles increased, as shown in figure 4. This result agreed with that of fatigue test in pointing to cyclic softening. The trend toward decreasing degradation rate with increasing number of cycles was also in accord with cyclic softening; in general, cyclic softening is usually rapid at first, but the change from one cycle to the next decreases with increasing numbers of cycles [20].

298

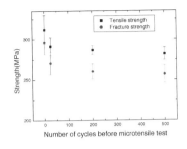 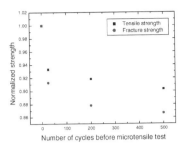

Figure 4. Variation of tensile and fracture strength with number of cycles before microtensile test.

CONCLUSIONS

Fatigue and tensile properties of LIGA Ni were investigated. Cyclic softening was observed and fatigue lifetime decreased with displacement amplitude. Evaluation of tensile properties after cyclic loading showed that tensile and fracture strength decreased with number of cycles. These results suggest that changes in mechanical behavior under cyclic loading must be considered in designing LIGA Ni MEMS. This cyclic softening phenomenon should be investigated further.

ACKNOWLEDGMENTS

This research was supported by grant M1040300001304001000610 from the Electrical Component Reliability Design Technology Program supported by the Ministry of Science and Technology, Korea.

REFERENCES

1. O. Kraft , L. B. Freund, R. Phillips, A. Arzt, *MRS Bulletin* **27**, 30 (2002).
2. S. Hong and R. Weil, *Thin Solid Films* **283**, 175 (1996).
3. D. T. Read, *Int. J. Fatigue* **20**, 203 (1998).
4. G. Cornella, R. P. Vinci, R. S. Iyer, R. H. Dauskardt, and J. C. Bravman, in: S.B. Brown, C. Muhlstein, P. Krulevitch, G.C. Johnston, R.T. Howe, J.R. Gilbert (Eds.), *Microelectromechanical Structures for Materials Research*, San Francisco, CA, U.S.A., April 13-17, 1998, Materials Research Society Symposium Proceeding **518** (1998) 81.

5. N. B. Barbosa III, P. El-Deiry, and R. P. Vinci, in: S. G. Corcoran, Y. –C. Joo, N. R. Moody, and Z. Suo (Eds.), *Thin Films-Stresses and Mechanical Properties X*, Boston, MA, U.S.A., December 1-5, 2003, Materials Research Society Symposium Proceeding **795** (2004) U11.39.1.

6. E. Mazza, S. Abel, and J. Dual, *Microsyst. Tech.* **2(4)**, 197 (1996).

7. J. Dual, E. Mazza, G. Schiltges, and D. Schlums, in: C.R. Friedrich, A. Umeda (Eds.), *Micromachining and Microfabrication*, Austin, TX, U.S.A., September 29-30, 1997, Proceedings of SPIE **3225** (1997) 12.

8. W. N. Sharpe Jr., D. A. LaVan, and R. L. Edwards, *IEEE 1997 International Conference on Solid-state Sensors and Actuators*, Chicago, IL, U.S.A., June 16-19, 1997, Proceedings of Transducers '97 (1997) 607.

9. W. N. Sharpe Jr., D. A. LaVan, and A. McAleavey, *Micro-Electro-Mechanical Systems (MEMS), ASME DSC, 62/HTD*, **354**, 93 (1997).

10. W. N. Sharpe Jr. and A. McAleavey, in: C. R. Friedrich, Y. Vladimirsky (Eds.), *Micromachining and Microfabrication*, Santa Clara, CA, U.S.A., September 20-22, 1998, Proceedings of SPIE **3512** (1998) 130.

11. T. R. Christenson, T. E. Buchheit, D. T. Schinale, and R. J. Bourcier, in: S. B. Brown, C. Muhlstein, P. Krulevitch, G. C. Johnston, R. T. Howe, J. R. Gilbert (Eds.), *Microelectromechanical Structures for Materials Research*, San Francisco, CA, U.S.A., April 13-17, 1998, Materials Research Society Symposium Proceeding **518** (1998) 185.

12. K. J. Hemker and H. Last, *Mat. Sci. Eng.* **A319–321**, 882 (2001).

13. K. P. Larsen, A. A. Rassmussen, J. T. Ravnkilde, M. Ginnerup, and O. Hansen, *Sens. Actuators* **A 103**, 156 (2003).

14. H. S. Cho, K. J. Hemker, K. Lian, J. Goettert, and G. Dirras, *Sens. Actuators* **A 103**, 59 (2003).

15. B. L. Boyce, J. R. Michael, and P. G. Kotula, *Acta Mater.* **52**, 1609 (2004).

16. S. M. Allameh, J. Lou, F. Kavishe, T. Buchheit, and W. O. Soboyejo, *Mat. Sci. Eng.* **A371**, 256 (2004).

17. D. W. Kim and D. Kwon, *Int. J. Modern Physics* **B17**, 1534 (2003).

18. I. Chasiotis and W. G. Knauss, in: Y. Vladimirsky, P. J. Coane (Eds.), *Micromachining and Microfabrication*, Santa Clara, CA, U.S.A., September 17-20, 2000, Proceedings of SPIE **4175** (2000) 96.

19. R. W. Hertzberg, *Deformation and fracture mechanics of engineering materials*, Third edition (John Wiley & Sons, 1989), 489.

20. N. E. Dowling, *Mechanical Behavior of Materials*, Second edition (Prentice Hall, New Jersey, 1999), 586.

Mater. Res. Soc. Symp. Proc. Vol. 875 © 2005 Materials Research Society   O11.6/B7.6

Effect of Microstructure and Dielectric Materials on Stress-Induced Damages in Damascene Cu/Low-k Interconnects

Young-Chang Joo, Jong-Min Paik and Jung-Kyu Jung
School of Materials Science and Engineering, Seoul National University, Seoul 151-744 Korea

ABSTRACT

The line width dependence of stress in damascene Cu was examined experimentally as well as with a numerical simulation. The measured hydrostatic stress was found to increase with increasing line width. The larger stress in an interconnect with large dimension is attributed to the larger grain size, which induce higher growth stress in addition to thermomechanical stress. A stress model based on microstructure was constructed and the contribution of the growth and thermal stress of the damascene lines were quantified using finite element analysis. It was found that the stress of the via is lower than that of wide lines when both the growth stress and thermal stress were considered. This stress gradient between via and line, which is the driving force of vacancy diffusion, is larger when the low-k with lower stiffness and higher thermal expansion is used for dielectric layer. For this reason, the Cu/low-k can be more vulnerable to stress-induced voiding.

Key words: Damascene Cu, microstructure, stress, low-k dielectric, finite element analysis, x-ray diffraction

INTRODUCTION

Since the RC delay and crosstalk of the interconnects are anticipated to be the limiting factors in high performance integrated circuits, copper and low-k dielectrics are now used in place of aluminum and silicon oxide. In contrast to Al interconnects, whose reliability limitations have been well characterized through 30 years of experience, Cu interconnect reliability is still a relatively new technology. Recent progress has shown that stress voiding in Cu interconnects is now considered an important reliability concern [1-2].

Stress in the interconnects can be divided into two classes according to its origin: thermal stress, which is generated by thermal expansion mismatch between metal lines and surrounding materials, and growth stress, which is induced by the grain growth of metal lines. Grain growth is a process of grain boundary elimination. Therefore, it can cause considerable volume shrinkage of the metal lines. Volume shrinkage of the metal lines with confinement by the surrounding materials causes tensile stress.

For the conventional Al interconnects process, lines are formed by an etching process from continuous thin films. Generally, grain growth is completed during film deposition and annealing prior to passivation, thus the growth stress of the lines is insignificant. As a result, the major stress component is thermal stress only.

On the other hand, considerable growth stress as well as thermal stress can develop in the Cu interconnects [3]. The initial grain size of the electroplated Cu is as small as few tens of nanometers. During subsequent high temperature annealing and passivation layer deposition, grain growth occurs under confinement by the predefined trenches. The final grain size of the Cu lines is known to be approximately the same as line width [4]. As a result, considerable strain due to the volume shrinkage develops during grain growth. For example, Lee et al. [5] reported that the growth stress in their electroplated Cu films was as large as 120MPa. Therefore it is believed that growth stress does affect the stress state of Cu lines in addition to thermal stress.

In order to quantify the growth stress in the damascene lines, a detailed understanding of the grain structure is required. The grain structure of Cu is a complicated 3-dimensional structure rather than a 2-dimensional columnar structure, which is the case of Al interconnects [6]. This means that the grain sizes along the line direction, across the line width, and through the line thickness are different from one another. Therefore, the in-plane grain structures of the top surfaces as well as the vertical structures through the thickness need to be analyzed.

The line width dependence of stress is important in determining chip reliability [7]. The measured hydrostatic stress of the Al interconnects are known to increase as the line width decreases. Hosoda et al. [8] explained that the reason for this behavior is that yielding is difficult in the narrow lines due to the dimensional restrictions. Due to these width dependencies of hydrostatic stress, stress induced damage is more problematic in narrow lines [7]. However, for the Cu interconnects, the line width dependence of stress and related reliability issues have not yet been studied. Furthermore, the intrinsic stress such as the "growth stress" is expected to play a significant role in the line width dependence of stress, which would be more complicated.

In this study, the line width dependence of stress in damascene Cu was investigated. The grain structures and stresses of lines with various widths were measured by transmission electron microscopy (TEM) and x-ray diffraction (XRD). The contribution of growth stress and thermal stress of the damascene lines were estimated based on the experimental data and finite element analysis (FEA). Using this method, the growth and thermal stresses were quantified, and their effect on the line width dependence and stress-related reliability is discussed.

EXPERIMENTAL DETAILS

For the stress measurement of damascene Cu lines, periodic line patterns with around 1 cm^2 area were fabricated. The line width was varied from 0.13 μm to 2 μm. The line widths in the mask were 0.18, 0.25 and 2 μm, and lines with various widths and spacings were obtained by changing the lithography dose. The thickness of the lines was 0.5 μm for all line widths. PECVD (plasma enhanced chemical vapor deposited) silicon oxide and low-k dielectrics (SiOC: CORAL^{TM}) were employed as an IMD (intermetal dielectric) layer to comparatively study the effect of the dielectric material on the stress in the Cu lines. After damascene trench formation, a Ta/TaN bi-layer with a thickness of 350 Å was used as a diffusion barrier for Cu. And then, seed Cu layer of 1500 Å thickness was deposited, which was followed by the electroplating for the Cu filling into the trench. Then, the samples were annealed at 200°C for about 30 min in a high vacuum furnace. After CMP of the overburden Cu, a silicon nitride capping layer of 700 Å thickness was applied and then 3000 Å thick layers of SiO$_2$ and silicon nitride were deposited sequentially for the passivation the structure. The stresses were measured using a 4-circle goniometer with a synchrotron radiation source. The detailed calculation process can be found in reference [9].

To study the stress state of the via-line structures, a finite element analysis was performed using the commercial program, ABAQUS 6.2 [10]. The width and thickness of via and line were fixed to be 0.2 μm and 0.5 μm, respectively. The thickness of the TaN barrier layer was 20 nm at the bottom and 10 nm at the sidewall. A 500 Å thick SiN etch stop layer was considered between IMD1/IMD2 and IMD3/IMD4. The boundary condition was applied in such a way that the via-lines are arrayed periodically along the x- and y-directions. A detailed description of the boundary conditions can be found in our previous work [11]. The properties of the materials used in this calculation are summarized in Table I.

Table I. Thermo-mechanical properties of the materials used in the finite element calculations.

Materials	CTE (ppm/°C)	Modulus (GPa)	Poisson's Ratio
Cu	17.7	111.5	0.343
SiO$_2$	0.51	71.7	0.16
CVD SiOC	20.0	9.5	0.3
Ta	6.5	185.7	0.342
SiN	3.2	220.8	0.27

RESULTS AND DISCUSSION

Microstructure of Damascene Cu Lines

Figure 1 shows the average grain size of the 0.18, 2 µm lines and the pad structures. Parameters d_x, d_y and d_z represent the grain sizes along the length, width and thickness, respectively. The grain size along the three principal directions of the 0.18 µm line was comparable to its width; however, d_x is slightly larger than d_y or d_z. This implies that grain growth is limited in the width and thickness but this effect is not substantial for grain growth along the line direction (Note that the value of d_x of the narrow line is even larger than its 0.18 µm width as shown in figure 1). The in-plane grain sizes of the 2 µm line and pad structure were comparable to their full thickness from the bottom of the trench to the top of the overburden (0.8 µm). Since the grain structure of the wide line or pad structure is columnar, d_z is the same as the trench depth. From this result, it is inferred that the minimum dimension restricts grain growth in the trench; for narrow lines (aspect ratio>1) grain growth is limited by its width, while it is limited by its thickness for wide lines (aspect ratio<1). If this is the case, it is expected that there exists a transition point of the grain size increasing rate as line width where the line width exceeds its height. The grain sizes of the narrow lines with various widths were also quantified and plotted with respect to the line width in figure 2. The average grain sizes along the three directions increase with increasing line width. It can be found that the grain size increases more rapidly in the narrow lines. Further detailed analysis was made by the theoretical approach based on grain growth Monte-Carlo simulation [12, 13].

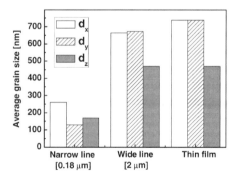

Figure 1. Median grain sizes as a function of the geometry of the interconnects; d_x, d_y and d_z are the grain size along the length and width and thickness, respectively.

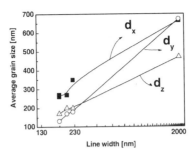

Figure 2. Median grain sizes of the set of narrow lines; d_y increases most rapidly and becomes similar to d_x for the 2000 nm line.

The Stress of Damascene Cu Lines

Figures 3 (a), (b) and (c) show the room-temperature principal stresses in the narrow lines (132~224 nm) and wide lines (2 μm) fabricated with the dielectric material, for the SiO_2 and the CVD low-k. As the line width increases, the stress along the lines (σ_{xx}) and across the lines (σ_{yy}) increase, while the stress normal to the surface (σ_{zz}) decreases. For the wide line (2 μm), the stress state and its magnitude of the Cu lines in SiO_2 dielectric are similar to those of the Cu lines in low-k dielectric. In the sub-micron lines, however, all the stress components of the Cu lines in low-k dielectric are lower compared to those of the Cu lines in SiO_2 dielectric, and this is especially prominent for σ_{zz}. This is because the elastic modulus of low-k dielectric (E=9.5 GPa) is much lower than that of SiO_2 (E=71.7 GPa). When Cu is embedded with a compliant material, the stress can be easily relaxed by the deformation of the surrounding material. Deformation along in-plane directions (x- and y- direction) is restricted by the rigid substrate rather than dielectric materials. Therefore, the differences in in-plane stresses do not reflect the differences in compliance of dielectric materials. On the other hand, deformation of Cu along the thickness direction is mainly affected by mechanical properties of dielectric materials because deformation through the thickness is free from mechanical constraint of the Si substrate. Furthermore, the poissonian deformation, which also should interact with surrounding dielectric materials, makes a considerable contribution to the stress through the thickness. For this reason, σ_{zz} of Cu line in low-k dielectric is relatively small with the value being below 200 MPa.

In figure 3 (d), the hydrostatic stress which was obtained from principal stresses was plotted versus line dimension. The level of hydrostatic stress increased with increasing interconnect dimension. The pad structures, which have a similar dimension to thin films, displayed the highest values. This is very interesting behavior because it is well-known that the hydrostatic stress is higher for Al lines with smaller dimension. Details on the effect of the line width dependence on reliability are discussed in the following section.

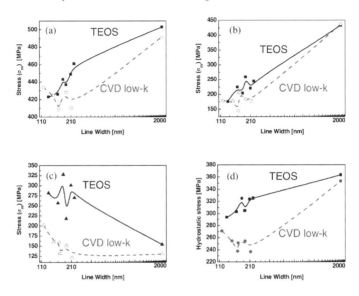

Figure 3. The principal stresses versus line dimension of Cu/SiO2 and Cu/Low-k interconnects measured by x-ray diffraction; (a) stress along the line direction, (b) stress along the width direction, (c) stress through the thickness and (d) hydrostatic stress.

Relationship between stress and microstructure

The thermal stresses were calculated separately using a 3-dimensional finite element method (FEM) with a linear elastic model in order to understand why structures with the larger dimension have a higher hydrostatic stress. Thermal stress was considered by thermal expansion of component materials as temperature changes from stress free temperature, T_0, to the temperature concerned, T. It was assumed that the virtual stress free temperature (T_0) is 100°C and the thermal stresses are developed during cooling from 100°C to 25°C. Figure 4 (a) shows the principal stress components and hydrostatic stress of damascene Cu lines with TEOS as a function of the line width. As the line width increased, the stress along the lines (σ_{xx}) did not vary, while the stress across the lines (σ_{yy}) increased, and the stress normal to the surface (σ_{zz}) decreased. As a result, the hydrostatic stress remained constant with increasing line width, which indicates that the line width dependence of the hydrostatic stress observed in these experiments is not caused by thermal stresses.

Grain growth is a process of grain boundary elimination, which can lead to the evolution of stress in Cu lines as well as thin films. The growth stress was modeled by introducing a set of stress-free homogeneous strains of the interconnect, namely, ε_x^T, ε_y^T and ε_z^T. Based on the initial grain diameters of d_x^0, d_y^0 and d_z^0 along the lines, across the lines and through the thickness, respectively, as the initial condition, and assuming that the grains grow to the final size d_x, d_y, and d_z, the grain growth induced principal strain components are given by

$$\varepsilon_x^T = \delta\left(\frac{1}{d_x^0} - \frac{1}{d_x}\right), \quad \varepsilon_y^T = \delta\left(\frac{1}{d_y^0} - \frac{1}{d_y}\right) \quad \text{and} \quad \varepsilon_z^T = \delta\left(\frac{1}{d_z^0} - \frac{1}{d_z}\right) \tag{1}$$

where δ is the excess volume per unit area. A reasonable assumption for δ was 1 Å, while the initial grain size was assumed to be 50 nm in all cases. Although the initial grain size of electroplated Cu have been not reported definitely, it can be estimated as few tens nanometer based on several references [14, 15] in which the grain size was measured after several hours from deposition. From the measured grain sizes (figure 2), values for d_x, d_y and d_z were entered into equation (1) to obtain the principal strain components by volume shrinkage. The grain size of 0.13 μm line was extrapolated from that of 0.18~0.22 μm lines assuming that the grain size of the narrow lines varies linearly with line width. Subsequently, the growth induced stresses of the Cu lines were calculated using the FEA.

Figure 4 (b) shows the line width dependence of the grain growth induced stress of the Cu lines. With the exception of the stress component through the thickness (σ_{zz}), the other stress components (σ_{xx}, σ_{yy}) increased considerably with increasing line width. Because both the measured line width and spacing were considered in the FEM model in order to compare the experimental data, the change in stresses due to the line spacing effect can be found as experimental data show (figure 3). Therefore, it can be shown that the hydrostatic stress increased with increasing line width. The total stress of the damascene line is a sum of the thermal stress (figure 4 (a)) and growth stress (figure 4 (b)). Figure 5 shows the experimentally measured stress and calculated stresses (thermal stress and growth stress) versus line width. Due to the larger grain size, wider lines displayed higher growth stress. The level of hydrostatic growth stress increased from 137 to 205 MPa with increasing line width, while the thermal hydrostatic stress at approximately 150 MPa, remained independent of the line width.

During the calculation of thermal stress, the stress free temperature was assumed to be 100°C to fit the calculated stress to measured counterpart. Although it is usually assumed that the stress free temperature (T_0) is the same as the deposition temperature (T_d), the measured stresses do not become zero at T_d [16]. For example, the experimentally determined T_0 varies in the wide ranges such that $T_0 = T_d$-(100~200°C) [17, 18]. Therefore, these calculations and its adjustment to compare the experimental results with the FEM are believed to be physically reasonable.

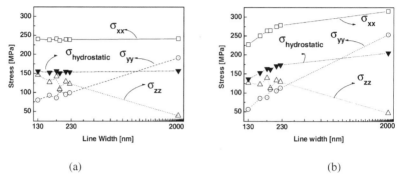

(a)   (b)

Figure 4. Calculated stress of damascene Cu lines with line width: (a) thermal stress and (b) growth stress.

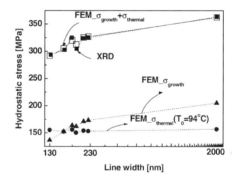

Figure 5. Hydrostatic stress as a function of the line width: experimentally measured stress (XRD), thermal stress by FEM, and grain growth stress from the grain size analysis and FEM.

The results of the individual calculation of the thermal stress and growth stress clearly show the effect of growth stress on the overall stress in the damascene Cu lines. Similarly, because the line width dependence of stress originates from the grain structures, control of the grain structure is expected to be the key factor in reducing the stress gradient between the wide and narrow lines.

It is well known that the hydrostatics stress increases with decreasing line width in Al interconnects. Therefore, the narrower Al lines have shorter stress-migration life times. However, in the case of the Cu interconnects, it was found that growth stress, which increases with increasing line width, makes the overall stress larger for the wider lines. These results indicate that stress-voiding in the Cu interconnects can be formed more easily in the wider lines than in the narrow lines, which is in contrast to that in the Al interconnects. In addition, it can be inferred that a strong stress gradient where the wide lines and narrow lines (or narrow via) meet may result. Since the stress gradient is the driving force of vacancy diffusion, it is probable that vacancies accumulate and voids form at the sites with dimensional transition in interconnects structure, for example, where either a small via or narrow line meets with a wide line.

Based on analysis for periodic line structures, we expand our growth stress model into via-line structures. Figure 6 shows the volume averaged hydrostatic stress for via-line structures with SiO_2 of M1, M2 and Via. The line width of M1 and M2 is 2000 nm and via width is 200 nm. It was assumed that the grain size was of M1 and M2 is 670 nm (which is measured), and the grain size of via is the same as that of 200 nm lines. When only the thermal stress is considered, the hydrostatic stress of via was larger than that of both M1 and M2. On the other hand, in the case where both the growth stress and thermal stress are applied to the structure, the stress of the via was lowest. Therefore, the vacancies are likely to diffuse toward the via from the M1 and M2 along the gradient of growth stress. Indeed, the stress induced voiding is frequently observed in the cross points of via and lines or via and pads in the damascene Cu interconnects [1, 2]. Ogawa et al. [2] proposed a model for stress voiding of Cu interconnects, in which a supersaturated vacancy migrates along the thermal stress gradient and accumulates below the via. In their model, thermal stress is the main contributor to the local stress gradient. In this study, it is believed that the growth stress does play at least a comparable role in the development of a wide-ranging stress gradient between a small structure (narrow lines or via) and a larger structure (wide lines or pads). Therefore, the site where a wide line meets with a via is vulnerable to stress-voiding. Because the stress gradient originates from dependence of grain structures on interconnect dimension, the control of the grain structure is expected to be an important factor in controlling the stress gradient between the wide and narrow lines.

The stress of Cu/low-k was also calculated considering growth stress and thermal stress, and is compared with that of Cu/SiO$_2$ in figure 7. During the analysis, all conditions were fixed and only the mechanical properties of dielectric materials were changed to that of low-k. Although the stress of Cu/low-k is lower than that of Cu/SiO$_2$, the differences in stress between M2 and via, however, is larger for the case of Cu/low-k. This indicates that vacancies in Cu/low-k can easily migrate from M2 to via, contrary to the case of Cu/SiO$_2$. The stress itself is the driving force of void nucleation; however, in the practical sense, the stress-voiding is not dominated by nucleation but by vacancy migration because it is known that a hydrostatic tension stress of the order of 5 GPa would be needed to cause void nucleation to occur at a significant rate [19]. Therefore, the stress gradient would be more important factor to determine the stress migration behavior. The observed experimental results [20, 21], which demonstrate that the Cu/low-k is weaker for stress-voiding, can be explained by this stress gradient effect.

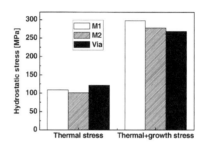

Figure 6. The volume-averaged hydrostatic stress for the via-line structures with SiO$_2$ of M1, M2 and Via.

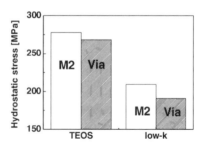

Figure 7. The volume-averaged hydrostatic stress for the via-line structures with SiO$_2$ of M2 and Via.

SUMMARY

The stress voiding in Cu interconnects was investigated by stress analysis using x-ray diffraction as well as finite element analysis. The stress variation as dimension resulted in stress gradient at the dimensional transition points. The stress gradient of Cu/low-k was higher than that of Cu/TEOS although the opposite was true for the stress level. Therefore, it is suggested that the stress reduction is important for Cu/TEOS while the stress gradient should be suppressed for Cu/low-k in order to improve interconnect reliability.

REFERENCES

1   K. Y. Y. Doong, R. C. J. Wang, S. C. Lin, L. J. Hung, S. Y. Lee, C. C. Chiu, D. Su, K. Wu, K. L. Young and Y. K. Peng, *IRPS*, 2003, pp. 156-160.

2   E. T. Ogawa, J. W. McPherson, J. A. Rosal, K. J. Dickerson, T. –C. Chiu, L. Y. Tsung, M. K. Jain, T. D. Bonifield, J. C. Ondrusek and W. R. McKee, *IRPS*, 2002, pp. 312-331.

3   P. R. Besser, Ehrenfried Zschech, Werner Blum, Delrose Winter, Richard Ortega, Stewart Rose, Matt Herrick, Martin Gall, Stacye Thrasher, Mike Tiner, Brett Baker, Greg Braeckelmann, Larry W. Zhao, Cindy Simpson, Cristiano Capasso, Hisao Kawasaki and Elizabeth Weitzman, *J. Electron. Mater.* **30**, 320 (2001).

4   Q. T. Jiang, M. Nowell, B. Foran, A. Frank, R. H. Havemann, V. Parihar, R. A. Augur, and J. D. Luttermer, *J. Electron. Mater.* **31**, 10 (2001).

5   H. Lee, S. S. Wong, and S. D. Lopatin, *J. Appl. Phys.* **93**, 3796 (2003).

6   J. M. Paik, K. C. Park, and Y. C. Joo, *J. Electron. Mater.* **33**, 48 (2004).

7   T. D. Sullivan, *Annu. Rev. Mater. Sci.*, **26**, 333 (1996).

8   T. Hosoda, H. Niwa, H. Yagi and H. Tsuchikawa, *IRPS*, 1991, p. 77.

9   S. H. Rhee, Ph. D. Dissertation, The University of Texas at Austin, Austin, TX, 2000.

10 ABAQUS, Version 6.2, general purpose finite element program, Hibbit, Karlson, and Sorensen, INC., Pawtucket, RI., 2001.

11 J. M. Paik, H. Park and Y. C. Joo, Microelectronic Engineering **71**, 348 (2004).

12 J. K. Jung, N. M. Hwang, Y. J. Park and Y. C. Joo, *Japanese J. Appl. Phys.* **43(6A)**, 3346 (2004).

13 J. K. Jung, N. M. Hwang, Y. J. Park and Y. C. Joo, *J. Electron. Mater.* **34**, 559.

14. S. P. Hau-Riege and C. V. Thompson, Appl. Phys. Lett. **76**, 309 (2000).

15. H. Lee, S. S. Wong, and S. D. Loptain, J. Appl. Phys. **93**, 3796 (2003).

16 H. Okabayashi, *Mater. Sci. Eng.* **R11**, 191 (1993).

17 A. Tezaki, T. Mineta, H. Egawa and T. Noguchi, *IRPS*, 1990, p. 202.

18 H. Yagi, H. Niwa, T. Hosoda, M. Inoue, H. Tsuchikawa and M. Kato, Stress-Induced Phenomena in Metallization, American Vac. Soc. Series 13, American Institute of Physics, New York, 1992, p. 44.

19 W. D. Nix and E. Arzt, Met. Trans. A 23A, 2007 (1992).

20 C. J. Zhai, H. W. Yao, P. R. Besser, A. Marathe, R. C. Blish II, D. Erb, C. Hau-Riege, S. Taylor and K. O. Taylor, "Stress Modeling of Cu/Low-k BEoL- Application to Stress Migration," *IRPS*, 2004, pp. 234-239.

21 W. C. Baek, P. S. Ho, J. G. Lee, S. B. Hwang, K. K. Choi, and J. S. Maeng, "Stressmigration Studies on Dual Damascene Cu/Oxide and Cu/Low-k Interconnects," in *Materials, Technology and Reliability for Advanced Interconnects and Low-k Dielectrics— 2004,* edited by R.J. Carter, C.S. Hau-Riege, G.M. Kloster, T.-M. Lu, and S.E. Schulz (Mater. Res. Soc. Symp. Proc. 812, Warrendale, PA , 2004), F 7.8.

Mater. Res. Soc. Symp. Proc. Vol. 875 © 2005 Materials Research Society   O11.7/B7.7

Comparison of Line Stress Predictions with Measured Electromigration Failure Times

Rao R. Morusupalli[1], William D. Nix[1], Jamshed R. Patel[1,2] and Arief S. Budiman[1]
[1]Materials Science and Engineering, Stanford University, Stanford, California,
[2]Advanced Light Source (ALS), Lawrence Berkeley National Laboratory (LBNL), Berkeley, California.

ABSTRACT

Reliability of today's interconnect lines in microelectronic devices is critical to product lifetime. The metal interconnects are carriers of large current densities and mechanical stresses, which can cause void formation or metal extrusion into the passivation leading to failure. The modeling and simulation of stress evolution caused by electromigration in interconnect lines and vias can provide a means for predicting the time to failure of the device. A tool was developed using MathCAD for simulation of electromigration-induced stress in VLSI interconnect structures using a model of electromigration induced stress. This model solves the equations governing atomic diffusion and stress evolution in one dimension. A numerical solution scheme has been implemented to calculate the atomic fluxes and the evolution of mechanical stress in interconnects. The effects of line geometries and overhangs, material properties and electromigration stress conditions have been included in the simulation. The tool has been used to simulate electromigration-induced stress in pure Cu interconnects and a comparison of line stress predictions with measured electromigration failure times is studied. Two basic limiting cases were studied to place some bounds on the results. For a lower bound estimate of the stress it was assumed that the interface can be treated like a grain boundary in Cu. For an upper bound estimate it was assumed that the interface can be treated like a free surface of Cu. Existing data from experimental samples with known structure geometries and electromigration failure times were used to compare the electromigration failure times with predicted stress build-up in the interconnect lines.

INTRODUCTION

Reliability and performance are equally important in the microelectronics industry of today. Product lifetimes are expected to be as high as 10 to 15 years of service. A typical microprocessor has millions of interconnect lines and the probability of line failure under operating conditions can be quite high.

In an interconnect line with blocking boundaries, the effect of electromigration is to deplete atoms on the cathode side while causing the atoms to accumulate at the anode. This results in the build-up of tensile stress at the cathode and compressive stresses at the anode end [1]. When the tensile stress at the cathode exceeds the critical stress necessary for void nucleation, a void will nucleate and begin to grow as illustrated in figure 1. Eventually the size of the void increases and leads to a resistance increase of the interconnect line leading to failure [2].

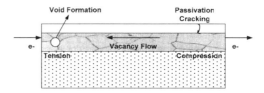

Figure 1. Illustration of electromigration, void formation and passivation cracking

MODELING OF LINE STRESS

The line structure is idealized as illustrated in the figure 2. The length of the line is L, the width of the line is 2a and the thickness of the line is h. Neither the passivation on top of the line nor the substrate below the line is shown. Atomic transport is assumed to occur along the top (and/or bottom) surface of the line. The thickness of the interfacial region in which transport occurs, denoted as δ, is greatly exaggerated in the Figure 2. [3].

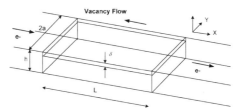

Figure 2. Schematic diagram of a simple interconnect line structure.

For the present analysis, the transport along the edges of the line is ignored. The atomic flux along the top and/or bottom interfaces in the presence of both a stress gradient and electron wind may be expressed as equation 1 [4]:

$$J^i_a = \frac{D_i}{kT}\left[\frac{\partial \sigma^i_n}{\partial x} + \frac{F_e}{\Omega}\right]_,$$

(1)

where i stands for either the top or bottom surface of the line. The atomic diffusivity is given by

$$D_i = D^o_i \exp\left(-\frac{Q_i}{kT}\right)_,$$

(2)

where Q_i is the activation energy for diffusion along the interface and D^o_i is the pre-exponential constant. As discussed below, we consider the interfacial diffusion to be bounded by the limits of grain boundary diffusion at the lower end and surface diffusion at the upper end. Also, F_e is the electron wind force in the direction of the current flow given by

$$F_e = -eZ^* \rho j \tag{3}$$

where eZ^* is the effective charge on the ions, ρ is the resistivity of the metal line and j is the current density [4,5].

In the present treatment the mass and/or vacancy flow into or out of the crystal lattice is ignored. The accumulation of mass at the interface may be computed using a mass balance. This leads to the expression for the volume of mass accumulation V at a given point along the interface

$$\frac{\partial V}{\partial x} = -2a\Omega\delta\frac{\partial J_d^i}{\partial x} \tag{4}$$

where Ω is the atomic volume. The accumulation of matter at any point along the interface causes the two sides of the interface (the line and the passivation, for example) to be displaced away from each other, leading to a pressure in the interface. The insertion of matter at an interface can be treated like a pressurized crack of length 2a [5]. According to this treatment the form of the normal displacements at the interface is

$$u_z(y) = 4\left(\frac{p}{M}\right)\sqrt{a^2 - y^2} \tag{5}$$

Where p is the pressure in the interface and M is the plane strain elastic modulus of the surrounding material. Also, the corresponding area of the opened crack is

$$A = 4\pi a^2\left(\frac{p}{M}\right) \tag{6}$$

The gradient in the rate of volume accumulation V at a given point along the interface is simply the rate of change of the opened area given by

$$\frac{\partial \dot{V}}{\partial x} = -2a\Omega\delta\frac{\partial \dot{J}_d^i}{\partial x} = \dot{A} \tag{7}$$

where the area displacement rate is,

$$\dot{A} = 4\pi a^2 \frac{1}{M}\frac{\partial p}{\partial t} \tag{8}$$

But since the normal stress in the interface is $\sigma_n^i = -p$ (tension is negative pressure) it follows that

$$\dot{A} = -4\pi a^2 \frac{1}{M}\frac{\partial \sigma_n^i}{\partial t} \tag{9}$$

Thus, combining equations 6 and 9 the governing equation is

$$4\pi a^2 \frac{1}{M} \frac{\partial \sigma_n^l}{\partial t} = 2a\Omega\delta \frac{\partial J_a^l}{\partial x},$$ (10)

and combining with equation (2),

$$\frac{2\pi a}{\Omega\delta M} \frac{\partial \sigma_n^i}{\partial t} = \frac{\partial}{\partial x} \frac{D_i}{kT} \left[\frac{\partial \sigma_n^i}{\partial x} + \frac{F_e}{\Omega} \right].$$ (11)

Equation 11 is the partial differential equation (PDE) for computing the stress evolution in the line. This PDE is solved numerically in the MathCAD tool that has been developed. The MathCAD and numerical implementation of the PDE are not discussed.

Initially the stress in the line is assumed to be zero. Then, as current flows in the line, a stress develops according to equation 11. Naturally the stresses develop first at the ends of the line where diffusion is blocked. Gradually a stress gradient develops everywhere in the line. Eventually, at long enough times, a linear, steady state, stress gradient develops. Typical results obtained for a simple interconnect structure are illustrated in figure 3.

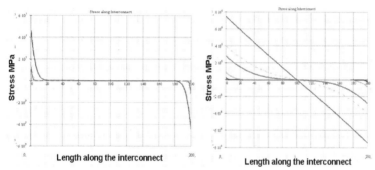

Figure 3. Stress (MPa) evolution in an interconnect line 70 microns long (at T= 573K and times of 30, 6, 3, 0.3 and 0.03 hours at a current density of 25 mA per square micron.

The most uncertain parameter in the model is the diffusivity in the interface. To estimate the stresses it is assumed that the interfacial diffusivity is bounded by grain boundary and surface diffusivity. Naturally, the stresses develop more quickly if surface diffusion is assumed and more slowly if grain boundary diffusion is assumed. The interfacial diffusivity might be lower than grain boundary diffusivity but it is not likely any higher than surface diffusion.

EXPERIMENTAL DATA

Existing electromigration Time to Fail (TTF) data from previous studies conducted by other groups [6] is used. Electromigration TTF data was available for experimental samples with known structure geometries and test conditions. Calculated line stresses were compared with available TTF data for the interconnect lines. TTF to line stress correlation was investigated for various sample geometries and test conditions.

RESULTS

Simulation runs were performed using the developed numerical scheme using MathCAD. Stress calculations were simulated for units with known geometry, electromigration stress conditions and TTF. For the calculations it was assumed that the interface can be treated like a free surface of Cu.

Table I. Line stress calculations for units with known geometries, test conditions and TTF data. (Measured TTF data by A.V. Vairagar et al, Microelectronics Reliability **44** (2004) 747–754.)

Geometry (microns)			Test Conditions		Experimental Data	Calculated Data
Type	L	W	J 10^9 A/m^2	Temp C	Measured TTF (Hrs)	Calculated Stress (MPA) at measured TTF
A Narrow	800	0.28	0.8	300	532.9	166.0
	800	0.28	0.8	325	246.1	161.0
	800	0.28	0.8	350	136.2	167.0
	800	0.28	1.2	350	79.33	190.0
	800	0.28	1.5	350	62.56	210.0
B Wide	800	0.7	0.8	300	510.7	101.0
	800	0.7	0.8	325	271.1	105.0
	800	0.7	0.8	350	120.5	98.0
	800	0.7	1.2	350	73.33	112.0
	800	0.7	1.5	350	54.56	120.0

Effect of geometry and temperature on line stress

For the narrow geometry samples stressed at the same current density, the calculated line stress at TTF is independent of electromigration temperature.
Wider geometry units also show a similar trend in line stress build up that does not vary with temperature for a particular stress current density. However, compared to the narrow units, the line stress at TTF is much lower for the wider units for electromigration under similar conditions. One explanation for the lower simulated line stress build up at TTF for the wider units could be that wider units undergo a much larger mass transport in less time when compared to the narrow units and therefore the line stress levels need not build up to the same levels at TTF.

Effect of geometry and current density on line stress

For a fixed geometry and temperature, higher electromigration current densities result in lower TTF, as expected, but the calculated line stress at TTF is higher for higher electromigration current densities. However, compared to the narrow geometry samples, the line stress at TTF is again lower for wider units with the same electromigration temperature and current densities. Although we can explain the lower stress for wider lines by making the mass transport argument again, it is not readily clear as to why the stress at TTF is higher for higher electromigration current densities and with fixed geometry and temperature. More analysis is needed. The effect of geometry, temperature and current density on line stress is summarized in table III.

Table II. The effect of geometry, temperature and current density on line stress

Geometry A Narrow	In general stress at TTF is higher than Type B	Same J Different T	Stress at Fail is independent of T
		Same T Different J	Stress at Fail is higher for higher J
Geometry B Wide	In general stress at TTF is lower than Type A	Same J Different T	Stress at Fail is independent of T
		Same T Different J	Stress at Fail is higher for higher J

CONCLUSIONS

In this work we investigated the correlation of measured electromigration TTF with calculated line stress. Line stress models are physically based, in contrast to the empirically based Black's law. If EM TTF correlates with line stress, we have a means for predicting line TTF through stress calculations.

In general, for a particular line geometry, the calculated line stress at TTF is constant and independent of electromigration temperature and furthermore line stress levels at TTF seem to be less for wider geometries. Line stress at TTF is higher for higher electromigration current densities and with fixed geometry and temperature. Further investigation is needed to understand the effect of current density on line stress at the point of failure.

ACKNOWLEDGEMENTS

We gratefully acknowledge Professor David M. Barnett of Stanford University for his help with the analysis of stability of our numerical implementation.

REFERENCES

1. C. Herring, J. Appl. Phys. **21**, 437 (1950).
2. Le, H.A.; Ting, L.; Tso, N.C.; Kim, C.-U, Analysis of the reservoir length and its effect on electromigration lifetime; Journal of Materials Research, **vol.17**, no.1, Jan. 2002. p. 167-71. Journal Paper.
3. Nix, W.D. Arzt, E , On void nucleation and growth in metal interconnect lines under electromigration conditions; Metallurgical Transactions A (Physical Metallurgy and Materials Science), **vol.23A**, no.7, July 1992. p. 2007-13. Conference Paper.
4. H. B. Huntington and A. R. Grone, J. Phys. Chem. Solids **20**, 76 ~1961
5. R.J. Gleixner and W.D. Nix, A physically based model of electromigration and stress-induced void formation in microelectronic interconnects, J. Appl. Phys., **86**, 1932-1944 (1999)
6. A.V. Vairagar, S.G. Mhaisalkar, and Ahila Krishnamoorthy, Electromigration behavior of dual-damascene Cu interconnects–Structure, width, and length dependences, Microelectronics Reliability **44** (2004) 747–754.

Stress-induced Void Formation in Passivated Cu Films

Dongwen Gan, Bin Li and Paul S. Ho
Laboratory for Interconnect and Packaging, The University of Texas at Austin,
Austin, TX 78712-1063, U.S.A

ABSTRACT

In this paper, we investigated void formation in passivated Cu films focusing on the kinetics of void formation under isothermal annealing as a function of temperature. Interestingly, we found that the kinetics of void formation in Cu films is consistent with that observed in Cu lines, which is driven by the combined effect of thermal stress and mass transport resulting in a peak growth rate at about 250°C. To analyze the observed results, we have calculated the stress state at the <111>/<200> Cu grain boundary to demonstrate the existence of a localized triaxial stress state as a result of elastic anisotropy. To account for void density, x-ray analysis was performed to measure the grain texture using inverse pole figure plot and the result can account for the void density observed. A kinetic model was used to analyze void growth under isothermal annealing. A threshold stress of about 40MPa was deduced for void growth in passivated Cu films with an activation energy of 0.75 eV.

INTRODUCTION

Thermal stress induced by thermal expansion mismatch is commonly observed in metal films and line structures. In interconnects, the confinement of the metal lines by surrounding dielectrics and the Si substrate yields a triaxial stress state. Relaxation of the hydrostatic stress leads to mass transport and void formation, raising serious reliability concern for interconnects [1]. Stress-induced void formation has been observed in Cu lines and interconnect structures under thermal cycling and isothermal annealing [2-4], where void formation is driven by time-dependent relaxation of the hydrostatic tensile stress. Recently, Shaw *et al* reported void formation in passivated Cu lines with line width ranging from 10 to 30 microns where the stress state comprised of predominantly biaxial tension [5]. Void formation under a biaxial stress state is unexpected and has seldom been observed in Al films or wide line structures. The phenomenon was attributed to the elastic anisotropy of <111> and <200> Cu grains which gives rise to local stress concentration at the grain boundary. While this study revealed an interesting observation, the kinetics of void formation has not been reported.

In this paper, we extended the study to focus on the kinetics of void formation under isothermal annealing as a function of temperature. Interestingly, we found that the kinetics of void formation in Cu films is consistent with that observed in Cu lines driven by the combined effect of thermal stress and mass transport reaching a peak rate at an intermediate temperature. To analyze the observed results, we used finite element analysis to calculate the stress state at the <111>/<200> Cu grain boundary and showed the existence of a localized triaxial stress state as a result of the elastic anisotropy. To account for void density observed, x-ray analysis was performed to measure grain texture using inverse pole figure plots.

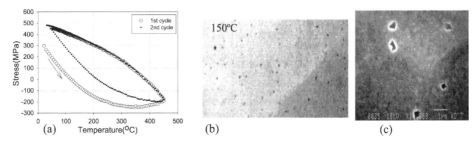

(a)   Temperature(°C)   (b)   (c)

Figure 1. (a) The thermal stress in the film in two thermal cycles; (b) Optical micrograph of SIVs; (c) The SEM image of SIVs after the removal of the passivation.

EXPERIMENTAL DETAILS

The copper film was prepared on a 720-micron thick (100) silicon wafer. A layer of SiN (50nm) was deposited on the wafer by chemical vapor deposition (CVD), followed by subsequent depositions of a diffusion barrier layer and a Cu seed layer. The rest of the Cu film was deposited using an electroplating (EP) process to yield a total thickness of 0.6μm. The film was finally passivated with a bi-layer of 50nm SiN and 200nm TEOS.

An optical microscope (OLYMPUS AH-2) equipped with a camera and a heating stage was employed for in-situ observation of void formation and to measure the void density in the film. SEM was used for void size measurement and topography analysis after etching off the passivation layer by HF acid. A bending beam system with a heating stage in a vacuum chamber [6] was used for thermal stress measurements during thermal cycling and isothermal annealing of the film. An ABAQUS finite element program was used to evaluate the local stress gradients in Cu films with an elastic model set up using 20-node quadratic brick elements. X-ray diffraction (XRD) was used to get texture information in Cu films. An inverse pole figure of the Cu film was obtained based on the (111), (200) and (220) pole figures.

RESULTS AND DISCUSSION

In-situ observation of void formation during thermal cycling

Void formation in the passivated Cu film was recorded under the microscope during thermal cycling from room temperature to 450°C with a ramping rate of 5°C/min. Typical film stress during thermal cycling is shown in Figure 1(a). Before cooling down the temperature to 315°C from 450°C in the first thermal cycle, no void was observed. After that, voids began to appear and the void density increased until the temperature decreased below 150°C. The optical micrograph in Figure 1(b) shows the voids observed at 150°C. Upon heating in the second cycle, the void density remained unchanged until the temperature reached 280°C. Upon further temperature increase, the voids gradually became faint and disappeared at 330°C, where the stress became compressive, indicating that void formation is indeed driven by tensile stresses. SEM observation shown in Figure 1(c) revealed that the voids have multiple edges, suggesting that void growth may be microstructure dependent. We noted that the voids did not close as soon as the stress became compressive, reflecting the kinetic nature of mass transport during void formation.

Kinetic analysis of void growth during isothermal annealing

The *in-situ* observation under thermal cycling showed that the void formation depends on stress and temperature. To study the kinetics of the void formation in passivated Cu films, isothermal annealing of the Cu film was performed at different temperatures from 100°C to 300°C at intervals of 50°C. To induce a tensile stress in the Cu film, the samples were first heated up to 400°C or 450°C then held it at a temperature of interest under isothermal annealing for void formation study. We found that the amount of voids and the growth kinetics depended on the maximum heating temperature and decreased considerably when the sample was heated up to 400°C instead of 450°C. That can be attributed to a decrease in the stress driving force for void formation due to a lower end-point temperature, which can be seen by comparing the stress levels in Fig.1a and Fig.2b. After cooling down from 400°C, the film was held at various temperatures to measure void growth kinetics under isothermal annealing. The void density was determined using an optical microscope and the void size was measured using SEM after 20 hrs. of isothermal annealing and results are shown in Fig.2a. Both the void density and void size were found to peak at an intermediate temperature, at about 150°C for void density and 250°C for void size.

During void growth under isothermal annealing, atoms diffuse away from the void with the atomic flux driven by a stress gradient as given by the Nernst-Einstein equation:

$$J = -\frac{D_{eff}}{kT}\nabla\sigma \qquad (1)$$

where D_{eff} is the effective diffusivity and $\nabla\sigma$ is the stress gradient driving mass transport for void formation. McPherson and Dum first proposed a power-law creep as the mechanism for stress relaxation driving void growth in Al lines [7]. Accordingly, the void growth rate can be expressed as:

$$R = C(T_0 - T)^N \exp(-\frac{Q}{kT}) \qquad (2)$$

where R is the void growth rate, T_0 the stress-free temperature, T the annealing temperature, N the "creep exponent", k the Boltzmann's constant and Q is the activation energy. This model was used to study stress voiding in Cu interconnect structures where the creep rate was deduced by measuring the resistance increase of the interconnect structures [4].

In our study, we applied a finite element analysis to evaluate the stress gradient at the void vicinity and found a linear dependence on the film stress (see next section). We propose therefore a linear diffusional creep instead of a power-law creep as a driving force for void formation in Cu films. Accordingly, Eq.2 can be expressed to give the following void growth kinetics:

$$V \propto d^3 = A\frac{(\sigma - \sigma_0)}{T}\exp(-\frac{Q}{kT}) \qquad (3)$$

where V is the volume of the void, which is proportional to the cube of d, the measured one-dimensional size. A is a material constant, σ_0 is the threshold stress for void formation and σ is the film stress. Equation 3 is similar to the Nabarro-Herring or Coble rate equations for the diffusional creep, but with different meaning of the material constant A. Using the stress determined in Fig.2b and the size in Fig.2a, equation 3 was deduced to be

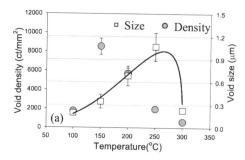
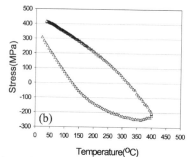

Figure 2. (a) Temperature dependence of the void density and void size in isothermal annealing cooling down from 400°C. (b) The correlative thermal stress of the film in a thermal cycle.

$$d^3 = 8.838 \times 10^7 \frac{(\sigma - 40)}{T} \exp(-\frac{0.75}{kT}) \qquad (4)$$

where σ is in MPa and d is in μm. By curve fitting, we found a threshold stress of 40MPa and an activation energy of 0.75eV for void growth. The growth kinetics as described by Eq.4 is plotted in Fig.2a as a solid line which agrees well with the observed void size d as a function of annealing temperature. The activation energy of 0.75eV for void growth suggests that the interface diffusion may play an important role in contributing to void growth in passivated Cu films. In Figure 2(a) the void density is shown to dramatically decrease at 100°C. It is possible that the voids were too small to be seen under the microscope due to the small growth rate at a low temperature. That is consistent with the observation that void formation became very slow after cooling down to below 150°C. If the product of the void density and the void size is used to obtain the total amount of void formation in the Cu film, the maximum "damage" is found to occur at an annealing temperature of ~ 240°C.

Local tri-axial stress state and stress gradients in Cu films

The *in-situ* observation of the void formation indicated that the voids formed under tensile stress and closed under compressive stress. As mentioned above, Cu crystal shows mechanical anisotropy that might result in the local tri-axial stress state and was believed to be one of the reasons for the void formation in Cu wild lines [5]. As an example, the FEA result of the stress in the vicinity of the boundary between a <111> and a <100> Cu line under a thermal loading of -100K from a stress-free temperature is shown in Figure 3, indicating the presence of a localized tri-axial stress state and stress gradients at the boundary. The stiffness tensors in the <100> and the <111> crystal coordinate systems [8,9] were used in this model.

In Figure 4, we show the inverse pole figure of the Cu film after thermal cycling from room temperature to 400°C. In addition to the strong (111) texture, the film has other texture components, i.e. (100) and (511). Considering the difference in modulus in different orientations, the texture components near <100> pole can give rise to a larger stress concentration at the grain boundary with <111> grains. The fraction of the <100> and the <511> grains to the <111> grains is estimated to be ~5%. Assuming each <100> or <511> grain provides a potential nucleation site for the void formation, the density of potential nucleation sites can be estimated to

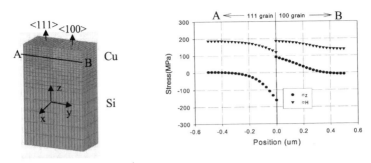

Figure 3. FEA result of the local stress in the vicinity of the grain boundary between a <111> and a <100> grain.

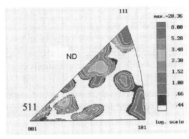

Figure 4. Inverse pole figure of the Cu film.

be ~5.5×10³ counts/mm² using a grain size of 3μm. This density is comparable with the observed void density of 2000 counts/mm² to 8500 counts/mm². This indicates that certain special grain boundaries can be favorite nucleation sites for void formation as a result of the localized stress concentration. Another mechanism causing the tri-axial stress state in passivated Cu films is related to the stress relaxation controlled by the interface diffusion. The relaxation of a tensile stress by the diffusion of atoms from the Cu/cap layer interface to grain boundaries will result in the tri-axial stress state at the grain boundary/interface junction [10]. This makes the junction to be a favorite site for void formation. Finally, we performed a finite element analysis to calculate the stress gradient near a void in a passivated Cu film. Considering a rectangular void going through the passivated Cu film, the hydrostatic stress in the void vicinity as a function of the void size was evaluated and the result is shown in Figure 5. The hydrostatic stress can be seen to increase from the void surface into the film but the stress gradient near the void surface decreases upon void growth.

CONCLUSIONS

Stress-induced void formation was studied in passivated Cu films during thermal cycling and isothermal annealing and correlated to the thermal stress behavior measured using a bending beam system. Voids were observed to form under tensile stress and close under compressive stress in Cu films. A kinetic model, similar to the diffusional creep model was found to account well with the observed void growth rate. Interestingly, we found that the kinetics of void

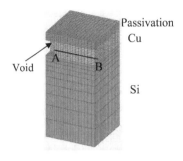 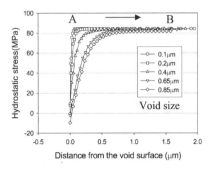

Figure 5. FEA results of the local stress in the vicinity of voids with different sizes.

formation in Cu films is consistent with that observed in Cu lines, which is driven by the combined effect of thermal stress and mass transport resulting in a peak growth rate at about 250°C. To analyze the observed results, we have calculated the stress state at the <111>/<200> Cu grain boundary to demonstrate the existence of a localized triaxial stress state as a result of elastic anisotropy. To account for void density, x-ray analysis was performed to measure the grain texture using inverse pole figure plot and the result can account for the void density observed. A kinetic model was used to analyze void growth under isothermal annealing. A threshold stress of about 40MPa was deduced for void growth in passivated Cu films with an activation energy of 0.75 eV.

ACKNOWLEDGMENTS

This work was supported in part by the Advanced Technology Program of Texas Higher Education Coordinating Board and the Semiconductor Research Corporation.

REFERENCES

1. M.A.Korhonen, P.Borgesen, and Che-Yu Li, MRS Bulletin/July 1992, P61-68.
2. J.A.Nucci, Y.Shacham-Diamand, and J.E.Sanchez, Jr., Appl. Phys. Lett. 65, 3585 (1995).
3. T.Oshima, T. Tamaaru, K. Ohmori, H. Aoki, H.Ashihara, T. Saito, H.Yamaguchi, M.Miyauchi, K.Torii, J.Muarata, A.Satoh, H.Miyazaki and K.Hinode, IEDM, 123 (2000).
4. E.T.Ogawa, J.W.McPherson, J.A.Rosal, K.J.Dickerson, T.-C.Chiu, *et al*, International reliability physics symposium, Piscataway, NJ, P312-321 (2002).
5. T.M.Shaw, L.Gignac, X-H.Liu, and R.R.Rosenberg, Sixth International Workshop on Stress-induced Phenomena in Metallization. Ithaca, NY, USA, 25-27 July 2001.
6. J.H Zhao, T. Ryan, P.S. Ho, *et al*, J. Appl. Phys. Sept. 88(5), 3029-38 (2000).
7. J.W. McPherson and C.F. Dunn, J. Vac. Sci.& Tech, B5(5), 1321-25(1987).
8. R.W. Hertzberg, Deformation and fracture mechanics of engineering materials; John Wiley and Sons, 1989.
9. J. Kasthurirangan, Ph. D thesis, The Univ. of Texas at Austin, 1998.
10. R.Huang, D.W.Gan, P.S.Ho, J. Appl. Phys. (accepted for publishing, May 2005)

Mater. Res. Soc. Symp. Proc. Vol. 875 © 2005 Materials Research Society

Stress Generation in PECVD Silicon Nitride Thin Films for Microelectronics Applications

M. Belyansky, N. Klymko, A. Madan, A. Mallikarjunan, Y. Li, A. Chakravarti, S. Deshpande, A. Domenicucci, S. Bedell, E. Adams, J. Coffin, L. Tai, S-P. Sun[1], J. Widodo[2] and C-W Lai[2]
IBM Semiconductor R&D Center, Hopewell Junction, NY 12533, USA
[1]Advanced Micro Devices, Inc; Hopewell Junction, NY 12533, USA
[2]Chartered Semiconductor Manufacturing; Hopewell Junction, NY 12533, USA

ABSTRACT

Thin SiN films deposited by plasma enhanced chemical vapor deposition (PECVD) have been analyzed by a variety of analytical techniques including Fourier Transform Infrared Spectroscopy (FTIR), X-ray reflectivity (XRR), and Rutherford Backscattering Spectrometry/Hydrogen Forward Scattering (RBS/HFS) to collect data on bonding, density and chemical composition respectively. Both tensile and compressive SiN films have been deposited and analyzed. Mechanisms of stress formation in SiN thin films are discussed. It has been found that amount of bonded hydrogen as detected by FTIR is higher for compressive films compared to tensile SiN films. Amount of bonded hydrogen in a film is correlated well with tensile stress. Effect of deposition temperature and other process parameters on stress have been studied. Exposure of SiN films to elevated temperature after deposition lead to increase in tension and degradation in compressive stress. New approaches to stress generation in thin films like creation of multilayer film structures have been delineated.

INTRODUCTION

Strain generation in silicon is becoming one of the major knobs in boosting performance of the leading edge metal-oxide-semiconductor field effect transistor (MOSFET) technology [1]. Tensile strain in the silicon channel is beneficial for electron mobility, while compressive stress increases hole mobility. While the basic physics behind stress induced mobility enhancement has been known for many years [2], only recently has stress engineering been incorporated into state of the art microprocessor logic technology. Strain in the silicon channel can be achieved by various techniques. For example, biaxial strain could be applied by depositing silicon on relaxed SiGe substrate, which is beneficial for both electron and hole mobility [3]. However, due to substantial technological difficulties biaxial strain has not been implemented yet in any logic technology. Another way of creating compressive strain in the silicon channel is introduction of an embedded SiGe layer close to the channel. The embedded SiGe approach has been successfully implemented [1], however it is technologically complex and is limited to hole mobility enhancement only.

This work explores ways of increasing the intrinsic stress level (both tensile and compressive) in the PECVD Silicon Nitride (SiN) films using various analytical techniques. Substantial increase in device speed has been achieved by application of these highly stressed SiN films, which in turn has been successfully incorporated into dual stress liner (DSL) technology benefiting both nFET and pFET type of transistors [4].

EXPERIMENTAL DETAILS

Commercially available semiconductor equipment has been used for deposition of PECVD SiN thin films with basic ammonia and silane chemistry. Films have been deposited on 300mm silicon wafer substrates with typical deposition parameters as follows: pressure 1-10 torr, silane flow 100-500 sccm, ammonia flow 1000-5000 sccm, high frequency plasma (13.56Mhz) and low frequency plasma (~ 400 kHz) within the 100-1000W RF power range. All films have been deposited at 400^0C. Film thickness is about 500A as measured by optical ellipsometry and verified by XRR. Stress is measured by a commercial laser beam scanning tool to determine silicon wafer bow before and after film deposition. After that film stress is calculated using Stoney's equation [5]. Conventional notation of tensile stress being positive and compressive stress as negative is used.

FTIR analysis is performed on Nicolet Model 870 Spectrometer. X-ray reflectivity (XRR) data was carried out using a BEDE High-Resolution X-ray diffractometer using a monochromatic CuKa radiation source. Data was collected from 600" to 10,000" in step sizes of 20" with a count time of 5 secs. Data was fitted using a multilayered model and simulated using commercially available BEDE Mercury software. The density and thickness of the layers were obtained from the best fits. RBS/HFS analysis was performed on a NEC Model 5SDH Spectrometer. Transmission Electron Microscopy (TEM) was done on JEOL 2010 field emission microscope. TEM samples were prepared using industry standard focussed ion beam technique.

DISCUSSIONS

A variety of compressive and tensile PECVD SiN films have been deposited at 400^0C using various silane/ammonia flow ratios and plasma powers. The main difference in compressive and tensile SiN film deposition conditions is the use of a high frequency plasma component only for tensile films, and for compressive films, mixing both low and high frequency plasmas. Films were analyzed with X-ray Photoelectron Spectroscopy (XPS), Auger Electron Spectroscopy (AES), Rutherford Backscattering (RBS), FTIR, XRR and TEM.

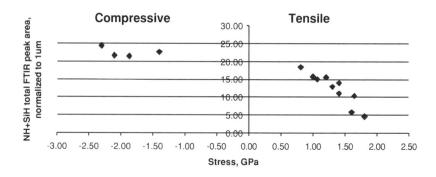

Figure 1. FTIR total bonded hydrogen correlation with film stress of as deposited films

Among those techniques only FTIR, XRR and RBS provided meaningful information on film composition and structure, while XPS and Auger could not detect any measurable difference between compressive and tensile films. Average film composition according to AES and XPS is very close to that of stoichiometric Si_3N_4. RBS was used for Si and N composition and HFS for hydrogen content. It was found that all as deposited SiN films contain about 10-15% of hydrogen similar to the estimate done by FTIR using published absorption coefficients [6]. The data suggests that all or most of the hydrogen in PECVD films is bonded.

The FTIR spectra of compressive and tensile films show significant differences. Both films exhibit NHx stretching and bending modes, however, the tensile film has relatively more NH2 bending intensity while the compressive film has more has overall higher NHx stretching intensity. Additionally, the tensile film shows significant SiH bonding, while the compressive film shows significantly lower SiH content. Also, the compressive film shows significantly higher SiN peak area than the tensile film.

Figure 1 shows the correlation between total bonded hydrogen measured from the combined NHx and SiH FTIR peak areas, with stress. Peak assignment of SiN FTIR spectra is based on the previous work of Yin and Lucovsky [7-9]. Compressive PECVD films have more hydrogen than tensile films and the most tensile films have the lowest amount of bonded hydrogen (combination of NHx and SiH). In compressive films the SiH content is generally very low so that the bonded hydrogen is dominated by NHx groups. In tensile films the SiH content generally tracks with the NHx content, in following a downward trend with increasing tensile stress.

While the SiN peak area in compressive films is relatively constant as a function of stress (similar to the bonded H content), in tensile films the SiN peak area increases with increasing stress, with concurrent decrease in total bonded hydrogen. This trend is shown in Figure 2.

Effectively, changing deposition parameters at 400^0C produce highly tensile films with

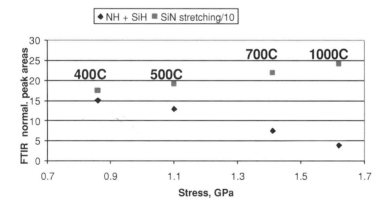

Figure 2. Change in bonded hydrogen (N-H+Si-H) and Si-N stretch peak areas after film anneal

less hydrogen content. Loss of hydrogen is typically associated with high temperature anneal of SiN films, so few compressive and tensile films have been exposed to temperatures much higher than original deposition temperature (in argon ambient) in order to amplify the effect of a hydrogen loss. Since the amount of bonded hydrogen is directly related to tensile SiN stress, high temperature anneal increased tension and decreased SiN film compression as expected. Figure 2 shows change in bonded hydrogen and Si-N peak area in annealed films. A significant decrease in Si-H and N-H stretch peak intensity has been observed. At the same time the intensity of Si-N peak increased suggesting a rearrangement of Si and N atoms concurrent with loss of H, resulting in the formation of new Si-N bonding. Ring like structures have been suggested previously for annealed tensile SiN films [10] or dense stoichometric Si_3N_4 framework for films deposited at high temperature [11]. Our initial data can not support either structural model. Work is in progress to build structural models of compressive and tensile PECVD SiN.

Increase in tension and loss of hydrogen during anneal was accompanied by film thickness shrinkage and subsequent increase in density (as shown in Table I). At the same time, compressive stress decreases linearly with an increase in anneal temperature suggesting that breaking of N-H and Si-H bonding is detrimental for compression.

Table I. Density and shrinkage data for annealed tensile films

Temperature (^{0}C)	Stress (GPa)	Density (g/cc)	Shrinkage (%)
400	1.0	2.4	0 (as deposited)
500	1.3	2.49	2
700	1.6	2.57	10

Mainstream optimization of SiN deposition process parameters did not produce further increase in tensile or compressive stress. New ideas are needed in order to increase intrinsic stress in SiN. One of the approaches is to deposit a multilayer SiN film, which according to recent report by Goto [12] produces a higher silicon channel strain.

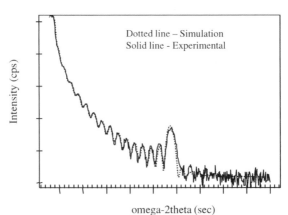

omega-2theta (sec)

Figure 3. XRR spectrum and simulation of a 16 layer SiN PECVD film

Multilayer films have been deposited by interrupting the deposition about one minute (switching of plasma and gases). Each individual layer is on the order of 10-50Å thick. It was found that blanket film stress increases linearly with the number of layers. Hence, a multilayer film produces higher stress than an identical film deposited without interruption, as a single layer. The presence of an interface increases stress level.

XRR technique was found to be most useful for measuring the number of layers as well as thickness and density of an individual layer. Figure 3 shows a sample XRR spectrum of a 16 layer film. The simulation profile accurately predicted the number of layers in the film. Figure 3 shows the experimental (solid) and simulated (dotted) data for an as-deposited multilayered nitride film. The widely spaced fringe spacing corresponds to the total thickness of the film. The peak between 5000" and 6000" is an indication of the presence of both: interfaces and film layers in the multilayered structure. The thickness of each layer is estimated from the position of this peak. This thickness when multiplied by the number of layers, matches the total thickness of the film.

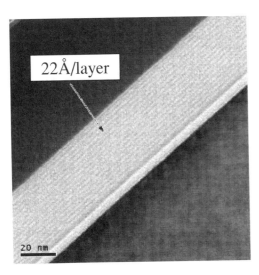

Figure 4. TEM of a 16 layer SiN film.

It was found that multilayer structures increase both tensile and compressive stress. For example a 5 layer SiN film showed about 10% increase in stress level, while a 20 layer film showed almost 20% increase comparing with a single layer film of the same thickness. The example of a multilayer compressive film is shown in Figure 4. Sixteen layers and interfaces are clearly seen on this TEM image, each of about 20-25Å thick. The very thin SiN layer relaxes and a subsequent layer deposition seem to produce a more strained film compared to an in-situ deposited single layer film of the same combined thickness. Single and multilayer films exhibit very similar FTIR bonding trends as shown in Figures1 and 2.

CONCLUSIONS

The study shows that the tension level in PECVD SiN films is inversely proportional to the amount of Si-H and N-H bonding. Decrease in film hydrogen content and formation of new Si-N bonds is associated with increase in tension level. Such an increase in film tensile stress could be achieved by process parameter optimization at fixed temperature or by annealing a film above the initial deposition temperature. Mutlilayer films are found to increase both tensile and compressive stress. While the exact reason for such an increase is not determined, it is believed that film growth on an interface produces a more strained film than a single layer film of the same thickness.

ACKNOWLEDGMENTS

This work is supported by the independent two alliance programs for SOI technology development and Bulk CMOS technology development.

REFERENCES

1. S. E. Thompson, M. Armstrong, C. Auth et al., *IEEE Electron Device Letters*, **25**, 4, 191 (2004); *IEEE Trans. Electron Devices*, **51**, 11, 1790 (2004)
2. C. J. Smith, *Phys. Rev*, 94, 42 (1954)
3. K. Rim, J. Chu, H. Chen et.al., *VLSI Symposyum Tech. Digest*, 98, (2002)
4. S. Yang, R. Malik, S. Narasimha, Y. Li, R. Divakaruni, et al., *IEEE IEDM Conference Proceedings*, **28.8**, 1075 (2004)
5. P. Ambree, F. Kreller, R. Wolf and K. Wandel, *J. Vac. Sci. Technol.*, **B11**, 614 (1993)
6. W.A. Lanford and M.J. Rand, *J.Appl. Phys.*, **49**, 2473 (1978)
7. Z. Yin and F.W. Smith, *Phys. Rev. B.*, **42**, 3666 (1990)
8. D.V. Tsu, G. Lucovsky and M.J. Mantini, *Phys. Rev. B.*, **33**, 7069 (1986)
9. P.D. Richard, R.J. Markunas, G. Lucovsky, GG. Fontain, A.N. Mansour and D.V. Tsu, *J. Vac. Sci. Technol.* **A3**, 867 (1985)
10. Y. Saito, T. Kagiyama and S. Nakajima, *Jpn. J. Appl. Phys.*, **42**, L1175 (2003)
11. Y.Toivola, J. Thurn and R.F. Cook, *J. Appl. Phys*, **94**, 6915 (2003)
12. K. Goto, S. Satoh, S. Fukuta., et al., *IEEE IEDM Conference Proceedings* 7803 (2004)

Deformation, Growth and
Microstructure in Thin Films

Mater. Res. Soc. Symp. Proc. Vol. 875 © 2005 Materials Research Society   O12.2

The Effect of Microstructural Inhomogeneity on Grain Boundary Diffusion Creep

Kanishk Rastogi and Dorel Moldovan
Department of Mechanical Engineering, Louisiana State University, Baton Rouge, LA 70803

ABSTRACT

Stress concentration at grain boundaries (GB), a phenomena arising from microstructural inhomogeneity, is an important factor in determining the mechanical properties of polycrystalline materials. In this study we use mesoscopic simulations to investigate characteristics of the deformation mechanism of grain-boundary diffusion creep (Coble creep) in a polycrystalline material. The stress distribution along the grain boundaries in a polycrystalline solid under externally applied stress is determined and the mechanism of how topological inhomogeneities introduce stress concentrations is investigated. Microstructures with inhomogeneities of various sizes and distributions are considered and their effect on the stress distribution and creep rate is quantified.

INTRODUCTION

In high temperature Coble creep deformation of fine-grained materials atoms are transported by grain boundary diffusion from boundaries in compression to those in tension [1]. The macroscopic creep behavior of a polycrystal undergoing diffusional creep is complex and generally governed by overall characteristics of the microstructure in which inhomogeneities play a significant role [2-7]. Stress concentration at grain-boundaries (GBs), a phenomenon arising from microstructural inhomogeneity, is an important factor in determining the mechanical properties of polycrystalline materials, such as crack nucleation or creep fracture in metals at elevated temperatures [8,9]. Although such stress concentrations may be of secondary importance for propagating the failure mechanism, their role is critical, for example, when a preferred site is selected for dislocation emission from a GB or for the nucleation of cavities or cracks along the GBs. At elevated temperatures, the very prominent GB sliding process prevents stress concentration at the triple junctions (the preferred site at low temperature). Instead, as illustrated below, there is a shift of the regions with higher stress away from the triple junctions into the GBs. In the present study we use a mesoscale simulation approach pioneered by Cocks [6] to simulate the stress distribution and the evolution of two-dimensional model microstructures subject to uniaxial tensile stress. Our focus is on elucidating the effect of microstructural inhomogeneity on stress distribution and on the creep rate.

SIMULATION METHODOLOGY

Following the work of Pan and Cocks [6] and Cocks and Searle [9], we briefly summarize the equations and the concepts used in our mesoscopic simulations. The diffusion of atoms along GBs is driven by the gradient of the chemical potential, which is induced by the gradient of the stress, σ, acting along each boundary. The chemical potential in the GB plane is related to the stress, σ, by the relation: $\mu = \mu_0 - \sigma\Omega$, where μ_0 is the chemical potential of an atom in a stress-free system and Ω is the atomic volume. From Fick's first law, the diffusive flux along each GB can be shown to depend on the gradient of the normal-stress component, σ_n, along the GB as follows:

$$J = \frac{D_{GB}\delta\Omega}{kT}\frac{\partial\sigma_n}{\partial s} \quad . \tag{1}$$

Here s is the local spatial coordinate along the diffusion path, D_{GB} the GB self-diffusion coefficient, δ the diffusion width of the GB, k Boltzmann's constant and T the absolute temperature. The atoms diffusing along the GBs can be either deposited at the GBs or removed from them. These processes cause grains on either side of the boundary to move with respect to each other in a direction normal to the GB at a rate v_n. Conservation of matter requires that

$$\frac{\partial J}{\partial s} + v_n = 0 \quad . \tag{2}$$

Using Eq. (1), this can be written as:

$$v_n = -\frac{D_{GB}\delta\Omega}{kT}\frac{\partial^2\sigma_n}{\partial s^2} \quad . \tag{3}$$

Assuming also that no microcracks are allowed to open up at the GBs or triple junctions, the fluxes J_i along the three GBs meeting at a triple junction must satisfy the condition

$$\sum_{i=1}^{3} J_i = 0 \quad , \tag{4}$$

where J_i is considered positive for the fluxes flowing into and negative for those flowing out of the triple junction. Equations (1)-(4), together with specified border conditions imposed on the simulation cell, describe the deformation process of a polycrystal with any grain topology and may be implemented in a computational framework. Our computational implementation follows the approach described by Cocks [9] and is based on the Needleman and Rice [10] variational functional. Cocks demonstrated that the solution is given by the velocity field, $\{v_n\}$, and flux field, $\{J\}$, along each GB in the system. These fields minimize the functional

$$\Pi = \int_S \frac{kT}{2D_{GB}\delta\Omega} J^2 ds - \int_\Gamma T_i V_i d\Gamma \quad , \tag{5}$$

subjected to the constraint of equation (4). Here S is the total length of the GBs through which the diffusional flow takes place, Γ is the length of the border of the simulation cell moving with velocity V_i when an external stress, σ_0, is applied. $T_i = \sigma_0\Gamma_i$ is the value of the traction exerted on the segment of length Γ_i located on the simulation-cell border.

The contribution to equation (5) from each GB segment can be written explicitly in terms of the following local degrees of freedom [9]: the velocity components of the center of mass of the two grains determining the GB, the rates of rotation of the grains, the value of the diffusional flux in the middle of the GB segment and the values of the normal stresses at the ends of the GB segment. The numerical method used here to minimize the variational functional follows the standard finite-element procedure [6]. This leads to the following matrix equation:

$$[G][U] = [F] \quad , \tag{6}$$

where [G] is the generalized stiffness matrix governed by the material properties, conservation law and simulation-border conditions. [U] is the global column vector of degrees of freedom and [F] is the external force vector which contains information on the external forces applied along the border of the simulation cell. The solutions of equation (6) are obtained by using a linear solver.

SIMULATION RESULTS

Using the method described above, we determine the stress distributions along various GB paths in three polycrystalline microstructures, subjected to uniaxial tensile stress and in the presence of topological inhomogeneities consisting of individual or clustered larger grains. All three microstructures were obtained by performing abnormal grain growth on regular hexagonal microstructures in which the GB properties (energies and mobilities) of the GBs surrounding specific grains were "biased" accordingly in order to promote a larger growth rate for those grains. The effect of the overall inhomogeneity size and of the actual cluster arrangement of the larger grains on the stress distribution is analyzed.

Figure 1(a) shows a microstructure consisting of 460 regular uniform sized grains, and one equiaxed larger grain in the middle. The uniform sized grains consist of 430 regular hexagonal grains and 30 smaller four or five-sided grains bordering the central large grain. The larger grain has an area about 66.46 times larger that the area of a hexagonal grain, A_h, and it accounts for about 13.5% of the total area of the simulation system. Figures 1(b) show the stress distributions along five different GB paths. Four paths labeled A1, B1, C1 and D1 are crossing the structure perpendicular to the straining direction and one labeled E1, runs parallel to the applied stress. As expected, the stress distribution along all five paths shows both regions that are in tension and regions in compression. Comparing the five stress distributions one can see the following: i) the amplitudes of the stress distributions along the median paths A1, B1 and E1, that intersect the larger grain are larger than the amplitude along the paths C1 and D1 (those that do not touch the larger grain) ii) the compressive stresses along paths C1 and D1 are significantly smaller and act on significantly fewer GBs than along the paths A1, B1 and E1 iii) the normal stress distributions converge to the characteristic periodic distribution of the normal stress in a uniform regular structure [7] (characterized by a maximum stress which is twice the value of the externally applied stress, σ_0) faster along the paths A1, B1 and E1 than along paths C1 and D1 and iv) the maximum normal stress in the microstructure, $\sigma_n = 3.25 \, \sigma_0$, is found along path C1 at the center of a GB located about one grain diameter away from the surface of the large grain. One can rationalize the last finding by noticing that the large central grain is surrounded by smaller grains (smaller than the hexagonal matrix grains) which contribute directly to lowering the stress concentration on the surface of the larger grain.

To explore further the effect of clustering of inhomogeneities on the stress distribution we investigated the same microstructure in which the larger grain is broken up in two and in three grains respectively. Figure 2 (a) shows the microstructure in the presence of two larger grains located close to the center of the system. Each of the two larger grains has an area of about 34.1 A_h, and their combined area is about 15% of the total system area. Fig. 2(b) shows the stress distribution along five GB paths. Similar to the system presented in Fig. 1, the amplitude of the normal stress is larger along the median path A2. Moreover for this system the maximum normal stress located along a GB on path B2 is only slightly larger than the stress at the tip of one of the larger grains (see path A2) and it is smaller than the maximum stress found in Fig. 1. More details on the clustering effect can be inferred from the system presented in Fig. 3 in which a triangular cluster of three larger grains is considered. Each of the three larger grains in Fig. 3(a) has an area of about 25.4 A_h. It is interesting to notice that of the three inhomogeneous microstructures investigated the system with triangular grain cluster in Fig. 3 shows the largest normal stress $\sigma_n = 3.75 \, \sigma_0$. This is again located at the center of a GB situated one grain away from the surface of the large grain sitting at the tip of the triangular cluster (see path E3 in Fig. 3(b).

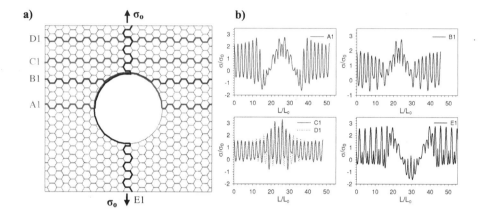

Figure 1. a) Model inhomogeneous microstructure consisting of 430 regular hexagonal grains, 30 smaller grains and one larger circular grain in the middle (A_{large}/A_{hex} = 66.5) subjected to the uniaxial tensile stress σ_0. The 30 smaller grains border the larger central grain **b)** The normalized stress profiles for the five distinctive paths highlighted and labeled: A1, B1, C1, D1, and E1 in **a)**.

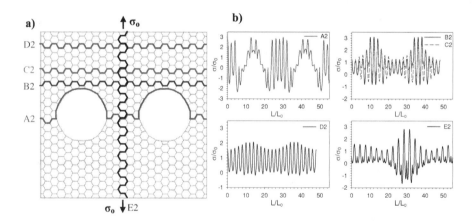

Figure 2. a) The inhomogeneous microstructure consisting of 428 regular hexagonal grains, 24 smaller grains and a two-grain cluster in the middle (A_{large}/A_{hex} = 34.3) subjected to the uniaxial tensile stress σ_0. **b)** The normalized stress profiles for the five distinctive paths highlighted and labeled: A2, B2, C2, D2, and E2 in **a)**.

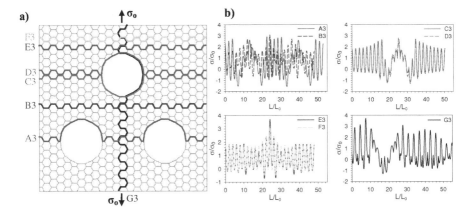

Figure 3. a) The inhomogeneous microstructure consisting of 426 regular hexagonal grains, 24 smaller grains and a three-grain cluster in the middle (A_{large}/A_{hex} = 25.4) subjected to the uniaxial tensile stress σ_0. **b)** The normalized stress profiles for the seven distinctive paths highlighted and labeled: A3, B3, C3, D3, E3, F3 and G3 in **a)**.

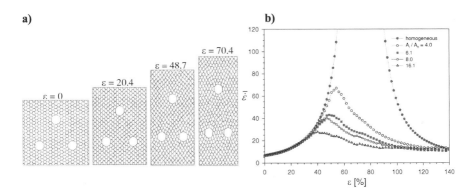

Figure 4. a) Four snapshots of the evolution of a regular hexagonal structure containing a cluster of three larger grains each with the area 8.0 times that of the regular hexagonal grains. **b)** Strain rate vs. strain for the homogeneous regular hexagonal structure and four non-uniform microstructures each containing a similar three larger-grains cluster.

Next we examine the effect of microstructural inhomogeneity on the deformation rate. Figure 4(a) shows the creep deformation of an inhomogeneous microstructure in the presence of a cluster of three large grains located at the center of the simulation system. The normalized strain rates versus strain for four such non-uniform microstructures are given in Fig. 4(b). For reference and comparison Fig. 4(b) also includes the results for the homogeneous hexagonal micrustructure [6]. The areas of each of the three larger grains, A_l, in the four non-uniform microstructures considered are: 4.0; 6.1; 8.1 and 16.1 times respectively larger than the area of a hexagonal matrix grain A_h. Interestingly, Fig. 4(b) shows that the presence of the inhomogeneity removes the singularity in the strain rate that is characteristic for homogeneous microstructures. Basically the micrustructural non-uniformity breaks the synchronism of grain-switching events, responsible for the high strain rate peaks in homogeneous microstructures. The larger the relative area, A_l/A_h, of the inhomogeneity the smaller the peak value of the strain rate. Moreover, the strain-rate peak position also shifts towards lower values of the strain with the increase of the area of the inhomogeneity.

CONCLUSIONS

Our simulations reveal that during Coble creep, in inhomogeneous microstructures, the stress acting on the grain boundaries surrounding larger grains are in general higher than the stress acting on the smaller matrix grains. Moreover, the study also reveals that the high-stress regions move away from the larger grains (inhomogeneities) if these are bordered by grains smaller than the matrix grains.

The simulations show that the presence of non-uniformity in the microstructure may also play an important role in determining the material response to creep deformation. Namely, we found that the larger the area of the inhomogeneity the smaller the peak value of the creep rate. One can rationalize this by noticing that the presence of inhomogeneity eliminates the synchronous neighboring grain-switching, typical for homogeneous microstructures.

ACKNOWLEDGMENTS

The work was supported in part by a grant from Louisiana Board of Regents under the contract number LEQSF 2003-06–RD–A–13

REFERENCES

[1] R.L Coble, J. Appl. Phys. **34**, 1679 (1963).
[2] R. Raj, M.F. Ashby, Metall Trans. **21**, 149 (1973).
[3] J.H. Schneibel, R.L. Coble, R.M. Cannon, Acta Metall. **29**, 1285 (1981).
[4] P.M. Hazzledine, J.H. Schneibel, Acta Metall. Mater. **41**, 1253 (1993).
[5] J.M. Ford, J. Wheeler, A.B. Movchan, Acta Mater. **50**, 3941 (2002).
[6] J. Pan and A.C.F. Cocks, Comput. Mater. Sci. **1**, 95 (1993).
[7] D. Moldovan, D. Wolf, S.R. Phillpot, A.K. Mukherjee, H. Gleiter, Phil. Mag. Lett. **83**, 29 (2003).
[8] A.S. Argon, in "Recent Advances in Creep and Fracture of Engineering materials and Structures" eds. Wilshire, B. and Owen, D.R.J., Pineridge Press, Swansea, 1-52 (1982).
[9] A.C.F. Cocks and A.A. Searle, Mech. of Mater., **12**, 279 (1991).
[10] A. Needleman and J.R. RICE, Acta Metall. **28**, 1315 (1980).

O12.6 Fracture and Deformation of Thermal Oxide Films on Si (100) Using a Femtosecond Pulsed Laser

Authors: Joel P. McDonald[1,4], Vanita R. Mistry[2], Katherine E. Ray[2], Steven M. Yalisove[3,4]

[1]Applied Physics Program, University of Michigan, 2477 Randall Laboratory, Ann Arbor, MI 48109-1120
[2]College of Engineering, University of Michigan, Lurie Engineering Center, 1221 Beal Avenue, Ann Arbor, MI 48109-2102
[3]Department of Materials Science and Engineering, University of Michigan, 2200 Hayward Ave., Ann Arbor, MI 48109
[4]Center for Ultrafast Optical Science, University of Michigan, 1006 Gerstacker Building, 2200 Bonisteel Avenue, Ann Arbor, MI 48109

Abstract: Femtosecond pulsed laser damage of Silicon (100) with thermal oxide thin films was studied in order to further understand the optical and electrical properties of thin films and to evaluate their influence on the damage of the substrate. The damage threshold as a function of film thickness (2 – 1200 nm) was measured. The damage morphology produced by single laser pulses was also investigated. Two primary morphologies were observed, one in which the oxide film is completely removed, and the other in which the film is delaminated and expanded above the surface producing a bubble feature.

Introduction: In this work, the mechanical properties of $Si - SiO_2$ structures are probed in an extreme fashion using high intensity femtosecond (fs) laser pulses. As the electrical and mechanical properties of Silicon are fairly well understood, Si has been a material of choice for fs laser damage studies [1-3]. Recently our work has shown that the naturally occurring ~2 nm thick native oxide layer supported by Si(100) influences both fs laser damage morphology and single shot damage threshold. In this work we have studied fs laser ablation of Si(100) with oxide films ranging in thickness from the ~ 2 nm native oxide up to the 1200 nm of thermal oxide (SiO_2). Our goal was to determine if the damage threshold and damage morphology of Si(100) with a surface oxide thin film was dependent on the oxide film thickness.

The following focuses on two results obtained during fs laser ablation studies of oxide thin films on Si(100) substrates. First, the damage threshold of Si(100) with a thermally grown oxide is presented as a function of thermal oxide film thickness (with laser incident perpendicular to sample surface, i.e. normal incidence). Second, the surface damage morphology resulting from single fs laser pulses incident on Si(100) with thermal oxide films is presented. Two different laser fluence dependent morphology types were observed, with variations within these types observed for different oxide layer thickness and laser beam angle of incidence. For laser fluences above a thickness dependent threshold, the oxide layer was removed in a nearly discrete fashion. For laser fluences below this threshold, uniform dome structures were produced, where the glass film appears to "bubble" out from the substrate. The domes are thought to result from a combination of thin film delamination and subsequent ablation plume expansion at the film/substrate interface. Similar dome features have been observed in other works addressing nanosecond laser ablation and modification of thin metal films [4]. However the data we present are distinctly different in both their nature and underlying mechanisms responsible for their formation. Atomic force microscopy (AFM) and optical microscopy (OM) data are presented showing both the discrete removal of the oxide film and the observed bubble phenomenon.

EXPERIMENT

Si(100) wafers with thermally grown oxide films and a Si(100) wafer with native oxide were used for this study. The thickness' of the thermal oxide films were found via ellipsometry to be 19.5 nm, 54 nm, 147 nm, and 1200 nm, while the thickness of the native oxide was determined (also by ellipsometry) to be 2.3 ± 0.3 nm. Prior to exposure to the laser, samples were prepared by an initial Triton degreasing scrub, followed by acetone, methanol, and deionized water baths.

A diagram of the experimental setup is shown in Fig. 1. All samples were laser machined at normal incidence with laser pulses from a Ti:sapphire laser. Select samples were machined at non-normal incidence and are the focus of future work. The incident laser pulse had a temporal width of ~150 fs with a central wavelength of 780 nm, and was delivered to the sample at a repetition rate of 125 Hz. The laser was focused onto the sample with a 35 cm focal length lens, producing a nearly circular spot at focus with a diameter of ~ 50 ± 5 μm (spot size at focus determined by fitting routine assuming a Gaussian beam profile [5]). Neutral density filters were used to control the intensity of the incident beam. The sample was positioned using a three-axis motorized translation stage and translated through the laser beam with constant velocity to obtain single shot features at normal incidence. In the grazing incidence beam geometry, the sample was positioned at the desired angle with respect to the beam and translated through the focus of the beam, using a fast shutter (Uniblitz) to control the exposure to the laser beam. For damage threshold determination at non-normal incidence, only the features produced nearest the focus were compared. The sample was rotated by 90° about the optic axis to change the polarization of the laser beam at the sample surface.

OM images were obtained for all samples and were used to determine the damage threshold. The damage threshold was determined from a Gaussian beam intensity profile fitting routine performed on the measured damage radius as a function of incident pulse energy [5]. For samples with native oxide and 19.4 nm of thermal oxide, the microscope images were obtained in Nomarski mode in order to obtain the contrast necessary to measure the damage radius. AFM

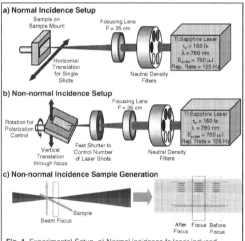

Fig. 1. Experimental Setup. a) Normal incidence fs laser induced damage threshold and morphology setup. b) Non-normal incidence setup (future work). Sample is scanned vertically through the beam focus to obtain features produced at focus, as well as before and after focus. c) Schematic of sample produced at non-normal incidence.

was performed on select features in order to obtain surface morphology information.

RESULTS

Femtosecond laser induced single shot damage threshold of Si(100) as a function of thermal oxide film thickness
The fs laser induced damage threshold is defined as the minimum incident laser pulse fluence (J/cm^2) for which damage to the Si(100) substrate is apparent in a 40x OM image. It should be noted that in the case of thicker films (147 and 1200 nm), the damage observed via OM for fluence near the

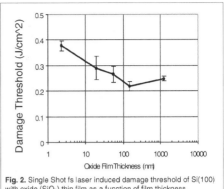

Fig. 2. Single Shot fs laser induced damage threshold of Si(100) with oxide (SiO_2) thin film as a function of film thickness.

damage threshold is only present at the oxide film, Si(100) substrate interface. AFM analysis indicates that near the damage threshold, no morphological modification to the upper surface of the SiO_2 film is present for films of 157 nm and 1200 nm. The damage threshold as a function of SiO_2 film thickness is shown in Fig. 2.
The damage threshold shows an apparent decrease with SiO_2 film thickness increase, with a slight variation observed for the sample with 1200 nm of thermal oxide. Certain measurement considerations were made to assure the measured damage threshold was as accurate as possible. First, only samples with similar beam waists (obtained from the fitting routine of the damage radius as a function of incident pulse energy) were compared. This assured that the damage threshold was measured as close to the beam focus as possible. Second, the nature of the damage from sample to sample was carefully considered to assure that the radial damage measurements were made with respect to the same characteristic damage, independent of film thickness.

Morphology of fs laser induced damage of Si(100) with surface oxide films at normal incidence
Two different damage morphologies were observed. For laser fluence above a particular thickness dependent threshold, the SiO_2 film was found to be removed in a nearly discrete fashion, with limited damage present in the underlying Si(100) substrate. This phenomenon is termed "oxide lift off," and was found to occur for all thermal oxide thicknesses. For laser fluence below the afore mentioned thickness dependent threshold for oxide lift off, the oxide thin film is delaminated from and expanded upward from the underlying substrate, forming a bubble or dome like feature. Complete (i.e. uncollapsed) dome features were observed for samples with 54 nm, 147 nm, and 1200 nm of thermal oxide. For film thicknesses of 19.5 nm and 54 nm, a collapsed bubble was often observed in which it appears as if the film was initially expanded from the substrate but collapsed once the internal pressure provided by the ablation plume subsided (see Fig. 7).

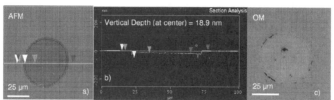

Fig. 3. AFM and OM of single shot features made in Si (100) with 19.5 nm of thermal oxide at normal incidence peak fluence = 1.05 J/cm^2. a) AFM image top view. b) AFM profile of single shot damage in a) showing depth of 18.5 nm. c) OM image of similar feature produced at same laser fluence.

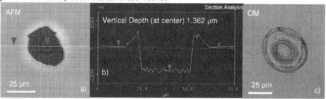

Fig. 4. AFM and OM of single shot features made in Si (100) with 1200 nm of thermal oxide at normal incidence peak fluence = 1.32 J/cm^2. a) AFM image top view. b) AFM profile of single shot damage in a) showing depth of 1.362 μm. c) OM image of similar feature produced at same laser fluence.

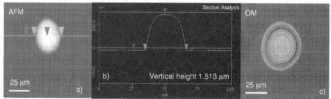

Fig. 5 AFM and OM of single shot features made in Si (100) with 147 nm of thermal oxide at normal incidence peak fluence = 0.41 J/cm^2. a) AFM image top view. b) AFM Profile of single shot dome in a). c) OM image of similar feature.

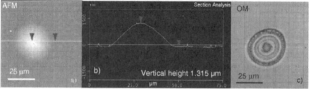

Fig. 6. AFM and OM of single shot features made in Si (100) with 1200 nm of thermal oxide at normal incidence peak fluence = 0.76 J/cm^2. a) AFM image top view. b) AFM profile of single shot dome in a). c) OM image of similar feature.

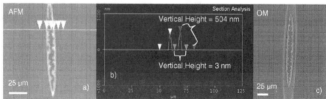

Fig. 7. AFM and OM of single shot features made in Si (100) with 54 nm of thermal oxide at grazing incidence (~6.4° relative to sample surface) with peak fluence = 0.341 J/cm^2 illustrating phenomenon of bubble collapse. a) AFM image top view. b) AFM profile of single shot damage in a) showing depth change of 3 nm between the interior and exterior of the damage feature, and a peak uncollapsed bubble height of 504 nm. c) OM image of similar feature produced with same laser fluence.

DISCUSSION

The decrease in the fs laser induced damage threshold as a function of SiO_2 film thickness is thought to be a direct result the thickness dependence of the optical properties of thin SiO_2 films [6]. Further investigation into the optical qualities of thin films (beyond damage threshold measurements) is necessary and will likely illuminate the trend in the damage threshold observed here.

The observed fs laser induced damage morphology of Si(100) with thin oxide films may provide many interesting insights into the interaction of fs laser pulses with materials, a topic of considerable interest to those in the fields of optics and materials science. A brief description of the interaction of the fs laser pulse with material may illuminate the nature of the damage. The SiO_2 film is largely transparent to the incident laser pulse, centered at 780 nm. As a result, the laser pulse passes through the oxide and interacts with the material at the film-substrate interface. As the Si melts, the interfacial bonds are severed between the film and substrate and the ablation plume comprised mostly of Si ions from the substrate expands upward against the oxide film. The oxide film is heated by the hot expanding plasma and is likely left in a low viscosity state. If the impulsive force from the ablation plume exceeds the fracture strength of the film, the film is removed, or popped off from the substrate. If the pressure from the plume is insufficient to remove the film, the film may be left in a bubble or dome shape. It should be noted that once the film is delaminated, a bubble could form simply due to the relaxation of compressive stress in the film. However, the stress relaxation alone does not completely explain the presence of bubble features in thick films (> 1000 nm) where residual stress is very low, nor does it explain the heights of the bubbles that we observe. Furthermore, we can continue to "pump" up the bubbles by hitting it with subsequent laser pulses. Additional investigation yielding the time scales of the observed oxide lift off or bubble formation may shed light on the dynamics of pressure produced by the interaction of intense laser pulses with materials.

CONCLUSIONS

The fs laser induced damage threshold of Si(100) with thin oxide films was found to decrease with increased oxide thickness. The damage morphology resulting from fs laser pulses directed at normal incidence onto Si(100) with oxide thin films was found to take two distinct forms. For laser fluence above a particular thickness dependent threshold, the oxide film was removed in a nearly discrete fashion. For laser fluence below this threshold, bubble like features were observed for films of thickness 54 nm, 147 nm, and 1200 nm. The bubble is thought to result from an initial delamination of the film from the substrate followed by an expansion of the film up above the substrate by the force of the ablation plume.

REFERENCES

[1] J. Bonse, *et. al.,* Appl. Phys. A - Mat. Sci. & Proc. **74**, p19 (2002).
[2] C. B. Schaffer, *et. al.,* Meas. Sci. & Tech. **12**, p1784 (2001).
[3] D. Du, *et. al.,* Appl. Phys. Lett. **64**, p3071 (1994).
[4] Xiao, Ke, *et. al.,* Appl. Phys. Lett., **85**, p1934 (2004).
[5] J. Bonse, *et. al.,* Appl. Surf. Science, **154-155**, p659 (2000).
[6] A. Kucirkova, *et. al.,* Appl. Spectrosc., **48**, p113 (1994).

Mater. Res. Soc. Symp. Proc. Vol. 875 © 2005 Materials Research Society   O12.7

Effects of Humidity History on the Tensile Deformation Behaviour of Poly(methyl –methacrylate) (PMMA) Films

Chiemi Ishiyama, Yoshito Yamamoto and Yakichi Higo
Precision and Intelligence Laboratory, Tokyo Institute of Technology
4259 Nagatsuta-cho, Midori-ku, Yokohama, 226-8503, JAPAN

ABSTRACT

Tensile testing of Poly(methyl methacrylate)(PMMA) films has been conducted to clarify the effects of humidity history upon their tensile properties.   Prior to testing, PMMA film specimens were kept under three different humidity conditions for 3 days to adjust the water content to each condition.   The conditions used in this study were 11%, 54% and 98% relative humidity (RH) respectively.   The tensile strength of PMMA films at these storage humidities tends to decrease with increasing humidity.   The tensile deformation behaviour is ductile at 11% and 54% RH although brittle at 98% RH.   When tensile tests were performed for different storage terms at 98% RH using specimens which had been equilibrated to 11% RH before testing, the tensile strength decreased within a short storage term.   The tensile deformation behaviour, however, remains ductile.   The tensile deformation behaviour drastically changes to a brittle mode after a 3 day storage term at 98% RH.   This suggests that the tensile strength is affected by moisture around the specimen surfaces.   This is attributed to a reduction in the crazing stress as a result of surface moisture.   Brittle failure, however, occurs after a long storage term at the high humidity levels tested.   This suggests that the ductile manner is caused by both absorbed water and the moisture around the specimen surfaces.

INTRODUCTION

In recent decades, the molding and processing technology of polymer materials has progressed remarkably, making polymer materials applicable to micro devices and the field of Micro-Electro-Mechanical Systems (MEMS).   The surfaces of micro-sized materials can be extremely sensitive to environmental conditions such as humidity.   Moreover, most polymers sorb water so that the content changes with humidity.   In Poly(methyl methacrylate)(PMMA) this water content strongly affects tensile [1], creep [2, 3] and other mechanical properties.   It is, therefore, necessary to investigate the relationship between humidity history and mechanical properties in polymers.   There are currently very few studies clarifying effects of humidity upon the mechanical properties of polymer thin films or micro-sized polymers.

In this study, tensile tests were performed on PMMA films (with thicknesses of 100 μm) under different humidity conditions in other to establish the relationship between the humidity history and mechanical properties of PMMA.   It is first necessary to clarify the relationship between water content and tensile properties of PMMA films under constant humidity conditions.   This is achieved using film specimens that have been equilibrated with the testing humidity.   Tensile tests are then conducted to evaluate the storage term under a high humidity condition (98% RH) using specimens equilibrated with a low humidity condition (11% RH) prior to testing.

EXPERIMENTAL PROCEDURE

The material used in this study was in the form of PMMA pellets of 1 x 10⁶ weight-average molecular weight (Mw), supplied by Sumitomo Chemical Co. Ltd. The pellets were cast into PMMA films by dissolving them in chloroform and then spreading the solution onto mercury, which was run into a laboratory dish. The solution was left for approximately one day, until a PMMA film was obtained.

Tensile specimens, as shown in Fig. 1, were stamped out from the PMMA films using a cutter, prior to heat-treatment at 348K for 1 hour. The glass transition temperature (T_g) of the PMMA pellets, PMMA film without heat-treatment and heat-treated film were measured using Differential Scanning Calorimetory (DSC). Specimens were kept under three different humidity conditions (11% RH, 54% RH and 98% RH) for three days to adjust the water content of the material. Specimen thickness at the gauge length was measured before testing using both a micrometer and a scanning laser microscope with an accurate Z axis controller.

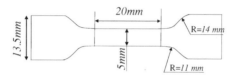

Figure 1. Schematic diagram of a tensile test specimen (Thickness: 100 μm).

Tensile tests were performed under three different humidity conditions (11% RH, 54% RH and 98% RH) using specimens which had been equilibrated with the test humidities. Stress-strain curves were obtained using a Shimazu EZ-test machine with a 100 N load cell, at a crosshead speed of 0.5 mm/min to provide a strain rate of approximately 1.1 x 10⁻⁴ s⁻¹.

Tensile tests were also performed at 98% RH using PMMA specimens that had been kept at 11% RH for three days prior to testing. These specimens were then kept at the testing humidity for time periods of 1 hour, 1 day, two days or three days respectively.

Specimens surfaces were observed using a scanning laser microscope and an optical microscope after testing in order to analyze plastic deformation behaviour on the specimen surfaces.

RESULTS AND DISCUSSION

Glass Transition Temperature (T_g)

The measured T_g values are shown in Table 1. T_g of the 100μm thick PMMA film is approximately fifty degrees lower than that of the PMMA pellets from which it originated. In addition, the T_g of PMMA film having been heat-treated at 348K for 60 min is ten degrees higher than that of the untreated film. This suggests that the mechanical properties of PMMA films are influenced by the chloroform solvent. The solvent remains in PMMA films leading to reduction in the T_g. Heat-treatment is therefore believed to have increased the T_g by causing the desorption of the solvent from the polymeric material.

Table 1. Glass Transition Temperature (T_g) of PMMA pellets and PMMA films

Materials	Thickness	Heat Treatment	Tg
PMMA pellets	-	none	388K
PMMA film	100 μm	none	338K
		at 348 K for 60 min	343K

The effect of water content on tensile properties of PMMA films

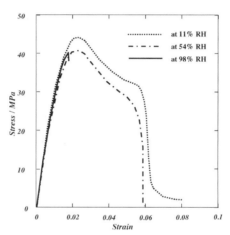

Figure 2. Typical stress-strain curves of PMMA films under each humidity condition at a strain rate of 1.1 x 10^{-4} /s. Test humidities are 11% RH, 54% RH and 98% RH. Water content in each specimen had been equilibrated with test humidity.

Table. 2. Average tensile strength and elongation to failure, which were obtained from the tensile testing of PMMA film, under three different humidity conditions at a strain rate of 1.1 x 10^{-4} /s.

Equilibrium Humidity (%RH at 20°C)	Test Humidity (%RH at 20°C)	Mean Tensile Strength (MPa)	Mean Elongation to Failure
11	11	45.1	0.053
54	54	40.7	0.048
98	98	39.0	0.022

Figure 2 indicates typical stress strain curves resulting from tensile testing using specimens equilibrated with the test humidity. The tensile strength data in a given humidity condition is scattered within approximately 8 % but the elongation to failure at both 11%RH and

98%RH is scattered by more than 30 percent due to necking. Table 2 therefore shows average tensile strength and average elongation to failure obtained from testing using more than 5 specimens at each humidity. The tensile strength of the films decreases with increasing humidity / specimen water content. This is similar to results previously obtained using a commercial cast PMMA sheet (1.5mm in thickness) [4]. It is considered that sorbed water in PMMA works as a plasticizer [1], decreasing the crazing stress [5], and therefore the tensile strength.

Figure 2 also shows that PMMA films at 11% RH and 54% RH behave in a ductile manner. This is in contrast to PMMA sheets that have behaved in a brittle manner previously [4]. The T_g of commercial PMMA sheet is around 388 K [6, 7], the same as that of the PMMA pellets. This suggests that the solvent present in the PMMA films makes them plastic at the low and middle humidities used. In contrast, PMMA films at 98% RH behave in a brittle manner, similar to that obtained using cast sheet in the previous study [8]

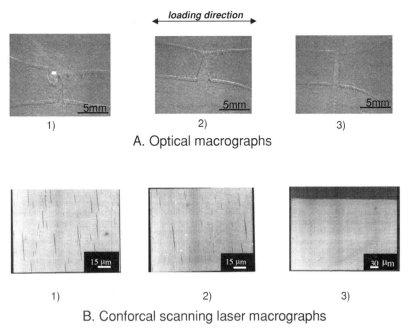

Figure 3. Optical (A) and Scanning Laser (B) micrographs after tensile testing at 1) 11% RH, 2) 54% RH, and 3) 98% RH: 1), 2) Necking areas visible, with a large number of crazes on the specimen surface. 3) No necking areas visible. A few crazes on the specimen surface.

Figures 3 A show optical micrographs around gauge lengths of specimens after failure. Necking areas are observed within gauge lengths of specimens that were tested under 11% RH and 54% RH conditions, see Fig 3A 1) and 2). This shows that shear deformation occurred under low and middle humidity conditions during tensile loading. Necking deformation is not usually observed under tensile loading in commercial PMMA cast sheets [8]. This suggests that the solvent is remaining in the films, making plastic deformation easier. In comparison, no necking is observed at 98% RH as shown in Figs. 3 A 3). This is similar to previous observations from commercial cast sheet [8]. This micrograph also shows that PMMA films at high humidity experience brittle failure.

Figure 3 B shows observations by scanning laser microscope. A large number of crazes are observed on the surfaces of specimens tested at 11% RH and 54% RH, see Figs 3 B 1) and 2). These crazes are located within a few microns of the specimen surfaces. Many crazes are specifically observed on necking areas with many short distortions, which are located between craze tips close to each other, appearing in the shear direction against tensile loading. It is suggested that craze tips may become the origin of shear deformation. Crazing behaviour at 98% RH does however, seem greatly different from that at other humidities, as shown Fig. 3 B3). Long creases were vaguely observed on film surfaces. This may be caused by brittle behaviour similar to that previously observed [4, 8]. It is considered that a few long crazes have rapidly occurred at the high humidity [8], leading to easier crack generation and failure.

The effects of humidity history in PMMA films on their tensile properties

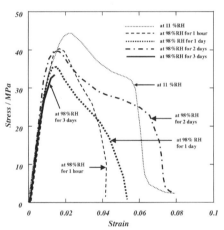

Figure 4. Typical stress-strain curves of PMMA films under different humidity history conditions: All the specimens had been kept under 11% RH condition prior to tensile testing.

Figure 4 illustrates typical stress-strain curves of PMMA films, that had been equilibrated with the low humidity (11% RH) prior to testing, after different storage terms at the high humidity(98% RH). Results of a specimen kept at 98% RH for 1 hour show that the tensile strength immediately decreases after a short term at the high humidity. This suggests that water in the vicinity of specimen surfaces strongly affects the tensile strength of PMMA films. The reduction in tensile strength is attributed to a decrease in crazing stress as a result of sorbed water at the specimen surfaces. In contrast, tensile deformation behaviour is largely unchanged after a short term exposure at 98% RH, although after 3 days the failure mode changes from a ductile to a brittle mechanism. This suggests that brittle behaviour may be caused by absorbed water inside the material as well as at its surfaces. Further studies are required to clarify the cause of this change in failure behaviours.

CONCLUSIONS

Tensile properties of PMMA films were investigated by tensile testing of water content adjusted PMMA film specimens, to clarify the effects of humidity history of the material upon the tensile properties. The following conclusions have been made;

1. Tensile testing of specimens equilibrated to their humidities showed a decrease in the tensile strength with increasing humidity. Sorbed water is believed to have decreased the crazing stress leading to a decrease in the tensile properties. At 98% RH, tensile deformation occurred in a brittle manner. This is attributed to crazing at the such a high humidity.
2. Tensile properties have been shown to alter depending upon the storage time at a high humidity condition. Tensile strength of PMMA films promptly decreases after a short time, although the deformation mode remains unchanged. Tensile strength appears to be strongly affected by an alteration in the necessary crazing stress at material surfaces. In comparison, brittle failure of PMMA film at high humidity may be affected by water absorbed into the material as well as on its surfaces.

REFERENCES

1. J. Shen, C. C. Chen and J. A. Sauer, *Polymer* 26, 511-518 (1985).
2. L. S. A. Smith and J. A. Sauer, *Plastics and rubber processing and applications* 6, 57-65 (1986).
3. D. G. Hunt, *Polymer* 19, 977-983 (1978).
4. C. Ishiyama and Y. Higo, *J. Polym. Sci. pert B: polym. Phys.* 40, 460-465 (2002).
5. Y. Bokoi, C. Ishiyama, M. Shimojo and Y. Higo, *J. Mater. Sci.* 35, 5001-5011 (2000).
6. E. V. Thompson, *J. Polym Sci.; Part A-2* 4, 199-208 (1966).
7. B. D. Washo, and D.Hansen, *J. Appl. Phys.* 40, 2423-2427 (1969).
8. C. Ishiyama, Y. Shiraishi, M. Shimojo and Y. Higo, *Proc of the 9th International Conference on Fracture* 2, 1029-1035 (1997).

Mater. Res. Soc. Symp. Proc. Vol. 875 © 2005 Materials Research Society   O12.9

Structural Control of Lithium Fluoride Thin Films

O.G. Yazicigil, D. Rafik†, V. Vorontsov† and A.H. King
School of Materials Engineering
Purdue University, West Lafayette, IN 47907-2044
† also at Department of Materials Science & Engineering
Imperial College, London, England

ABSTRACT

Polycrystalline lithium fluoride thin films have a number of existing and potential uses, but the optimization of their microstructure has not yet been addressed systematically. We have developed a means of measuring the porosity in LiF films, and a method for performing detailed electron-microscopical studies on this normally beam-sensitive material. These techniques have been applied to assess the structure of LiF films immediately after deposition from the vapor phase, and also after subsequent annealing.

INTRODUCTION

Lithium Fluoride thin films are often used as an "electron photoresist," and they have many other potential uses because of their unique optical and electrical properties [1]. A barrier to the application of LiF thin films, however, is the poor understanding of how the film structure is affected by the processing conditions. Structural "zone models" are widely used to describe how the structure of metallic thin films is affected by their processing [2,3,4] and Kaiser *et al.* [5] have attempted to apply a similar description to LiF films, but were only able to use C/Pt replicas of the film surfaces to study in the transmission electron microscope (TEM), presumably because of the beam-sensitivity of LiF. Nunzio *et al.* [6] have used scanning electron microscopy to study the microstructures of LiF films, and they report structures that differ significantly from those reported by Kaiser *et al.* In this paper, we demonstrate a technique for the direct study of LiF in the TEM, and use it in conjunction with atomic force microscopy (AFM) to elucidate the structures of LiF films deposited on unheated substrates, and also to study their microstrucural development on annealing.

MATERIALS AND METHODS

LiF films were prepared by thermal evaporation onto amorphous 50nm thick Si_3N_4 membrane substrates, and also onto glass microscope slides. A Quartz Crystal Thickness Monitor was used to measure the thickness of the films.

Films of nominal thickness 30 nm, 50 nm and 90 nm, were deposited on unheated substrates for TEM analysis, and additional LiF films of thicknesses of 100 nm and 600 nm were grown only on glass substrates. The films were studied in the as deposited condition, and also after 250°C anneals of 10 minutes and 6 hours duration. The films were annealed in a controlled atmosphere of Ar-5%H_2 flowing approximately at

100sccm, with a heating ramp rate of 1000°C/h followed by cooling to room temperature at a rate of 600°C/hr.

A Digital Instruments Multimode Atomic Force Microscope (AFM) was used in tapping mode to study the surface topography of LiF thin films deposited onto glass substrates. Film thickness measurements were made with the AFM.

A JEOL JEM 2000FX Electron Microscope, equipped with a LaB_6 cathode, was used to collect bright field images, dark field images and diffraction patterns. The sensitivity of LiF to electron beams makes it an excellent electron photo-resist material, but is also a serious impediment to performing electron microscopy on the films [7]. We found that we were able to make our films stable in the TEM by carbon coating them prior to observation, to allow charge dispersal. Gatan Digital Micrograph software Version 3.65 was used to acquire and analyze the images. Two sets of axial dark field images were taken using segments of the {220} and {111}+{200} diffraction rings, respectively, in order to reveal any texture evolution.

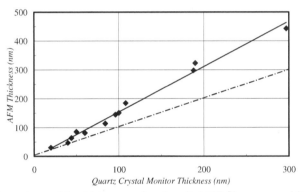

Figure 1: Comparisons of the thicknesses of as-deposited LiF films measured by the quartz-oscillator thickness monitor, and AFM. The broken line has a slope of unity, and is the expected relationship if the films are fully dense.

RESULTS AND DATA ANALYSIS

Film Thickness and Porosity

Thickness measurements were made on films deposited on glass substrates, using the AFM to compare with the thickness monitor readings. The films were scratched to expose the bare substrate in order to measure the thickness of the film directly, using the AFM. The thicknesses of the LiF films measured by the AFM and thickness monitor are shown in Figure 1. The AFM measurements were consistently larger than those reported by the thickness monitor, and the difference between the measurements can be understood when it is recognized that the AFM directly measures the thickness of the film, while the quartz crystal oscillator measures its mass per unit area, and then

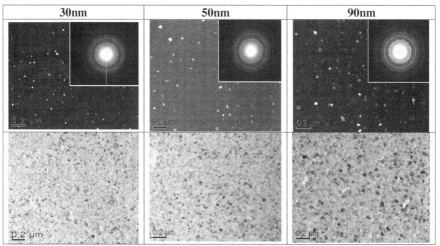

Figure 2: Dark field images shown for {111}+{200} diffracting grains for 30nm ,50nm, 90nm LiF films in as-deposited condition. Bright field images from the same films are also shown.

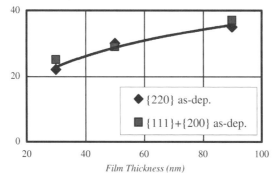

Film Thickness (nm)

Figure 3: Modal equivalent grain diameters for the as-deposited films, from dark field TEM.

calculates the thickness assuming the full theoretical density. The quartz crystal monitor is well-calibrated for other, fully dense films, so the discrepancy between the measurements can be ascribed to a reduced density in the films. If we assume that this is a result of porosity, it is straightforward to calculate the pore fraction in the film. The films, as deposited, exhibit a 33% increase of thickness, attributed to porosity. There is no evidence that this value varies with the thickness of the deposited film. After a ten-minute anneal at 250°C, the porosity is reduced to 23%.

Electron Microscopy and Grain Size Analysis

Our films produced characteristic FCC diffraction ring patterns that match the known lattice spacings for LiF, confirming the composition and phase of the material. Dark field images showed isolated bright grains which are suitable for automated grain size

Figure 4: Dark field images showing {111}+{200} diffracting grains for 30nm LiF film in the as-deposited condition (*a*), and after10 minute (*b*) and 6 hour anneals (*c*) at 250°C.

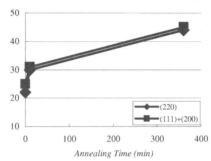

Annealing Time (min)

Figure 5: Grain diameter comparisons for {220} and {111}+{200} diffracting grains as a function of annealing time in a 30 nm thick LiF film. The points represent the mode values.

analysis. Grain sizes were obtained by counting from 300 to 1000 diffracting grains in each type of dark field image ({220} or {111}+{200}). Grain size measurements were also made from AFM images, for the as-deposited condition and the grain sizes were from the two techniques were found to be consistent. The grain sizes reported here are the modal values of the grain size distributions, in the form of the diameter of an equivalent cylindrical grain. The grain size increases with increasing thickness for the as-deposited condition, as shown in Figures 2 and 3.

The effect of annealing is illustrated in Figures 4 & 5. Grain growth is quite rapid in the first 10 minutes of annealing, but essentially halted between 10 minutes and 6 hours, except for the 30nm films.

Film Thickness	Modal Grain Size (equivalent grain diameter), nm					
	As-deposited		10 minutes		6 hours	
	{220}	{111}+{200}	{220}	{111}+{200}	{220}	{111}+{200}
30nm	22	25	30	31	44	45
50nm	30	29	44	40	44	40
90nm	35	37	52	55	50	50

Table 1: modal equivalent grain diameters for films of various thickness, as a function of annealing time at 250°C. The two columns in each set of data refer to grains that produce diffraction intensity of the specified condition.

The modal grain size values for the film in various conditions are given in Table 1. The reduction in grain size for the 90nm films, between 10 minutes and 6 hours is attributable to variance in the data, but is still worth further study, since it may also be related to the sintering of porosity in the material.

There is no significant difference between the diameters of the {220}-diffracting grains and the {111}- and {200}-diffracting grains.

Additional Observations

A number of additional observations have been made, along with the findings reported above. In one of our 50nm depositions, which was otherwise identical to the ones reported above, the films contained several larger, faceted crystals. These were identified from their diffraction patterns to be Li_2S. The origin of the sulfur contamination in this single fun is unknown. Grain growth in these films was severely retarded, by comparison with the results for the uncontaminated films, reported above.

In cases where our films were not sufficiently protected from the electron beam by carbon coating, we observed the formation of Li_2O, *in-situ* in the TEM, after some time of exposure. The oxygen source for this is not known, but the likely sources are residual water vapor in the vacuum, and SiO in the substrate.

DISCUSSION

The microstructures of clean LiF films, deposited onto unheated sustrates, consist of fine equiaxed grains and large amounts of porosity. They differ from the "zone 1" microstructures of polycrystalline metal films [1,2,3] in the significance of the porosity content. Montereali *et al.* [8] have deduced from their refractive indices that LiF films deposited on low-temperature substrates are porous, and our results confirm this by direct measurement.

The as-deposited grain size increases as the deposited film thickness increases. This is most likely an effect of competitive growth and shadowing effects, allowing some grains to grow to larger sizes as the film thickens. It can also be a result of grain growth during the deposition, since thicker films are grown for longer times than thinner ones. There is an uncontrolled rise in the temperature of the film during deposition, due to radiant heating from the LiF source, and also the heat of sublimation as the LiF condenses. Further experimentation would be required to assess the relative contributions of these two mechanisms.

There is no significant difference between the number-density or size of the {220} and {111}+{200} diffracting grains. Grains which appear bright in the {220} DF images can have surface normals of the form ⟨100⟩, ⟨110⟩ or ⟨111⟩, while those that are bright in the {111}+{200} images cannot include grains with ⟨111⟩ surface normals. Comparison of the images therefore yields some limited information about the preference for certain surface orientations, and we are able to conclude only that ⟨111⟩ grains neither form preferentially nor grow preferentially in these films. This conflicts with the findings of Bauer [9] but is not especially surprising, in our view, since the {111} surface is not a low-energy orientation in LiF, in contrast with the FCC metals. Kaiser *et al.* report the formation of ⟨100⟩ textures [5], while Nunzio *et al.* [6] report ⟨16 9 7⟩ at low temperatures and ⟨522⟩ at high temperatures.

Upon annealing at 250°C, the porosity of the films decreased, and the grains of 30nm thick LiF films grew to 1.5 times the film thickness while for the grains of 50nm and 90nm thick LiF films grew to 0.8 and 0.6 times the film thickness respectively, before stagnating. In contrast with metal films [10] the grain size in these LiF films thus tends to stagnate at a decreasing fraction of the film thickness, as the film thickness increases, rather than attaining a size that depends linearly on the film thickness. Interestingly, the stagnated grain size of the 50nm films is possibly smaller than the grain size attained by our 30nm films. This could be due to porosity in the films, which pins the motion of the grain boundaries: the removal of the porosity depends upon the diffusion of atoms into the film (or vacancies out) and should therefore be faster for thinner films.

CONCLUSIONS

We have developed practical techniques for measuring porosity, and performing transmission electron microscopy on lithium fluoride thin films. These have been used to show that films deposited onto cold substrates exhibit considerable porosity, accounting for as much as one third of their thickness. The grain structure of the films appears to be equiaxed. The porosity decreases, and grain size increases with annealing at 250°C, though it has not been possible to remove all of the porosity, or reliably obtain columnar-structured films by this means. There is no evidence of the formation of any preferred orientation in these films.

ACKNOWLEDGMENTS

The participation of Dena Rafik and Vassili Voronstov in this work was made possible by an undergraduate student exchange program between Purdue and Imperial College.

REFERENCES

1. G. Baldacchini, *Journal of Luminescence*, **100**, 333-343 (2002)
2. B.A. Movchan and A.V. Demchishin, *Physics of Metals and Metallography*, **28**, 83-90 (1969).
3. J.A. Thornton, *Ann.Rev. Mater. Sci.* **7**, 239-260 (1977).
4. C.R.M. Grovenor, H.T.G. Hentzell, and D.A. Smith, *Acta Metall.*, **32**, 73-78 (1984).
5. U. Kaiser, N. Kaiser, P. Weissbrodt, U. Mademann, E. Hacker and H. Muller, *Thin Solid Films*, **217**, 7-16 (1992).
6. P.E. Nunzio, L. Fornarini, S. Martelli and R.M. Montereali, *Phys. Stat. Sol (a)*, **164**, 747-756 (1997).
7. A. Murray, M. Scheinfein and M. Isaacson, *J. Vac. Sci. Tech. B* **3**, 367-372 (1985).
8. R.M. Montereali, G. Baldacchini, S. Martelli and L.C. Scavarda do Carmo, *Thin Solid Films*, **196**, 75-83 (1991).
9. E. Bauer, in *Single Crystal Films*, Eds. M.H. Francombe and H. Sato, Pergamon Press, Oxford (1964).
10. W.W. Mullins, *Acta Metall.*, **6**, 414-427 (1958).

Mater. Res. Soc. Symp. Proc. Vol. 875 © 2005 Materials Research Society　　　　　　　　O12.10

Role of Stress on the Phase Control and Dielectric Properties of (1-x) BiFeO₃ - x Ba₀.₅Sr₀.₅TiO₃ Solid Solution Thin Films

Chin Moo Cho, Hee Bum Hong, and Kug Sun Hong

School of Materials Science and Engineering, Seoul National University, Seoul 151-742, Korea

ABSTRACT

Dielectric properties and structure of (1-x) BiFeO₃ (BFO) - x Ba₀.₅Sr₀.₅TiO₃ (BST) (x = 0 ~ 1) solid solution thin films were investigated. All films were prepared at 600 °C on (111) oriented Pt / TiO₂ / SiO₂ / Si substrates by pulsed laser deposition (PLD) technique. Solid solution could be achieved in all composition ranges, evidenced by X-ray diffraction (XRD) and field emission scanning electric microscope (FE-SEM). The intermediate compositions ($0.4 \leq x \leq 0.8$) exhibited a distinct (111) oriented cubic perovskite structure, while rhombohedra symmetry was found in the x < 0.4 range. Dielectric constant and tunability of the (1-x) BFO – x BST films within this composition region ($0.4 \leq x \leq 0.8$) decreased from 1110 to 920 at 1 MHz, and increased from 28.34 % to 32.42 % at 200 kV/cm, respectively, while loss tangent remains constant. A systematic decrease in lattice parameter with BST addition reduced stress due to reduction of lattice parameter mismatch between film and the substrate. In that range, the improvement of the dielectric properties without a degradation of loss tangent is attributed to the presence of the stress relaxation, which was quantitatively confirmed by a surface profiler based on Stoney's equation.

INTRODUCTION

Recently, perovskite structure such as BiFeO₃ (BFO) [1] has attracted much attention as it shows coexistence of ferromagnetic (antiferromagnetic) and ferroelectric ordering simultaneously. These materials present opportunities for potential applications in information storage, the emerging field of spintronics and sensors [2, 3]. It has been reported that BFO has rhombohedrally distorted simple perovskite structure showing ferroelectric properties with high Curie temperature ($T_c \sim 850\,℃$) and antiferromagnetic behavior with Néel temperature ($T_n \sim 370$) simultaneously [4, 5]. BFO is interested in terms of practical applications, because both electrical and magnetic ordering temperatures of BFO are well above room temperature. Unfortunately, application of the BFO has been interrupted by leakage problems, likely a result of defects and nonstoichiometry. To overcome this obstacle, recent work has focused on solid solutions of BFO with ABO₃ materials, such as BaTiO₃ and Pb(Zr, Ti)O₃.

In this study, solid solution was formed by mixing a certain amount of BFO and $Ba_{0.5}Sr_{0.5}TiO_3$ (BST), perovskite structure material. The (1-x) BFO – x BST solid solution targets were made at the range of $0.4 \leq x \leq 0.8$, and then the thin films were deposited on (111) preferred Pt / TiO_2 / SiO_2 / Si substrates by pulsed laser deposition (PLD) technique. The thin films having orientations were studied in the view of stress and the characteristics changes of lattice mismatch between the substrate and thin film. The relation of stress and dielectric properties of the thin films with various thicknesses was also investigated.

EXPERIMENT

The stoichiometric sintered ceramic targets of (1-x) BFO – x BST ($0.4 \leq x \leq 0.8$) were prepared by conventional mixed oxide methods.

The BFO – BST thin films were grown on (111) preferred Pt / TiO_2 / SiO_2 / Si substrate (Inostek, Korea) by PLD with a laser flounce of 2 J/cm². All the films were deposited at 600℃ under oxygen partial pressure in the range of 50 to 100×10^{-3} Torr. Deposited films were, then, cooled down in an oxygen partial pressure of 350 Torr without any further thermal treatment.

The structure of BFO – BST target and thin film were observed using X-ray diffraction (XRD, Model MX18HF-SRA, Macscience Instrument, Japan). The lattice parameter of the bulk was measured with Si standard. Microstructures and thickness of the film were examined using field emission scanning electron microscope (FE-SEM, Model JSM-6330F, JEOL, Japan).

Platinum electrodes (200 ㎛ in diameter) were deposited using RF magnetron sputter onto the film surface using a metallic mask. The capacitance and the loss tangent (tan δ) of the Pt / BFO – BST / Pt capacitor were measured with a impedance analyzer (HP 4194, Hewlett-Packerd, USA) in the frequency range from 100 Hz to 10 MHz.

The residual stress in the thin films was determined by wafer-curvature measurements using a laser reflectance system. An accurate fit for the radius of curvature of the substrate was determined using a line scan consisting of 9 points. The radius of curvature of the bare substrate was determined as the reference point for all subsequent measurements. A change in radius of curvature after film deposition can then be used to calculate the stress in the film via the Stoney equation

RESULT AND DISCUSSION

1. Stress in the thin films with various compositions

Figure 1 shows X-ray powder diffraction profiles of bulk (1-x) BFO – x BST (0 ≤ x ≤ 1) solid solution. Solid solution could be achieved in all composition ranges without any secondary phase, confirmed by XRD analysis. In the range of 0.4 ≤ x ≤ 0.8, all patterns were indexable according to cubic perovskite unit cell. In this region, the lattice parameter increased linearly with the content of BFO from 3.9631 Å (x = 0.8) to 3.9778 Å (x = 0.4). However, in the range of x < 0.4, rhombohedral symmetry was observed. The reason is that as BFO content increases, BFO – BST solid solution system follows the BFO structure having rhombohedral symmetry.

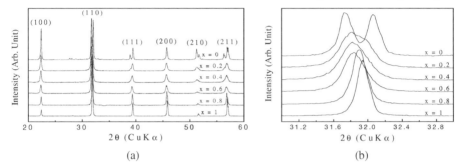

(a)   (b)

Figure 1.   **Powder XRD patterns of (1-x) BFO – x BST (0 ≤ x ≤ 1) samples sintered.**

Figure 2 shows the XRD patterns of BFO – BST thin films, nearly 200 nm thick at the composition of x = 0.4, x = 0.6 and x = 0.8, deposited on Pt / TiO$_2$ / SiO$_2$ / Si substrates under various oxygen partial pressure at 600℃. The XRD patterns demonstrate that the films have cubic perovskite unit cell and single phase without secondary phase.

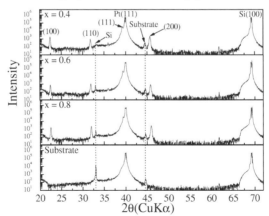

Figure 2.   **XRD patterns of (1-x) BFO – x BST (0.4 ≤ x ≤ 0.8) thin films and substrate.**

Table 1 shows the lattice parameters and residual stress (σ_s) of the BFO – BST ($0.4 \leq x \leq 0.8$) of bulk and thin film. The lattice parameters of the bulk and thin films were calculated using XRD. The residual stress in the thin films was determined by wafer-curvature measurements using a laser reflectance system. An accurate fit for the radius of curvature of the substrate was determined using a line scan consisting of 9 points. The radius of curvature (R_0) of the bare substrate was determined as the reference point for all subsequent measurements. A change in radius of curvature (R) after film deposition can then be used to calculate the stress (σ_s) in the film via the following equation

$$\sigma_s = \frac{E_s}{6(1-\upsilon_s)} \frac{t_s^2}{t_f} \left(\frac{1}{R} - \frac{1}{R_0} \right)$$

where, E_s, υ_s, and t_s are the Young's modulus, Poission ratio, thickness of the substrate and t_f is the thickness of the film, respectively [6]. In a substrate, the thickness of the Si, Pt and TiO_2 was 500 μm, 150 and 20 nm, that we regarded only Si single crystal. The Young's modulus and Poission ratio of the (100) oriented Si single crystal is 129 GPa and 0.28, respectively. The thicknesses of the films were verified to 200 nm by FE-SEM.

Table 1. Lattice parameter and residual stress of the (1-x) BFO – x BST thin films

Content of BST	Lattice of bulk (a_b)	Lattice of thin film (a_t)	Lattice difference (a_b - a_t)	Residual stress (MPa)
X = 0.8	3.963 Å	3.946 Å	0.0171 Å	745.0069
X = 0.6	3.967 Å	3.959 Å	0.0076 Å	624.7511
X = 0.4	3.978 Å	3.967 Å	0.0106 Å	675.0445

In the table 1, as more BFO was added, the lattice parameter of both bulk and thin film increased. However, in the case of thin films, lattice parameters were not coincident with those of bulky samples. It was thought that the lattice mismatch between thin film and Pt substrate (a = 3.923 Å) induced the distortion of the thin films, resulting in variation of lattice parameter. Therefore, the lattice difference between the bulk and thin film was considered to be the residual stress of the films. At x = 0.6 composition, it was thought that the relatively small residual stress was affected by the small lattice difference. In contrast to x = 0.4, the large lattice difference between the bulk and thin film enhanced the residual stress in the thin film, which is consistent with other reports [7].

2. Stress and dielectric properties of the thin films with the various thicknesses

(a) Changing of the stress according to the thickness of thin films

In order to investigate the thickness dependence on the stress and dielectric properties of BFO – BST system, x = 0.8 composition was deposited with 140 and 450 nm thickness, respectively. We verified that both films are single phase without secondary phase, evidenced by XRD analysis. The stress of the thin films with various thicknesses was also measured using a surface profile. According to the stress measurement, the stress of 1330 MPa was applied to the sample with a thickness of 140 nm. However, in the sample with a thickness of 450 nm, the stress dramatically decreased to 456 MPa. It can be explained that the decrease of stress was affected by the growth of the thickness [7, 8].

(b) Effects of stress on the dielectric properties

To investigate the effects of stress on the dielectric properties in thin film, dielectric constant (ε_r), loss (tan δ), and tunability were measured with various thicknesses. Table 2 shows the stress of the thin films, dielectric constant (ε_r), tan δ, tunability, and K-factor (tunability / tan δ) in accordance with the thickness.

Table 2. Stress and dielectric properties in accordance with the thickness

Thickness (nm)	Residual stress (MPa)	Dielectric Const. (ε_r)	tan δ	Tunability (%)	K-factor
140	1330	786.7	98.49×10^{-3}	69.5	7.06
450	456	825.6	51.34×10^{-3}	72	14.02

With increasing the thickness of the thin film, dielectric constant, tan δ, tunability, and K-factor were improved from 786.7 to 825.6 at 1MHz, from 98.49×10^{-3} to 51.34×10^{-3}, from 69.5 to 72 % at 600 kV/cm, from 7.06 to 14.02, and from 1330 to 456 MPa. Other studies show that misfit strain between thin films and substrates leads to a change of dielectric constant [7, 8]. Tunability is also reported to be affected by stress and / or strain in thin films. According to the study of Park. *et al.*, tunability of BST thin films decreased with increasing both tensile and compressive stress except for the case of small tensile stress [9]. In the study of J. Lu *et al.*, tensile stress caused by the thermal mismatch between thin films and substrates was proposed to be a possible reason for the high tunability [10]. In this study, the deposition temperature was tied to 600 ℃, however, the film thickness changed from 140 nm to 450 nm. Therefore, it is clear that the increased tunability was affected by thin film thickness

CONCLUSION

BFO – BST solid solution system was formed and thin films with the composition of $0.4 \leq x \leq 0.8$ were deposited on (111) preferred Pt / TiO_2 / SiO_2 / Si substrate by PLD technique. As a result of measurement, as a content of BFO increased, the lattice parameter of both bulk and thin film increased. However, lattice parameters in the case of thin films were not coincident with those of bulk samples. It can be induced by lattice mismatch between thin film and Pt substrate. The difference of the lattice parameter between the bulk sample and thin film has a linear relationship and was considered with the residual stress of the films. It is evident that the stress was originated from the difference of lattice parameter between the bulk sample and thin film.

In order to investigate the thickness dependence on the stress and dielectric properties of BFO – BST system, $x = 0.8$ composition was deposited with 140 and 450 nm thickness, respectively. According to the stress measurement, the stress with 1330 MPa was applied to the thin film with a thickness of 140 nm. However, in the thin film with a thickness of 450 nm, the stress dramatically decreased to 456 MPa. It can be explained that the decrease of stress was affected by the growth of the thickness. With increasing the thickness of the thin film, dielectric constant, tan δ, tunability, and K-factor were improved from 786.7 to 825.6 at 1MHz, from 0.0985 to 0.05134, from 69.5 to 72.0 % at 600 kV/cm, from 7.056 to 14.02, and from 1330 to 456 MPa.

REFERENCE

1. Y. E. Roginskaya, Y. N. Venevtsev, and G. S. Zhdanov, J. Exp. Theor. Phys. 44, 1418 (1963).
2. Y. E. Roginskaya, Y. N. Venevtsev, and G. S. Zhdanov, Sov. Phys. JETP 21, 817 (1965).
3. G. A. Smolenskii and I. E. Chupis, Sov. Phys. Usp. 25, 475 (1982).
4. J. G. Ismilzade, Phys. Status Solidi. B 46, K39 (1971).
5. G. A. Smolenskii and V. M. Yudin, Sov. Phys. JETP 16, 622 (1963).
6. Y. C. Zhou, Z. Y. Yang, and X. J. Zheng: Surf. Coat. Tech. 162, 202 (2003).
7. W. Chang, C. L. Gilmore, W. J. Kim, J. M. Pond, S. W. Kirchoefer, S. B. Qadri and D. B. Chirsey, J. Appl. Phys. 87, 3044 (2000).
8. J. S. Horwitz, W. C. Chang, W. Kim, S. B. Qadri, J. M. Pond, S. W. Kirchoefer, and D. B. Chrisey, J. Elec, 4:2/3, 357 (2000).
9. B. H. Park, E. J. Peterson, Q. X. Jia, J. Lee, X. Zeng, W. Si and X. X. Xi, Appl. Phys. Lett. 78, 533 (2001).
10. J. Lu and S. Stemmer, Appl. Phys. Lett. 83, 2411(2003).

Mater. Res. Soc. Symp. Proc. Vol. 875 © 2005 Materials Research Society   O12.13

Residual Stresses in TiO$_2$ Anatase Thin Films Deposited on Glass, Sapphire and Si Substrates

Ibrahim A. Al-Homoudi[1], Linfeng Zhang[2], D.G. Georgiev[2], R. Naik[3], V.M. Naik[4], L. Rimai[2], K.Y.Simon Ng[5], R.J. Baird[2], G.W. Auner[2], G. Newaz[1]
[1]Department of Mechanical Engineering, Wayne State University, Detroit, MI 48202.
[2]Department of Electrical and Computer Engineering, Wayne State University, Detroit, MI 48202.
[3]Department of Physics and Astronomy, Wayne State University, Detroit, MI 48201.
[4]Department of Natural Sciences, University of Michigan-Dearborn, Dearborn, MI 48128.
[5]Department of Chemical Engineering and Material Science, Wayne State University, Detroit, MI 48202.

ABSTRACT

Anatase-TiO$_2$ films (thickness 100-1000 nm) were grown on glass, sapphire, and Si(100) substrates using pulsed dc-magnetron reactive sputtering. By measuring the curvature of substrates before and after the thin film deposition, the residual stresses were determined. These results clearly show that the bi-axial stresses are compressive type and decreases with the increasing film thickness. The Raman spectra of these films were measured with two different excitation wavelengths (514 and 785 nm) and the thickness dependent shifts of E$_g$ phonon mode were studied. The dominant 144 cm^{-1} E$_g$ mode in TiO$_2$ anatase clearly shifts to a higher value by 0.45 to 17.4 cm^{-1} depending on the type of substrate and the thickness of the film. Maximum shift was seen for the films on glass substrate indicating a higher bi-axial compressive stress in agreement with the curvature measurements. The excitation wavelength dependent shift of E$_g$ mode clearly shows that the bi-axial stress increases along the film depth, being larger at the film/substrate interface.

INTRODUCTION

Titanium dioxide (TiO$_2$) thin films have been used extensively in many device applications such as gas and humidity sensors, protective coatings on optical elements, solar energy converters etc. Many different fabrication techniques have been developed to prepare TiO$_2$ thin films, however, the processing conditions have been found to strongly influence the structural properties of the resulting films. Often the films exhibit strain due to thermal stresses caused by differential thermal expansion between substrate and the film or by lattice constant mismatch between the two. It has been recognized that the residual compressive stresses may cause film delamination from the substrate whereas the tensile stresses may cause surface cracks in the films. Most of these stresses either compressive or tensile will eventually end up in severe failure problems in devices [1-6].

Most commonly used methods to measure the residual stresses in films are substrate curvature measurement, X-ray diffraction (XRD) and Raman spectroscopy. The XRD and the curvature measurements can be used directly to determine the average stress in the film. It is also known that zone center Raman phonon lines shift to higher/lower frequency under compressive/tensile stress [7-11]. However, relating the observed Raman spectra to the residual stress is more complicated due lack of information regarding the stress or the crystal orientation, particularly in a polycrystalline samples. Quantitative measurements of the residual stress in diamond films, grown by chemical vapor deposition, have been done using Raman scattering, assuming that the stress is primarily biaxial and is in the plane of the film [11, 12]. In this paper we present the residual stress measurements in polycrystalline TiO_2 anatase thin films of different thicknesses (100 to 1000 nm) deposited on glass, sapphire and Si(100) substrates. The residual stresses were determined using substrate curvature measurements, and are correlated to the results from Raman spectroscopy.

EXPERIMENT

TiO_2 anatase thin films were grown by pulsed dc-magnetron reactive sputtering using Ti source in Ar + O_2 gas mixture as described in our previous work [6]. The films of thickness ranging from 100 to 1000 nm were prepared on 20×10×0.5 mm on glass, sapphire and Si(100) substrates. The curvature (deflection) measurements were performed using 128 intelligent film stress measurement system manufactured by Frontier Semiconductor Measurements (FSM) Inc. The unpolarized-Raman spectra were recorded in backscattering geometry using Renishaw Raman-microscope. The Raman spectra were recorded under ambient conditions using 514 and 785 nm excitation wavelengths. Owing to the optical absorption of these films, these two excitation wavelengths have different penetration depth into the film and hence a comparison of Raman line shapes and peak shifts yields information on the strain distribution as a function of depth within the film.

RESULTS AND DISCUSSION

The residual film stress (σ) was calculated by using Stoney's equation:

$$\sigma = \frac{E_{sub}}{6(1-v_{sub})}\left(\frac{t_{sub}^2}{t_{fil}}\right)\left(\frac{1}{r_a} - \frac{1}{r_b}\right) \ (MPa) \quad (1)$$

where, E_{sub} is the Young's modulus and v_{sub} is the Poisson ratio of the substrate, t_{sub} and t_{fil} are the thickness of the substrate and the film, and r_a and r_b are the radii of curvature of substrate with and without the film.

In the curvature measurement, the use Stoney's equation is justified because the film thickness is less than 5% of the substrate thickness. Figure 1 shows the stress as a function of the TiO_2 film thickness.

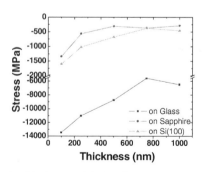

Fig. 1: Variation of residual stress with TiO_2 film thickness on different substrates.

Clearly, the stress is compressive and decreases with increasing the film thickness. Further, the magnitude of residual stress in the film grown on glass substrate is substantially larger than on sapphire or Si(100) substrates. The lower residual stress in films prepared on sapphire and Si(100) substrates can be explained based on the closeness of thermal expansion coefficient and lattice match between sapphire substrate and TiO_2 compared to Si(100). On the other hand glass is amorphous. It is interesting to note that TiO_2 films grown on sapphire and Si(100) exhibit a stress relaxation over a small thickness range compared with the films grown on glass substrate. Based on the data shown in Fig. 1 one can estimate a critical film thickness (i.e. thickness at which the compressive stress reduces to zero) of ~ 1200 nm for TiO_2 films grown on sapphire and Si(100), and perhaps much higher in films grown on glass or it is possible that the films remain strained. More data is needed to assess the critical thickness.

TiO_2 in anatse form has a tetragonal structure. It has six Raman active modes ($A_{1g}+2B_{1g}+3E_g$). Figures 2 and 3 show the Raman spectra of TiO_2-anatase films on various substrates obtained using 514 and 785 nm excitation wavelengths. Clearly, all the expected phonon modes are observed and the frequencies are consistent with the literature values [4,8,9]. The additional sharp peak observed at 520 cm^{-1} in Fig.2(c) and 3 (c) is due to Si, and it served as an internal standard for calibration. The strong peak observed ~ 144 cm^{-1} is an E_g mode of anatase TiO_2 and it shifts to lower peak energy with increasing film thickness as expected. The E_g mode intensity gets stronger and the peak width gets smaller with increasing film thickness.

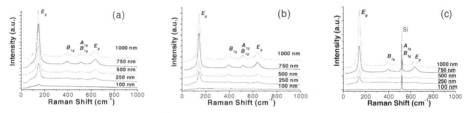

Fig. 2: Raman spectra of TiO_2 thin film with different thickness deposited on (a) glass, (b) sapphire, and (c) Si(100) with wave length λ = 514 nm.

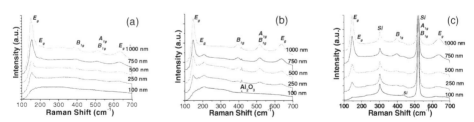

Fig. 3: Raman spectra of TiO_2 thin film with different thickness deposited on (a) glass, (b) sapphire, and (c) Si(100) with wave length λ = 785 nm.

The strong E_g mode observed in thicker TiO$_2$ films on Si or sapphire, with 514 nm excitation, appears at ~ 1.8 cm^{-1} higher and broader in the Raman spectrum obtained with 785 nm excitation wavelength. This may be due to the increased penetration of 785 nm probing more strained layers underneath the surface.

Raman spectroscopy is a very convenient method to measuring stresses in small area on samples. Fig. 4 shows the variation of $\Delta\omega$ (the Raman peak shift with respect to its position in the stress free sample), as a function of film thickness. Quantitative relationships between biaxial stress and the shift in the phonon frequencies exist for diamond films [5]. It is known that $\Delta\omega$ is proportional to the magnitude of residual stress in thin films [7, 11,12]. The stress of the thin films can be calculated by using the shift in the Raman peaks using a relation:

$$\sigma = I_{S\lambda} \cdot \Delta\omega \quad (MPa) \tag{2}$$

where; $I_{S\lambda}$ – proportionality constant is a substrate and excitation wavelength dependent constant for the TiO$_2$ thin film, and $\Delta\omega$ is the Raman peak shift.

However, no theoretical work describing such relations for tetragonal TiO$_2$ is reported in the literature and thus $I_{S\lambda}$ are not known. Fig. 4 shows that $\Delta\omega$ decreases with increasing film thickness for both the excitation wavelengths. However, the shifts are larger with 785 nm excitation compared with the 514 nm.

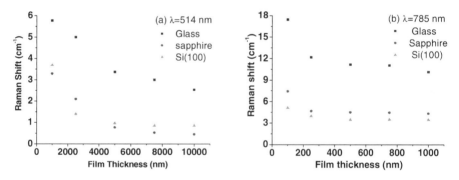

Fig.4: Variation of $\Delta\omega$ (for 144 cm^{-1} E_g mode) with film thickness for TiO$_2$ anatase thin films on various substrates.

Figures 5 and 6 show the Raman line shift vs. average residual stress values determined from curvature measurements obtained with 514 and 785 nm excitation. The Raman peak positions were obtained by fitting Gaussians to spectra shapes. The value of $I_{S\lambda}$ in Eq. 2 was obtained by fitting a straight line through the origin as shown in Fig 5 and 6. For TiO$_2$ films on the glass, $I_{G514} = -2055.8$ MPa/cm^{-1}, for the films on sapphire substrate, $I_{SP514} = -347.3$ MPa/cm^{-1}, whereas for the films on Si(100), $I_{Si514} = -445.5$ MPa/cm^{-1}. The $I_{S\lambda}$ constants were also measured in the same way but with the Raman peak shift using 785 nm laser light, where $I_{G785} = -343$ MPa/cm^{-1}, $I_{SP785} = -102$ MPa/cm^{-1} and $I_{Si785} = -196.4$ MPa/cm^{-1}, on glass, sapphire and Si(100) respectively.

The magnitude of the stress in the film changes depending on the film thickness and substrate. Depending on the penetration depth of the laser light in Raman measurements, stress estimated can also vary. The compressive stresses gradually increase with lower film thickness, and with higher excitation, which means that the stress gets higher closer to the film and substrate interface. It is clear that the contribution from deeper layers to the Raman spectrum become increasingly important with the increasing excitation wavelength due to decreased absorption.

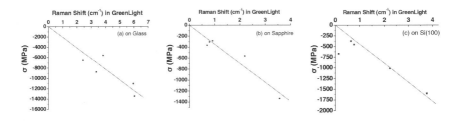

Fig. 5: Stress values from curvature measurements vs. Raman shifting of 144 cm^{-1} E$_g$ mode with 514 nm wavelength.

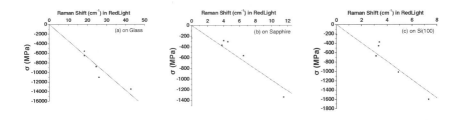

Fig. 6: Stress values from curvature measurements vs. Raman shifting of 144 cm^{-1} E$_g$ mode with 785 nm wavelength.

CONCLUSIONS

Measurements of residual stresses for crystalline TiO$_2$ anatase thin films of different thicknesses (100 to 1000 nm) deposited on glass, sapphire and Si(100) substrates using pulsed D.C. magnetron reactive sputtering have been studied by Raman spectroscopy and curvature measurements. It was evident that all TiO$_2$ thin films are under a compressive stress which decreases with increasing film thickness. The empirical relations using Raman shift and the measured stress from curvature have been established.

ACKNOWLEDGMENT

The authors wish to acknowledge the support of the Institute for Manufacturing Research at Wayne State University.

REFERENCES

1. S. Ikezawa, F. Mutsuga, T. Kubota, R. Suzuki, K. Baba, S. Koh, T. Yoshioka, A. Nishiwaki, K. Kida, Y. Ninomiya, K. Wakita, Vacuum **59**, 514 (2000).
2. F. Vaz, L. Rebouta, Ph. Goudea, J.P. Riviere, E. Schaffer, G. Kleer, M. Bodmann, Thin Solid Films **402**, 195 (2002).
3. Q. H. Fan, A. Fernandes, E. Pereira, J.Gracio, Diamond and Related Materials **8**, 645 (1999).
4. D. Bersani and P. P. Lotticia, X. Ding, Appl. Phys. Lett. **72,** 73 (1998).
5. J. W. Ager III and M. D. Drory, Physical rev. B, **48**, 2601 (1993).
6. I. A. Al-Homoudi, L. Zhang, E. McCullen, C. Huang, L. Rimai, R. Baird, K.Simon Ng, R. Naik, G.W. Auner, G. Newaz, Mater. Res. Soc. Symp. Proc. Vol. **848** (2005) FF3.20.5.
7. W. Zhang, Y. He, M Zhang, Z Yin, and Q Chen, J. Phys. D: Appl. Phys. **33**, 912 (2000).
8. G. Gu, Y. Li, Y. Tao, Z. He, J. Li, H. Yin, W. Li, Y. Zhaoa, Vacuum **71**, 487 (2003).
9. J. C. Paker and R. W. Siegel, Appl. Phys. Lett. **57**, 943(1990).
10. M. H. Grimsditch, E. Anastassakis, and M. Cardona, phys. Rev. B, **18**, 901 (1978).
11. Q. H. Fan, J. Gracio, E. Pereira, Diamond and Related Materials **9**, 1739 (2000).
12. V. Paillard, P. Puech, P. Cabarrocas, Journal of Non-Crystalline Solids, **299**, 280 (2002).

Mater. Res. Soc. Symp. Proc. Vol. 875 © 2005 Materials Research Society   O12.15

Influence of stress on structural and dielectric anomaly of Bi$_2$(Zn$_{1/3}$Ta$_{2/3}$)$_2$O$_7$ thin films

Jun Hong Noh, Hee Bum Hong, and Kug Sun Hong[1]
School of Materials Science & Engineering, College of Engineering, Seoul National University, Seoul, Korea
[1]Corresponding author, E-mail address: kshongss@plaza.snu.ac.kr

ABSTRACT

Bi$_2$(Zn$_{1/3}$Ta$_{2/3}$)$_2$O$_7$ (BZT) thin films were grown on the (111) oriented Pt/TiO$_x$/SiO$_2$/Si substrates using a pulsed laser deposition (PLD) technique. BZT thin films deposited at an oxygen partial pressure of 400 mTorr have the non-stoichiometric anomalous cubic phase despite the BZT target was the monoclinic phase. Compositions, the lattice mismatch, the interfacial layer and the residual stress in the film were investigated as the factors which may affect the formation of the anomalous cubic phase. Among them, the coherent interfacial layer which formed at high oxygen pressures resulted in the formation of the cubic phase by reducing the internal stress.

INTRODUCTION

Bi$_2$O$_3$-ZnO-Ta$_2$O$_5$ (BZT) pyrochlore ceramics have been developed for low firing temperature multilayer capacitors since they exhibit excellent dielectric properties of the large dielectric constants (k = 60~70) with low dielectric losses (tanδ < 0.001) and the small temperature coefficients of capacitance (TCC = -170 ~ 60 ppm/°C) [1-3]. The BZT ternary systems have two phases with different composition; one is monoclinic Bi$_2$(Zn$_{1/3}$Ta$_{2/3}$)$_2$O$_7$ phase and the other is cubic (Bi$_{1.5}$Zn$_{0.5}$)(Zn$_{0.5}$Ta$_{1.5}$)O$_7$ phase [4, 5]. The dielectric constant, the dielectric loss and the TCC of the monoclinic phase are 61.4, below 0.001 and 60ppm/°C, while those of the cubic phase are 71.4, below 0.005 and -172ppm/°C, respectively. The dielectric properties of BZT pyrochlores may be controlled by combining the monoclinic and cubic phases [1].

Dielectric thin films have many potential applications because of their outstanding advantage over bulk ceramics in several aspects such as the faster response, the lower operation voltage, and feasibility for epitaxial growth [6,7]. However, thin films of the BZT pyrochlores have been rarely reported. The pulsed laser deposition (PLD) technique can synthesize the thin films with multi components at high deposition rates with a precise control of the compositions, compared to other techniques such as rf-sputtering, sol-gel and metal organic chemical vapor deposition (MOCVD) processes [8].

In this study, we fabricated the BZT thin films on (111)-oriented Pt/TiO$_x$/SiO$_2$/Si substrates and SrTiO$_3$ (STO) single crystal substrates by using PLD technique. The residual stress varied with the oxygen partial pressure during the deposition and the influence of the internal stress on the evolution of the crystalline phase of BZT thin films and the resulting dielectric properties were discussed in terms of the interfacial layer between BZT films and substrates and the residual stress of the thin films.

EXPERIMENTAL DETAILS

BZT thin films were deposited on (111)-oriented Pt/TiOx/SiO2/Si substrates (Pt substrate) and (111)-oriented $SrTiO_3$ single crystal substrate (STO substrate) at 700°C using the pulsed laser deposition (PLD) technique. The target with $Bi_2(Zn_{1/3}Ta_{2/3})_2O_7$ composition was prepared by a conventional mixed oxide method. The Pt substrates were ultrasonically cleaned with acetone and ethanol for 2 min in each step, and then rinsed with DI water. The laser was operated at a pulse rate of 3 Hz and the energy density of the laser beam at the target surface was maintained at 2 J/cm^2. The target-substrate distance was 5 cm. The deposition chamber was initially evacuated to 3.0×10^{-6} Torr and oxygen gas was introduced into the chamber to maintain the desired pressure (50 to 400 mTorr) during deposition. After the deposition, the thin films were cooled down with a cooling rate of 10°C/min in an oxygen pressure of 350 Torr without any further thermal treatment.

Phase of the BZT films was observed using X-ray diffraction (XRD) and cross sectional microstructures were examined using transmission electron microscope (TEM) and high-resolution transmission electron microscope (HR-TEM). The chemical composition of the films was determined by electron probe microanalysis (EPMA) on several different points for each sample. The residual stress of the thin films can be calculated by measuring the radius of curvature of substrates with and without thin films from the Stoney's equation [9], which is expressed as followed.

$$\sigma_f = \frac{E_s t_s^{\,2}}{6(1-v_s)t_f}\left(\frac{1}{R_{post}} - \frac{1}{R_{pre}}\right) \qquad (1)$$

Where, R_{post}, R_{pre}, t_f, t_s, E_s and v_s are the radius of curvature of substrates with and without thin film, the thickness of the thin film, the thickness, the Young's modulus, and the Poisson's ratio of the substrate, respectively. In order to measure the dielectric properties, platinum top electrodes with a diameter of 250 μm were deposited by the sputtering method. The relative dielectric constant (k) and the dielectric loss (tanδ) of the films were measured with an impedance analyzer (HP 4194a) at 1MHz. Temperature coefficient of capacitance (TCC) was measured in the temperature range from -20 to 125°C.

RESULTS AND DISCUSSION

X-ray diffraction (XRD) patterns of the BZT films deposited on Pt substrates at various oxygen partial pressures from 50 mTorr to 400 mTorr are shown in Figs. 1 (a) and (b). As shown in Fig. 1(a), all of the diffraction peaks are indexed as the monoclinic or the cubic pyrochlore structure and the former is highly oriented along (221) direction and the latter is highly oriented along (222) direction. As shown in Fig. 1(b), the intensity of the monoclinic (442) peak decreases with increasing oxygen partial pressures until it starts to disappear at the oxygen partial pressure over 200 mTorr. It should be noted that the BZT film deposited at 400 mTorr is only composed of the anomalous cubic phase, which is not synthesized in target composition due to the lone electron pairs of Bi^{3+} ions.[1,10] There are some possible reasons which can affect the evolution of phases; changes in the composition due to the evaporation of the volatile components, internal stress due to the lattice mismatch and/or thermal strain, and the interfacial layer between thin films and substrates.

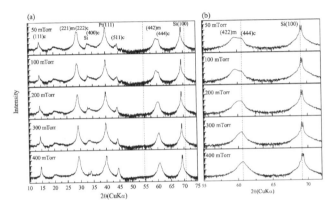

Figure1. XRD patterns of the BZT films deposited on Pt substrate over (a) wide and (b) narrow range.

Figure 2 (a) shows the composition of the BZT films as a function of the oxygen partial pressure, analyzed using electron probe microanalysis (EPMA). It reveals that the composition of the BZT films, deposited at oxygen partial pressures from 50 mTorr to 400 mTorr, is in accordance with that of the target, while the composition of the BZT films deposited at oxygen partial pressures lower than 50 mTorr deviates from that of the target. This implies that the thin films maintain the target composition during deposition at various oxygen partial pressures and the composition does not seem to affect the formation of the anomalous cubic phase.

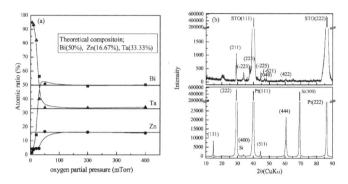

Figure 2. (a) Atomic ratio of Bi, Zn, and Ta of the BZT films, (b) XRD patterns of the BZT films deposited platinum and STO substrate.

Another possibility is the lattice mismatch between BZT thin films and substrates. In order to investigate the influence of the lattice mismatch, we deposited the BZT film on the (111)-oriented STO substrate which is almost the same in the lattice parameter (a = 3.905 Å) with (111)-oriented Pt substrate (a = 3.923 Å). XRD patterns of BZT thin films deposited on the different substrates at 400 mTorr are presented in Fig. 2 (b). In spite that the lattice parameters of the two substrates show very small difference (0.46 %), the diffraction patterns are quite different from each other. BZT thin film deposited on the STO substrate shows the pure monoclinic phase with a random orientation, while the counterpart shows the pure cubic phase with a highly preferred (222)-orientation. Thus, the lattice mismatch does not seem to affect the anomalous cubic phase, either.

Figure 3 (a) shows cross-sectional TEM image of the BZT thin film with cubic phase deposited at 400 mTorr on the Pt substrate. Although the interface between the BZT thin film and the Pt substrate looks clear in the image with low magnitude, the high resolution image in Fig. 3 (b) shows the interfacial layer, which is considered to be the composite of Pt and BZT, obviously. Although it is not easy to analyze the precise composition of the interfacial layer as thin as 5 nm, the interface layer is believed to be $Bi_2Pt_2O_7$. It has been reported for the bismuth-based layered thin films such as $SrBi_2Ta_2O_9$ and $CaBi_4Ti_4O_{15}$ that the interfacial layer between platinum and bismuth-based films might have the pyrochlore structures [10-12].

Cross-sectional TEM images of the BZT film deposited at 50 mTorr, which is composed of both cubic and monoclinic phases, are shown in Fig. 4(a). The monoclinic phase with a dark contrast and the cubic phase with a bright contrast grown on the Pt substrate are shown in the

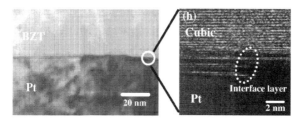

Figure 3. Cross section (a) TEM and (b) HR-TEM image of 400mTorr–deposited BZT film.

low magnitude image. As can be seen in the high-resolution image of the triple junction among the cubic phase, the monoclinic phase, and the Pt substrate (denoted by '2'), the cubic phase is grown on the interfacial layer, while the monoclinic phase is grown on the bare Pt substrate. The interfacial layer gradually disappears along the interface from the cubic phase to the monoclinic phase. Another high-resolution image denoted by '1' also shows that the monoclinic phase is grown directly on the bare Pt substrate without the formation of any interfacial layers. $Bi_2Pt_2O_7$, which is considered as the interfacial layer, has a cubic pyrochlore structure and its lattice parameter is $10.37\,\text{Å}$, which is very close to that of the bismuth-based cubic pyrochlore structure. Therefore the anomalous cubic phase with a non-stoichiometric composition forms on the coherent interfacial layer in order to reduce the internal stress.

    The residual stress of BZT thin films deposited at various oxygen partial pressures were also investigated quantitatively in order to confirm the effects of the internal stress on the formation of the anomalous cubic phase. Figure 4 (b) shows residual stress, measured by the substrate curvature method, of the BZT films as a function of oxygen partial pressure.

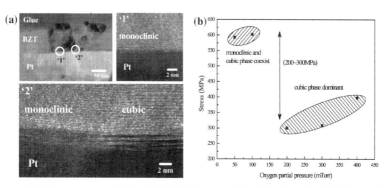

Figure 4. (a) Cross section TEM and HR-TEM image of 50mTorr-deposited BZT film, (b) residual stress of BZT films

As can be seen in Fig. 4 (b), all of the thin films are under tensile stress and the values for the BZT films deposited at high oxygen partial pressure above 200 mTorr with the anomalous cubic phase are ranging from 300 to 400 MPa, which are almost half that of the BZT thin films deposited at low pressure below 100 mTorr with the mixed cubic and monoclinic phases. Assuming that the other factors such as the thermal expansion and the thickness of the thin films for the two phases with the identical composition can be ruled out, the difference in the residual stress between the thin films with different phases can be attributed to the formation of the coherent interfacial layer in the anomalous cubic phase. In other words, while thin films with the stoichiometric monoclinic phase grown on the bare Pt substrate are under the large tensile stress, the non-stoichiometric anomalous cubic phase can form under the small tensile stress since the interfacial layer formed at high oxygen partial pressure plays an important role as a buffer layer to reduce the internal stress.

The dielectric properties of the BZT films deposited at 50 mTorr and 400 mTorr are summarized in Table 1. The BZT film deposited at 400 mTorr with the anomalous cubic phase has a high dielectric constant and a negative temperature coefficient of capacitance (TCC). This negative TCC of the cubic BZT film is consistent with other bulk cubic pyrochlore ceramics such as $Bi_{3/2}ZnTa_{3/2}O_7$ and $Bi_{3/2}ZnNb_{3/2}O_7$ with larger unit-cell volumes [2, 3]. $Bi_2(Zn_{1/3}Ta_{2/3})_2$ bulk ceramics with the monoclinic phase were reported to have a positive TCC and moderate relative dielectric constant; TCC = +60 ppm/°C, k = 63.6 [3]. The lower relative dielectric constant (k = 100) and the more stable TCC (-40 ppm/°C) of BZT film deposited at 50 mTorr result from monoclinic phase in low oxygen partial pressure.

Oxygen partial pressure (mTorr)	Phase	k	tanδ	TCC (ppm/°C)
50	monoclinic + cubic	100	0.010	-40
400	cubic	178	0.007	-170

Table 1. Dielectric properties of the BZT films deposited at 50 and 400 mTorr.

CONCLUSION

BZT thin film deposited at high oxygen partial pressure by PLD technique have the anomalous cubic phase, which is not in agreement with the target monoclinic phase. This growth of the anomalous cubic phase which has the same composition as the target results from the formation of the coherent interfacial layer. Since the lattice parameter of this interfacial layer is close to that of cubic phase, induced internal stress in the non-stoichiometric anomalous cubic phase grown on the coherent interfacial layer is less than that in the stoichiometric monoclinic phase grown on bare Pt substrate. The anomalous cubic phase can form under small tensile stress by the coherent interfacial layer. The anomalous cubic phase has a high dielectric constant and a negative TCC, which is similar to dielectric properties of other cubic pyrochlore ceramics.

REFERENCE

1. C. Ang, Z. Yu, H. J. Youn, C. A. Randall, A. S. Bhalla, L. E. Cross, M. Lanagan, Appl. Phys. Lett.., 82, 3734 (2003)

2. C. Ang, Z. Yu, H. J. Youn, C. A. Randall, A. S. Bhalla, L. E. Cross, J. Nino, M. Lanagan, Appl. Phys. Lett.., 80, 4807 (2003)

3. H. Youn, T. Sogabe, C. A. Randall, T. R. Shrout, and M. T. Lanagan, J. Am. Ceram. Soc..84, 2557 (2001)

4. I. Levin, C. A. Clive, J. Nino, M. T. Lanagan, J. Mater. Res.., 17, 1406 (2002)

5. H. B. Hong, D. Kim, and K. S. Hong, Jpn. J. Appl. Phys.. 42, 5172 (2003)

6. B. H. Hoerman, G. M. Ford, L. D. Kaufmann, B. W. Wessels, Appl Phys. Lett..73, 2248 (1998)

7. P. K. Petrov, E. F. Carlsson, P. Larsson, M. Friesel, and Z. G.Ivanov, J. Appl. Phys..84, 3134 (1998)

8. B. E. Watts, F. Leccabue, G. Bocelli, G. Calestani, F. Valderon, O. Demelo, P. P. Gonzalez, L. Vidal, and D. Carrillo, Mater. Lett..11, 183 (1991)

9. G. G. Stoney, Proc. R. Soc. London, ser. A 82, 172 (1909)

10. S. H. Kim, D. J. Kim, J. Maria, A. I. Kingon, S. K. Streiffer, J. Im, O. Auciello, A. R. Krauss, Appl. Phys. Lett..76, 496 (2000)

11. K. Kato, K. Suzuki, D. Fu, K. Nishizawa, and T. Miki, Appl. Phys. Lett.. 81, 3227 (2002)

12. X. Wang, H. Wang, and X. Yao, J. Am, Ceram. Soc..80, 2745 (1997)

Mater. Res. Soc. Symp. Proc. Vol. 875 © 2005 Materials Research Society   O12.19

Strain and Grain Size Effects on Epitaxial PZT Thin Films

Oscar Blanco and Jesus Heiras[1]
Centro de Investigación en Materiales, DIP-CUCEI, Univ. de Guadalajara
Apdo. Postal 2-638, CP. 44281, Guadalajara, Jal., México
[1]Centro de Ciencias de la Materia Condensada, UNAM
Apdo. Postal 2681, C.P. 22800, Ensenada, B.C., México

ABSTRACT

Epitaxial ferroelectric thin films of lead zirconium-titanium oxide, $Pb(Zr_{0.53}Ti_{0.47})O_3$ (PZT), were successfully grown on $SrTiO_3$, $LaAlO_3$, and $Sr(Nb)TiO_3$ single crystal substrates by a modified RF sputtering technique at high oxygen pressures. The structural properties of the films were evaluated by $\theta/2\theta$, ω and ϕ scans. From these data the crystalline orientation relationships may be extracted. For films grown on $SrTiO_3$ and $Sr(Nb)TiO_3$ substrates, the following orientation relationships were determined: PZT [001] parallel to [001] of the substrate, and PZT [100] parallel to [100] of the substrate. Films grown on $LaAlO_3$ substrates showed a bi-domain crystalline structure with orientation relationships as follows: PZT [100] parallel to [001] of the substrate and PZT [001] parallel to [001] of the substrate. This work was focused to the determination the strain and grain size coefficients, and the analysis of their contribution on the peak broadening in the XRD patterns, and in considering their effects over the ferroelectric behavior. From Williamson-Hall plots, it was possible to conclude that the enhancement of the crystalline film properties (epitaxy and single crystalline domains) reduce the short range strains contribution to peak broadening. On other hand, the grain size contribution to peak broadening was increased with the enhancement of the film cristallinity.

INTRODUCTION

Thin films of several ferroelectric materials have been studied by their potential use in numerous applications. Such applications included devices in the nano-scale sizes, for example, FRAM´s, NVRAM´s and electro-optical devices [1,2,3]. In this way, many research efforts are dedicated to studying and understanding the deviations from the bulk properties observed in thin films resulting from the lattice mismatch strain and the size effects (film thickness and grain size). In particular, epitaxial films deposited on lattice-mismatched substrates are often subject to the large coherence strain [4]. In these films, the broadening (FWHM) of the diffractions peaks is the sum of contributions from the grain (domains) size and the local strain. In this work, we studied the strain and grain size effects on the structural characteristics and ferroelectrics properties of the films.

EXPERIMENTAL DETAILS

Epitaxial ferroelectric thin films of lead zirconium-titanium oxide, $Pb(Zr_{0.53}Ti_{0.47})O_3$ (PZT), were RF sputtered onto $SrTiO_3$ (STO), $LaAlO_3$ (LAO), and $Sr(Nb)TiO_3$ (SNTO) single crystal substrates by a modified RF sputtering technique at high oxygen pressures as reported previously [5,6]. The relevant deposition conditions are summarized in the Table I.

The crystalline quality and structure features were evaluated by detailed XRD analyses, included $\theta/2\theta$, ω and ϕ scans. Calculations of the relative magnitudes of the lattice strain and crystalline size were performed by an analysis based on the Williamson-Hall plots. The ferroelectric characterization, polarization vs. applied field hysteresis loops, was obtained with an RT66A ferroelectric test system.

Table I. Relevant deposition conditions of epitaxial PZT films [5]

Target	Pb(Zr $_{0.53}$Ti$_{0.47}$)O$_3$, 1" diameter
Substrate	SrTiO$_3$, LaAlO$_3$ and Sr(Nb)TiO$_3$
Substrate temperature	550° C - 600°C
Oxygen pressure	1 Torr – 2 Torr

RESULTS AND DISCUSSION

<u>**Crystalline properties**</u>

The crystalline properties were evaluated by XRD analyses using a four-circle diffractometer. From the $\theta/2\theta$ and ω scans we determined the c-axis and a-axis lattice parameters and the out-of-plane crystalline orientation of the film (in the plane perpendicular to the substrate surface). The in-plane orientation of the film (in the substrate surface plane), and its mosaic spread, were investigated recording a ϕ-scan of the substrate and thin film from the (202) reflection.

Figures 1, 2 and 3 show the $\theta/2\theta$, ω and ϕ scans performed in the PZT/STO and PZT/SNTO structures. These analyses show that the PZT films grown on STO and SNTO substrates are single crystalline, c-axis oriented with small dispersion, and have an exact parallel orientation with the substrates: (001) PZT ∥ (001) STO or SNTO and [001] PZT ∥ [001] STO or SNTO.

a) 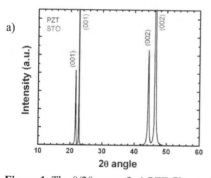 b)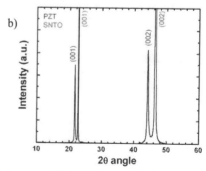

Figure 1. The $\theta/2\theta$ scan of: a) PZT film on STO substrate, and b) PZT film on SNTO substrate. Both show c-axis orientation.

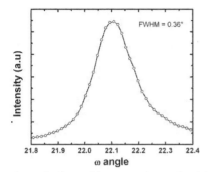

Figure 2. The rocking curve (ω-scan), of the (002) reflection form a PZT film on a SNTO substrate. The FWHM value proofs the low orientation dispersion.

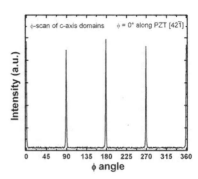

Figure 3. The φ-scan for a PZT thin film c-axis oriented on a STO substrate, with four peaks positioned at 90° relative to each other.

Figure 4, and 5 show the θ/2θ and φ scan performed on the PZT/LAO structure. These analyses show that PZT films grown on LAO substrates have a crystalline bi-domain structure with two crystalline orientations: a-axis oriented and c-axis oriented. Both domains are single crystalline and have an exact parallel orientation with the substrate: (100) PZT ‖ (001) LAO and (001) PZT ‖ (001) LAO.

In all cases, for the films grown on SrTiO₃, on Sr(Nb)TiO₃ and on LaAlO₃, the φ-scan of (202) crystalline planes shows a four-fold symmetry with peaks positioned at 90° relative to each other, without intermediate peaks (Figures 3 and 5). The fact that there are only 90° spaced peaks in the φ-scan of film grown on LAO, suggests that the spatial relationship between c-axis domains and a-axis domains adjust with the presence of 90° domains wall.

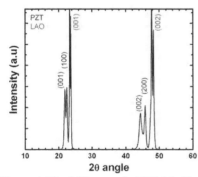

Figure 4. The θ/2θ scan of a PZT thin film on LAO substrate showing the c-axis and a-axis oriented crystalline domains.

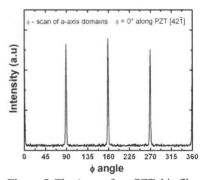

Figure 5. The φ-scan for a PZT thin film a-axis oriented on a LAO substrate, with four peaks positioned at 90° relative to each other.

Consistent with the PZT crystalline structure, the (202) crystallographic plane has a 45° inclination respect to the (001) crystallographic plane. Additionally, the basal axis of the (202) plane is the crystalline a-axis. In this form, the values of crystallographic angles implicated in the ϕ-scan are: $\theta = 33°$, $2\theta = 66°$, $\psi = 43.6°$ and $44.9°$, and $\phi = 0 - 360°$.

The calculated value of c-axis length of PZT films on STO and SNTO is 4.09 ±0.01 Å, and 4.07±0.01 Å for films on LAO. The same analyses reveal an a-axis lattice parameter of 4.01 ±0.03 Å for films on STO and SNTO, which gives a tetragonal distortion ratio (c/a)-1 of 0.019. The a-axis lattice parameter for films on LAO was 3.96±0.04 Å, with a tetragonal distortion ratio of 0.027. From these results, is possible deduce that the unitary cell of the PZT films on LAO substrates is subject at more complex combination of strain, lattice-mismatch and thermal strain, than films on STO.

Strain and grain size coefficients

Epitaxial films deposited on substrates witch large mismatch are generally are subject to large coherency strains. These strains are in some degree responsible for the diffraction peak broadening. Broadening of the diffraction peaks is due to the sum of the contributions arising from grain size and strains (tensile or compressive). The contribution of local strain to the FWHM of the $\theta/2\theta$ Bragg peaks is given by [4]:

$$\Gamma_S^2 = K_S \tan^2 \theta \quad (1)$$

$$\Gamma_G^2 = K_G \frac{\lambda^2}{\sin^2 2\theta} \quad (2)$$

where K_S and K_L are the coefficients due to strain and grain size effects, respectively, and λ is the wavelength of X-Ray radiation. Hence the broadening is given by [4]:

$$\Gamma_{\theta/2\theta}^2 \sin^2 \theta \cos^2 \theta = K_S \sin^4 \theta + K_G \quad (3)$$

where $\Gamma_{\theta/2\theta}^2$ is the FWHM measured in the peaks of the $\theta/2\theta$ scan. Figure 6 shows the Williamson-Hall plots built according to equation 3; from these plots the relative contribution of strain and grain size to the FWHM are determined from the slope and the intercept of the plots.

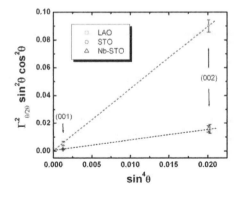

Figure 6. The Williamson-Hall plots built according to equation 3 and taking the XRD data's from the $\theta/2\theta$ scan of epitaxial PZT films deposited on LaAlO₃, SrTiO₃ and Sr(Nb)TiO₃ substrates.

Table II. Calculated values of strain (K_S) and grain size coefficients (K_G), used in the determination of the local strain in the diffraction peak broadening.

Substrate	Angle 2θ	$\Gamma_{\theta/2\theta}$	K_S	K_G ($\times 10^{-4}$)
SNTO	21.68	0.21 ±0.01	0.75	5.56
	44.21	0.36 ±0.02		
STO	21.74	0.20 ±0.02	0.77	1.31
	44.33	0.38 ±0.02		
LAO	21.74	0.41 ±0.01	4.48	0.18
	44.24	0.87 ±0.01		

The calculated values of the strain (K_S) and the grain size (K_G) coefficients are shown in Table II. In general, the local strain contribution to the FWHM is reduced by the presence of a unique crystalline orientation, c-axis domains with 180° walls. Then, the broadening effect is basically due to the size of these domains. In other hand, the grain size contribution to the FWHM showed a strong reduction when two crystalline orientations are present: a-axis domains and c-axis domains with 90° walls and 180° walls. In this case the local strains introduced basically for the lattice-mismatch and for thermal effects, are the principal contribution in the diffraction peak broadening. From the crystallographic analyses and the determination of strain coefficients, the proposed model for the crystalline domains organization in PZT films grown on SrTiO$_3$, Sr(Nb)TiO$_3$ and LaAlO$_3$ substrates are shown in the Figure 7.

a) 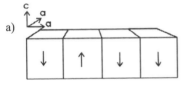   b)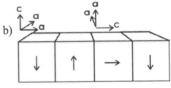

single c-axis domains   c-axis / a-axis domains

Figure 7. Proposed model for crystalline domains of epitaxial PZT film grown on: a) SrTiO$_3$ and Sr(Nb)TiO$_3$, b) LaAlO$_3$, single crystal substrates.

The domain-reorientation proposed in films on LAO, is possible when the films are under tensile strain, since c-axis of the PZT unit cell is longer than a-axis. From the point of view of the thermal strain, thermal expansion coefficient of PZT is larger than that of LAO. Therefore, in this case, the homogeneous strain induced by thermal expansion of the PZT film adhering to the LAO substrate gives rise to a tensile strain since all d spacings are elastically expanded. In other hand, the lattice misfit gives rise to compressive strain. Because of limitation of space in this paper the conflict of two effects will be analyzed elsewhere.

Ferroelectric properties

The characterization of ferroelectric properties made by the hysteresis loops measurements en film deposited on Sr(Nb)TiO$_3$ substrates, had been reported elsewhere [5,6]. Figure 8 shows a typical hysteresis loops obtained in this structure. The principal characteristics of this loops is the saturation in the polarization and the present of a slightly sloped square shape. These characteristic have been associated with two factors: the presence of the interfacial lead oxide layer that introduces a thin dielectric layer in the film – substrate interface and, as shown in this work, at the presence of 180° c-axis domains that are associated with the 180° ferroelectric domains. These types of domains are responsible for the reduction of the efficiency of the polarization inversion mechanism. Hysteresis loops measurements performed in films deposited on LaAlO₃ are in process and will be present in future reports.

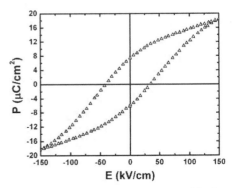

Figure 8. Hysteresis loop measured in the Pt/PZT/SNTO structure

CONCLUSIONS

Ferroelectric thin films of $Pb(Zr_{0.53}Ti_{0.47})O_3$ (PZT), were successfully grown on $SrTiO_3$ (STO), $Sr(Nb)TiO_3$ (SNTO) and $LaAlO_3$ (LAO) substrates. The structural properties showed epitaxial growth and two crystalline orientation: c-axis in film on $SrTiO_3$ and $Sr(Nb)TiO_3$ substrates, and c-axis and a-axis in film on $LaAlO_3$ substrates. Both crystalline domains show a four-fold symmetry. These suggest the presence of 180° and 90° ferroelectric domains, 180° are the unique domains in the structure PZT/STO and PZT/SNTO, and a mixing of both domains are present in the PZT/LAO structure. Based on the Williamson-Hall plots, it was possible to conclude that the enhancement of the crystalline film properties (epitaxy and single crystalline domains) reduce the short range strains contribution to peak broadening. On other hand, the grain size contribution to peak broadening was increased with the enhancement of the film cristallinity.

ACKNOWLEDGMENTS
Authors would like to thank D. Schlom and V. Vaithyanathan (Penn State Univ.) for their assistance on four-circle XRD analyses. This work was partially supported by Conacyt proj. 40604-F, PAPIIT-UNAM proj. IN116703-3 and PromeP-UdeG CA-379 and EXB-308.

REFERENCES
1. J.F. Scott, M. Ross, C.A. Paz de Araujo, M.C. Scott and M. Huffman, *MRS Bulletin* **21**(7), 33 (1996).
2. J.F. Scott, in *"Ferroelectric Memories"*, Ed. Springer (2000).
3. A.Y. Wu, F. Wang, Ch. Juang and C. Bustamante, *Mat. Res. Soc. Symp. Proc.*, **200**, 261-266 (1990)
4. J.G. Yoon and T.K. Song, "Fabrication and Characterization of Ferroelectric Oxide Thin Films" in *"Handbook of Thin Films Materials"*, ed. H.S. Nalwa (Academic Press, 2002) pp. 309-360.
5. O. Blanco, J. Heiras, J.M. Siqueiros, E.Martínez and A.G. Castellanos-Guzmán, *J. Matt. Sci. Lett.*, **22**, 449-453 (2003).
6. O. Blanco, J. Heiras, J.M. Siqueiros, E. Martínez and E. Andrade, *Mat. Res. Soc. Symp. Proc.* **784**, C3.23 (2004).

Mater. Res. Soc. Symp. Proc. Vol. 875 © 2005 Materials Research Society

Effects of Ru Vacancies and Oxygen Synthesis Pressures on the Formation of Nanodomain Structures in SrRuO$_3$ Thin Films

Y. Z. Yoo, O. Chmaissem, S. Kolesnik, B. Dabrowski, C. W. Kimball

Institute of NanoScience, Engineering and Technology (INSET), Physics Department, Northern Illinois University, DeKalb, IL 60115

L. McAnelly, M. Haji-Sheikh, A. P. Genis

INSET, Electrical Engineering Department, Northern Illinois University, DeKalb, IL 60115

ABSTRACT

SrRuO$_3$ (SRO) thin films were grown on SrTiO$_3$ (100) substrates using the pulsed laser deposition method. The films' growth properties widely changed in response to different working oxygen partial pressures. An island growth mode was dominant for low pressures up to 10 mTorr followed by a step flow growth mode at 60 mTorr and step flow plus 2 D growth at 200 mTorr then reverting back to island growth at 300 mTorr. Significant out-of-plane strains of SRO films were observed for low growth pressures (up to 10 mTorr) but became notably reduced at 60 mTorr and continued to decrease gradually with further pressure increases. Formation of Ru vacancies occurs regardless of the working pressure values and appears to be minimized at 60 mTorr. Highest T$_C$'s were obtained in films exhibiting the step flow growth mode. The role of Ru deficiencies in relation to strain, growth mode, and magnetic properties is discussed.

INTRODUCTION

SrRuO$_3$ (SRO), a ferromagnetic metal oxide with curie temperature of 163 K, crystallizes the perovskite structure of the GaFeO$_3$ orthorhombic family and has a pseudocubic lattice constant of ~ 0.393 nm that is nearly the same as that of cubic SrTiO$_3$ (a ~ 0.391 nm). The structural, magnetic and transport properties of SRO make the material suitable as an electrode candidate for perovskite and spin-based ferroelectronics.

SrRuO$_3$ films grown on SrTiO$_3$ (STO) substrates exhibit interesting properties that are different than the corresponding bulk properties. For example, SRO thin films grown on STO are compressively strained owing to its in-plane lattice matching to that of STO. The presence of such strains is known to negatively interfere with the spontaneous magnetization process, thus, resulting in a reduction of the ferromagnetic critical temperature T$_C$. In addition, SRO films typically show the evolution of unique terrace and step structures in a narrow region of growth conditions on STO substrates presenting surface step and terrace structures, which is also related to the magnetic properties and microstructure of the films.[1,2] These properties cannot be considered separately because they correlate to each other. Thus, in order to study the relationship among the various above-mentioned physical properties, SRO films with different surface structures should be fully investigated for a better understanding of their magnetic, transport and structural properties.

In this paper, SRO thin films have been grown with systemically modified physical properties by changing the working oxygen partial pressure. This work addresses a wide range of oxygen partial pressures employed in the growth of SRO films on STO substrates and the coupling between the surface, magnetic and structural properties.

EXPERIMENTAL DETAILS

SRO films were grown on nominal SrTiO$_3$ (100) substrates (MaTeck GmbH) using the Pulsed Laser Deposition method (PLD) and a KrF excimer laser (λ = 248nm, 10 Hz). A base pressure of 5×10^{-7} Torr was achieved in the chamber prior to deposition. A stoichiometric single phase SRO target was ablated for 30 minutes at a laser power of 250 mJ at a fixed growth temperature of 830 °C. Oxygen partial pressure was varied from 0.1 mTorr to 300 mTorr at the same growth temperature. Structural properties of the obtained films were examined using a Rigaku D$_{max}$ x-ray diffractometer (XRD). Surface profiling was carried out using atomic force microscopy (AFM) (Quescant Q-Scope) and film compositions were measured by energy dispersive x-ray analysis (EDX). To determine the exact Sr/Ru contents in the films, additional EDX measurements were carried out for SRO films on Sr-free LaAlO$_3$ substrates grown under identical growth conditions. Magnetic properties of the grown films were investigated using a Superconducting Quantum Interference Device (SQUID) (Quantum Design).

RESULTS AND DISCUSSION

Atomic force microscopy (AFM) images of SRO film surfaces that formed under various oxygen pressures are displayed in Fig. 1. As shown in parts (a), (b) and (c) of the figure, the surface morphologies for the 0.1 mTorr, 1 mTorr and 10 mTorr films exhibit a 3D island-like growth mode with a relatively low RMS roughness of less than 1.1 nm.

When the oxygen working pressure was raised to 60 mTorr, well developed terraces and step structures formed as can be seen in Fig. 1(d). A continuation of this trend is shown in Fig. 3(e), where the terrace width increases with increasing working pressure from 60 to 100 mTorr.

The SRO film grown at 200 mTorr in Fig. (f) shows a unique surface morphology in which terraces and two different types of steps are observed. The height of the small steps and the width of the terraces, along the line labeled B-B' shown in Fig. 1(f) are approximately 0.4 nm that correspond to one unit cell of SRO, and 430 nm, respectively. The height of the larger steps observed along the C-C'direction (Fig. 1(f)) is about 3 nm, with an estimated terrace width of ~ 1300 nm. SRO films grown at 300 mTorr display the same island growth mode characteristic of the low pressure grown films (up to 10 mTorr) but with a larger estimated island size, thus, indicating that coalescence begins to dominate at high pressures (300 mTorr).

SRO films grown on STO at 60 and 100 mTorr show distinct structural directionalities. However, as the oxygen partial pressure is increased to 200 mTorr, the step and terrace patterns lose directionality. This observation clearly indicates that step and terrace morphology in SRO films depends on a narrow range of oxygen pressure as well as on the surface structure of the STO substrates.

Fig. 2 shows the Sr/Ru intensity ratio, film strains, Curie temperature and growth mode as a function of oxygen partial pressure. The intensity ratio of Sr (K_α) to Ru (K_α) energy peaks from the EDX profiles for all the SRO films has a higher value than that of the bulk. Our analysis indicates the formation of Ru vacancies in all films when compared to bulk SRO over the entire working pressure range. From this result, it can be presumed that SRO film compositions cannot be directly controlled by the partial pressure parameter.

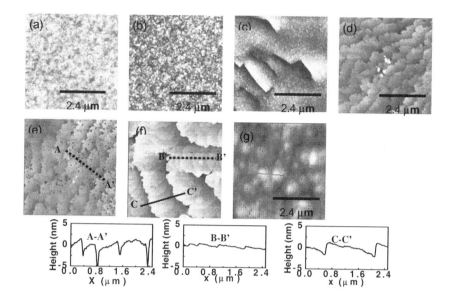

Figure 1. AFM images (5 μm x 5 μm) for SRO films grown at 0.1 mTorr ;(a), 1 mTorr; (b), 10 mTorr; (c), 60 mTorr;(d), 100 mTorr;(e) , 200mTorr(f) and 300 mTorr;(g). Surface profiles along the A-A', B-B' and C-C' lines are shown in bottom figures, respectively. SRO films at 60 mTorr and 100 mTorr show step flow growth mode. SRO film at 200 mTorr showed a combined growth mode of 2 D (main) and step flow.

This is in qualitative agreement with previous results showing that Ru deficiencies were most likely generated during growth in rich oxygen atmospheres.[3] All films were preferentially grown along the pseudocubic [001] orientation (not shown). SRO Films grown in the lower working pressure region (from 0.1 mTorr to 10 mTorr) have a significantly larger out-of-plane lattice constant when compared to that of the bulk (0.398-0.400 nm versus 0. 393 nm, respectively). As the working pressure increases, the out-of-plane lattice constants of SRO films decreased to 0.360 nm at 60 mTorr and neared the bulk value at pressures above 60 mTorr, although they were still somewhat larger (not shown here). The elongation of the out-of- plane lattice constant was explained as due to strains and Ru deficiency.[3,4] Based on this, the out-of-plane film strains (calculated from the lattice deviations) were determined and shown in Fig. 2. The plot reveals that SRO films grown in pressures up to 10 mTorr exhibit substantial strains exceeding 1% whereas films grown in pressures above 10 mTorr have weaker strains.

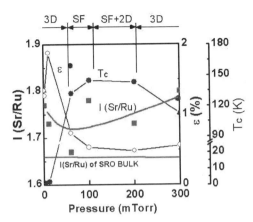

Figure 2. Sr/Ru intensity ratios obtained from EDX measurements (I), film strains (ε), Curie temperature (T_C) and growth mode as a function of oxygen partial pressure. Some SRO films show two T_C values: 139 K (major) and 163 K (minor) for the 60 mTorr film and 112 K (major) 133 K (minor) for the 300 mTorr. SF, 3D and 2D represent step flow growth, three-dimensional island growth and two-dimensional layer-by-layer growth, respectively.

The ferromagnetic Curie temperature, T_C, was determined by taking the derivative of magnetization with respect to temperature (dM/dT). No ferromagnetic properties are observed for any of the SRO films grown at pressures at 10 mTorr or below, irrespective of growth temperature.

A ferromagnetic behavior was consistently observed in films grown at oxygen partial pressures of 60 mTorr or above. The highest T_C values of 145 K and 143 K were obtained of the 100 and 200 mTorr films, respectively. The magnetization results are in good agreement with the XRD data which show that the 100 and 200 mTorr films have out-of-plane lattice constants closer to that of the bulk than the 60 mTorr films. The lower than expected T_C values (T_C = 163 K for the bulk $SrRuO_3$ materials) and the absence of ferromagnetism in the low pressure grown films may be explained by increased structural distortions involving RuO_6 octahedra as well as other interatomic structural features like the bond angles and bond lengths due to the significant elongation of the out-of-plane lattice constant.

The 300 mTorr film showed a significantly reduced T_C of 112 K. Concerning the 300 mTorr grown film, the low T_C cannot be explained based on strain effects alone, since this sample exhibits strains that are very similar to those of the 100 mTorr film. Also, the 60 mTorr SRO film showed two different T_C's, at 139 K (major one) and 163 K (minor one). Instead, we interpret the observation of two T_C values to result from some inhomogeneous distribution of the Ru vacancies over different domains. The 163 K transition is herein attributed to minority domains that are stoichiometric, while the reduced 137 K transition is explained by the combination of Ru deficient domains and compressive strain effects. As such, the results of both the 60 and 300 mTorr films are corroborated by the explanation that the generation of Ru vacancies, coupled with compressive strains, are two of the major mechanisms leading to a T_c reduction in SRO films.

Finally, the presence of Ru vacancies plays a critical role in the reduction of T_C, and results in the elongation of out-of-plane lattice constants in grown films. It is found that SRO films show high T_C's when the growth mode of SRO films is governed not by 3D growth mode but by either step flow or 2D growth modes. This indicates that the growth mode is closely associated with both film strain and stoichiometry deviation resulting from Ru vacancies.

ACKNOWLEDGMENTS

Work at NIU was supported by the Institute of NanoScience, Engineering and Technology - US Department of Education and by the NSF Grant No (DMR-0302617).

REFERENCE

1. E. Vasco, R. Dittmann, S. Karthäuser, and R. Waser, Appl. Phys. Lett. **82**, 2497 (2003).

2. Rao, R.A.; Gan, Q.C.B.Eom, Appl. Phys. Lett. **71**, 1171 (1997).

3. B. Dabrowski, O. Chmaissem, P. W. Klamut, S. Kolesnik, M. Maxwell, J. Mais, Y. Ito, B. D. Armstrong, J. D. Jorgensen, and S. Short, Phys. Rev. B **70**, 014423 (2004).

4. Y. Z. Yoo, O. Chmaissem, L. McAnelly, A. P. Genis, M. Haji-Sheikh, S. Kolesnik, B. Dabrowski, M. Maxwell, C. W. Kimball, J. Appl. Phys. **97**, 103525 (2005).

Mater. Res. Soc. Symp. Proc. Vol. 875 © 2005 Materials Research Society

Failure Analysis of Thermally Shocked NiCr Films on Mn-Ni-Co Spinel Oxide Substrates

Min-Seok Jeon, Jun-Kwang Song, Eui-Jong Lee, Yong-Nam Kim, Hyun-Gyu Shin and Hee-Soo Lee
Material Testing Team, Korea Testing Laboratory,
222-13 Guro-dong, Guro-gu, Seoul 152-848, Korea

ABSTRACT

NiCr films were thermally evaporated on the Mn-Ni-Co-O thick-film substrates. The NiCr/Mn-Ni-Co-O bi-layer systems were tested in a thermal shock chamber with three temperature differences of 150, 175 and 200°C. The systems were considered to have failed when the sheet resistance of NiCr films changed by 30% relative to an initial value. As the cyclic repetition of thermal shock increased, the sheet resistance of NiCr coatings increased. The Coffin-Manson equation was applied to the failure mechanism of cracking of NiCr coatings and the SEM observation of cracks and delamination in NiCr coatings due to thermal cycling agreed well with the failure mechanism.

INTRODUCTION

NiCr films are applied for humidity sensor, thin film resistor or infrared sensor [1-3]. In our case, the NiCr films represent an infrared absorbing coatings or electrodes because they have a high emissivity of 0.97 comparable to gold black and low cost in fabrication. Tailored Mn-Ni-Co-O thick film thermistor is one of promising candidates for commercial infrared sensing element. The output of thermistor is large in comparison with other infrared sensing element and an infrared sensor using it is able to operate at temperature of up 150°C. NiCr films absorb an infrared and transport its heat to the NTC (negative temperature coefficient) thermistor. The resistance of Mn-Ni-Co oxide decreases with its temperature and this change of the resistance is converted in the form of output voltage for sensing a infrared from objects.

NiCr/Mn-Ni-Co-O bi-layer sensing element experiences periodical temperature changes during long term operation. In addressing the issue of ceramic sensor reliability, an understanding of the failure mechanisms is required. Accelerated degradation testing is being used frequently to predict the failure and reliability of various materials [4]. In addition the temperature cycling test is extensively used in microelectronic industry to qualify new products [5]. Therefore the assessment of the durability and characteristics of the sensing element are needed by an accelerated degradation testing (ADT) of thermal cyclic method. One of important aspect of maintaining the stability of NiCr/Mn-Ni-Co oxide sensing element is the control of mechanical stresses [6,7]. These stresses are able to be generated and accumulated in the NiCr films by an operation condition of periodical temperature changes. Therefore the thermo-mechanical stability of NiCr films is required to be evaluated through the testing in an accelerated condition. Thermal shock was given to NiCr/Mn-Ni-Co oxide thick films and their characteristics were compared before and after the accelerated testing. This paper is aimed at analyzing failure mechanisms of NiCr/Mn-Ni-Co oxide sensing elements by thermal cycling.

EXPERIMENTAL DETAILS

Mn-Ni-Co-O green sheets were prepared using a tape caster. The green sheets were stacked using a one-axial pressing machine and sintered at 1200°C using a conventional furnace. NiCr films were coated on the 100μm-thick Mn-Ni-Co-O substrates at room temperature using a thermal evaporator. The ratio of Ni and Cr was 80:20 and the coating thickness was between 150nm and 200nm. The NiCr/Mn-Ni-Co-O systems were exposed to periodic changes in temperature in a thermal shock chamber (TSA-41L, ESPEC Corp., Osaka, Japan) and the sheet resistance of NiCr coatings was measured after 100, 250 and 500 cycles using a four-point probe (MCP-T600, Mitsubishi Chemical Corp., Tokyo, Japan) at room temperature. The temperature difference at thermal cycling was 150°C, 175°C and 200°C. NiCr/Mn-Ni-Co-O systems were exposed to an atmosphere of higher and lower temperature periodically by a hot and cold air. The lower temperature in the thermal cycling testing was fixed at -40°C and the dwell time at each temperature was 30 minutes. The micro-structure of the surface and cross-section of the NiCr/Mn-Ni-Co-O were observed using a field emission scanning electron microscope (S-4700, Hitachi Ltd., Tokyo, Japan) and the change in the surface roughness of the NiCr was analyzed using an atomic force microscope (XE-100, PSIA Corp., Sungnam, Korea).

DISCUSSION

Thermal cyclic method was used to accelerate the degradation and to address the reliability of NiCr coatings on Mn-Ni-Co-O substrates. Figure 1 shows the variation in sheet resistance of NiCr coatings with the number and temperature range of thermal cycling. As the thermal cycling was repeated, the sheet resistance of NiCr coatings was increased parabolically. Increasing rate in sheet resistance was gradually decreased as thermal cycling was increased. Moreover, when the temperature difference of thermal cycling was enlarged, the sheet resistance of NiCr coatings was increased. A sample size of 15 specimens was subjected to each condition and the NiCr coatings were considered to fail when the sheet resistance was varied by 30% relative to an initial value.

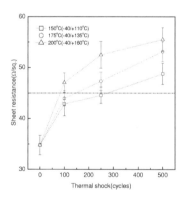

Figure 1. The variation in sheet resistance of NiCr coatings with the number of thermal cycling.

This was marked in figure 1 with a dotted horizontal line. The plot of variation in sheet resistance at each condition was fitted by a polynomial expression. The number of thermal cycling to the failure of the NiCr coatings at $\Delta T = 150°C$, $175°C$ and $200°C$ was found to be 260, 135 and 83 cycles, respectively.

The logarithmic relationship between ΔT and the number of cycle to failure (N) was represented in figure 2. The well-known Coffin-Manson equation is widely used for modeling a failure mechanism of low cyclic fatigue of a metal by a thermal cycling [8, 9]. Coffin-Manson(C-M) relationship is given below.

$$N_f = A(\Delta T)^{-B} \qquad (1)$$

where N_f is the number of cycles to failure, ΔT is the temperature range of thermal cycling, A is a constant related to the material type, testing method, etc, B is the Coffin-Manson coefficient determined experimentally. Because there was the same failure mechanism for all three conditions in this study, the Coffin-Manson coefficient could be determined by using a least square method of fitting N_f and ΔT to equation 1, the results are shown in figure 2. The Coffin-Manson coefficient was turned out to be 3.97. It was recently reported that the Coffin-Manson coefficient values are in the 1-3 range for ductile metal (e.g. solder), 3-5 for hard alloys or intermetallics (e.g. Al-Au) and 6-9 for brittle fracture mechanisms (e.g. Si, SiO_2, Si_3N_4) [5]. Therefore, it is concluded that the failure mechanism of the NiCr/Mn-Ni-Co-O systems is the cracking of the NiCr coatings which causes the change in their sheet resistance.

The scanning electron microscopy verifications were done with as-grown and failed NiCr coatings (figure 3). The surface of the NiCr coatings stressed at $\Delta T = 150°C$ showed many surface defects. After thermal shock testing of 500 cycles at $\Delta T = 200°C$, many cracking places were shown on the surface of the NiCr coatings. Figure 4 shows the cross-section of as-grown and failed NiCr/Mn-Ni-Co-O samples. It was shown that when NiCr/Mn-Ni-Co-O was stressed at $\Delta T = 150°C$, some delamination was formed at the NiCr-Mn-Ni-Co-O interface.

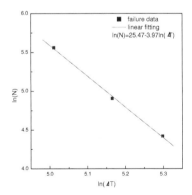

Figure 2. The logarithmic plot of the number of cycle to failure as a function of temperature difference.

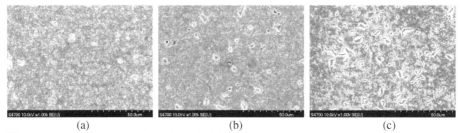

(a)   (b)   (c)

Figure 3. SEM micrographs of surface of NiCr coatings, (a) as-grown sample (b) after 500 cycles at ΔT=150°C and (c) after 500 cycles at ΔT=200°C.

The cross-section, which was stressed at ΔT=200°C, of NiCr/Mn-Ni-Co-O exhibited the surface cracking of NiCr coating including its delamination. This failure analysis confirmed again the observation from the Coffin-Manson model.

Figure 5 shows atomic force microscopy images of the surface of the NiCr coatings before and after the thermal cycling testing. The surface roughness of the NiCr coatings was found to increase with temperature range of thermal cycling. The surface defects of Figure 3 (b) were turned out to be pits from the atomic force microscopy image of Figure 5 (b). The NiCr coating stressed at temperature range of 200°C had rougher surface than that stressed at 150°C, which is consistent with the scanning electron microscopy photographs of the cross-section.

(a)   (b)   (c)

Figure 4. SEM micrographs of cross-section of NiCr/Mn-Ni-Co-O, (a) as-grown sample (b) after 500 cycles at ΔT=150°C and (c) after 500 cycles at ΔT=200°C.

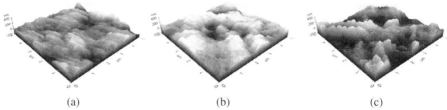

(a)   (b)   (c)

Figure 5. The atomic force microscopy images of NiCr coatings, (a) as-grown sample (b) after 500 cycles at ΔT=150°C and (c) after 500 cycles at ΔT=200°C.

CONCLUSIONS

A thermal cyclic method in accelerated testing has been presented that can be used for reliability testing and addressing the failure mechanisms of coating materials, NiCr coating on Mn-Ni-Co-O substrates. The accelerated degradation testing was performed to NiCr/Mn-Ni-Co-O bi-layer in three temperature ranges of 150, 175 and 200°C. As the cyclic repetition of thermal shock was increased, the sheet resistance of NiCr coatings was increased. The Coffin-Manson coefficient was turned out to be 3.97, which was consistent with a brittle fracture of hard alloys or intermetallics, as was confirmed by failure analysis indicating cracking and delamination of NiCr coatings after thermal cycling.

ACKNOWLEDGMENTS

This work has been carried out under the National Research Laboratory (NRL) Program supported by Ministry of Science and Technology (MOST), Korea.

REFERENCES

1. L.I. Belic, K. Pozun, and M. Remskar, *Thin Solid Films*, 317, 173 (1998).

2. S. Takeda, *Vacuum*, 41, 1769 (1990).

3. A. Peled, J. Farhadyan, Y. Zloof, and V. Baranauskas, *Vacuum*, 45, 5 (1994).

4. W. Nelson, *Accelerated Testing*, (John Wiley & Sons, New York, 1990).

5. EIA/JEP122B, "Failure mechanisms and models for semiconductor devices", EIA/JEDEC Pub. Aug. 2003.

6. J.G. Fagan, *American Ceramic Society Bulletin*, 72, 70 (1993).

7. J. Lesage, M.H. Staia, D. Chicot, C. Godoy, and P.E.V. De Mirande, *Thin Solid Films*, 377, 681 (2000).

8. L.F. Coffin, Jr., ASME Transactions, 76, 931 (1954).

8. S.S.Manson, *Experimental Mechanics*, 5, 193 (1965).

Mater. Res. Soc. Symp. Proc. Vol. 875 © 2005 Materials Research Society   O12.24

The Effect of Porogen on Physical Properties in MTMS-BTMSE Spin-on Organosilicates

B.R. Kim*, J. M. Son, J.W. Kang, K.Y. Lee, K.K. Kang, M.J. Ko
LG Chem. Ltd./Research Park, Corporate R & D, Daejeon Korea
D.W. Gidley
University of Michigan, Department of Physics, Ann Arbor, MI. 48109

ABSTRACT

Decreasing the circuit dimensions is driving the need for low-k materials with a lower dielectric constant to reduce RC delay, crosstalk, and power consumption. In case of spin-on organosilicate low-k films, the incorporation of a porogen is regarded as the only foreseeable route to decrease dielectric constant of 2.2 or below by changing a packing density. In this study, MTMS-BTMSE copolymers that had superior mechanical properties than MSSQ were blended with decomposable polymers as pore generators. While adding up to 40 wt % porogen into MTMS:BTMSE=100:50 matrix, optical, electrical, and mechanical properties were measured and the pore structure was also characterized by PALS. The result confirmed that there existed a tradeoff in attaining the low dielectric constant and desirable mechanical strength, and no more pores than necessary to achieve the dielectric objective should be incorporated. When the dielectric constant was fixed to approximately 2.3 by controlling BTMSE and porogen contents simultaneously, the thermo-mechanical properties of the porous films were also investigated for the comparison purpose. Under the same dielectric constant, the increase in BTMSE and porogen contents led to improvement in modulus measured by the nanoindentation technique but deterioration of adhesion strength obtained by the modified edge lift-off test.

INTRODUCTION

As integrated circuit-feature dimensions continue to shrink, there is a strong need for new low-k materials to be in use with copper as the metal line [1]. In parallel with an effort to seek a homogeneous alternative to incorporate atoms and bonds with lower polarizability, the introduction of porosity into base matrix materials has been attempted to reduce a dielectric constant. The porous films are generally synthesized by blending a thermally labile porogen with a thermally stable matrix resin. However, one of the fundamental dilemmas for developing low-k dielectrics is the fact that a low dielectric constant is incompatible with high mechanical stability, particularly for porous materials where the mechanical properties are generally deteriorated further by the presence of pores. Therefore, no more porosity than necessary to achieve the dielectric objective should be incorporated to survive the harsh semiconductor fabrication processes.

In the previous works [2], MTMS-BTMSE copolymers with five different compositions were synthesized and their physical properties were characterized. Network formation enhanced as the amount of BTMSE increased, which led to the improvement of mechanical properties and the increase in the dielectric constant. In the present study, the MTMS-BTMSE copolymers were added with sacrificial porogens. At first, MTMS:BTMSE=100:50 matrix was blended with up to 40 wt% porogen to systematically investigate the influence of porogen content on pore structure and physical properties. The pore structure was characterized by positron annihilation lifetime spectroscopy (PALS) using radioactive bata-decay positrons, and the variations of optical, electrical, and mechanical properties were investigated. In addition, mechanical properties of the five different MTMS-BTMSE porous films with the same dielectric constant of 2.3 were also recorded by adjusting porogen content in accordance with the variation of BTMSE content in the matrix resin.

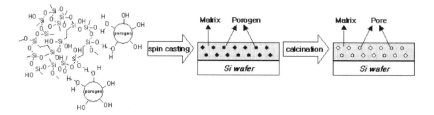

Figure 1. Schematic flowchart used to produce porous organosilicate low-k films.

EXPERIMENTAL

The porous organosilicate films were prepared by the following procedures. First, MTMS and BTMSE prepolymers were mixed with deionized water and solvent with stirring. 0.5 ~ 5.0 equivalent of deionized water per each alkoxy group was added and then the concentration of reaction mixture was adjusted to 1 M ~ 3 M by the supplement of water and solvent. Second, deionized water including a catalytic amount of HNO_3 was blended to this solution at room temperature and stirred for 30 minutes or longer. Acid catalyst was added in molar ratio of 5 × 10^{-3} ~ 5 × 10^{-2} relative to the moles of silicate precursor. Under these conditions, the turbid solution turned clear after a few minutes. This mixture was stirred at 50 ℃ ~ 80 ℃ over 6 hours and then cooled to room temperature. To this mixture was blended high boiling point solvents mentioned above. Low boiling point components were removed from the resulting reaction mixture through the vacuum evaporation method at 45 ℃, which yielded the final matrix solution containing the prepolymers with average molecular weights of 1,000 ~ 5,000. The decomposable polymers as pore generators were loaded to the matrix solution from 0 to 40 wt% before the solution was deposited on silicon wafers.

The coating solution was applied to a 4 inch silicon wafer with a spin coater at an acceleration rate of 1000 rpm/sec and a spinning rate of 2000 rpm/sec for 30 seconds. The deposited films were uniform, defect-free and reproducible after repeated experiments. After deposition, the films on wafer were dried for 2 minutes on a hot plate maintained at the temperature of 100 ℃ to thereby remove the solvent. Subsequent annealing was usually performed for 60 minutes at 430 ℃ under N_2 environment, which produced colorless transparent porous films. The thickness of the films ranged from less than 0.1 µm up to 2.0µm, depending on the spinning speed and the solid content of the coating solution. A schematic diagram used to produce porous low-k films is summarized in Figure 1.

The refractive indices and the thickness of the porous films were determined from the spectroscopic ellipsometry method [3]. The nine different points for a single film were measured and the results were averaged. The deviation of these nine points was normally less than 0.3%. The accurate measurement for thickness enabled the exact calculation of dielectric constant from the capacitance measured with metal-insulator-semiconductor (MIS) structures. Volume fraction porosity can be extracted from the refractive index and the dielectric constant using Lorentz-Lorentz equation and Maxwell-Garnett equation

$$\frac{n_0^2 - 1}{n_0^2 + 2} \times (1 - P) = \frac{n^2 - 1}{n^2 + 2}, \quad \frac{k_0 - 1}{k_0 + 2} \times (1 - P) = \frac{k - 1}{k + 2} \qquad (1)$$

where n_0 and n represent refractive indices of base matrix material and porous film, k_0 and k stand for dielectric constants of base matrix material and porous film, and P denotes volume fraction porosity.

In order to characterize the pore structures such as pore size and pore interconnectivity, PALS was employed at room temperature [4, 5]. PALS made use of an electrostatically focused beam of several keV positrons generated in a high vacuum system. When the beam was implanted into the sample, positronium (Ps), the bound state of a positron and electron, was formed in the void volume of pores by electron capture. Using the effect that Ps annihilation lifetime (2-142 ns) was shortened from vacuum lifetime due to the collisions with the pore walls, the average pore size could be determined by Ps annihilation lifetime with the extended Tao-Eldrup model [6].

Both elastic modulus and hardness often used as indications of mechanical stability for low-k candidate materials were also measured by the continuous stiffness measurement (CSM) technique on a MTS nanoindenter XP equipped with the DCM head [7]. Frequency was fixed to 75 Hz and Berkovich (three-side pyramid shape) indenter was adopted to carry out the test. Nine indentations were performed on each sample up to 200nm depth and the results were averaged. In addition, residual stress and fracture toughness were evaluated by the well-known beam bending method and the modified edge lift-off test (MELT) method, respectively [8].

RESULTS AND DISCUSSIONS

Figure 2 is thermal gravimetric analysis (TGA) data of porogen polymer under N_2 environment. Decomposition of porogen occurred between $270\,°C$ and $400\,°C$, and approximately 1.0 wt% of residue remained after $430\,°C$ annealing. Final solid contents of coating solutions were fixed to 20 wt% in proper coating solvents.

Figure 3(a) plots the measured refractive indices and the dielectric constants as a function of weight percent porogen loading for MTMS:BTMSE=100:50 base matrix. Both the properties decreased monotonically over the entire compositional range, which hinted that the

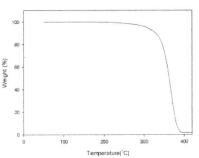

Figure 2. TGA measurement of porogen polymers.

pores were efficiently generated without pore collapse or film shrinkage upon the thermal cure. Figure 3(b) shows that volume fraction porosities extracted from the refractive index and the dielectric constant using equation (1) were somewhat different from porogen loading. The discrepancy seemed to happen because the matrix resin itself lost some of its initial weight upon curing as well as porogen loading was calculated on a weight percent basis.

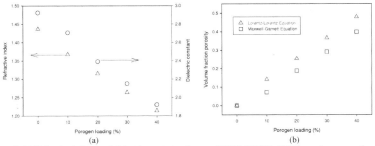

(a)   (b)

Figure 3. (a) Refractive indices and dielectric constants of porous MTMS:BTMSE=100:50 copolymers as a function of porogen loading, (b) volume fraction porosity calculated with Lorentz-Lorentz equation and Maxwell-Garnett equation.

Table 1. PALS results for the mesopores of the porous MTMS:BTMSE=100:50 films.

Porogen loading (%)	Ps lifetime (ns)	$D_{cylinder}$ (nm)	D_{sphere} (nm)	Mean Free path (nm)	Interconnection length (nm)
0 (matrix)	2.5-6.0	-	0.6-1.0	-	0(closed)
10	16	1.4	1.7	1.1-1.4	0(closed)
20	34	2.1	2.6	1.7-2.1	<30
30	58	3.2	4.2	2.8-3.2	>500

According to the experiments, not only the volume fraction of the pores but pore size, shape and distribution were also the critical factors that affected the properties of porous low-k films [4]. The increase in porosity was inclined to enlarge the pore size and when porosity exceeded about 20-30%, the pores became percolated and interconnected. These issues gave rise to reliability concerns such as moisture absorption, metal diffusion, and crack formation. Therefore, the porous low-k films with homogeneous and nanometer-sized closed pores were preferred to preserve physical properties.

Table 1 illustrates the mean pore diameter deduced from Ps lifetime by using both spherical and cylindrical pore models. Since the diffusion of Ps into the vacuum occurred for the films with an interconnected porous network, a 60 nm silicate cap was deposited to seal interconnected pores at the surface and to keep Ps confined within the films. Each sample was investigated at several implantation energies, 2.0, 3.0, and 5.0 keV, to search for any depth-related heterogeneity. In addition, the mean free path (4V/S, where V and S are the pore volume and surface area, respectively) range was set from the two pore models. The mean free path in a long cylinder was equal to the cross-sectional diameter, whereas that in a sphere was 2/3 of the sphere diameter. The interconnection length represents the effective depth from which Ps can diffuse through the porous network and escape into the vacuum.

The matrix film exhibited no mesoporous Ps signal but had a Ps signal annihilating in micropores. Two strong Ps components of 2.5 and 6.0 ns corresponded to spherical diameters of 0.6 nm and 1.0 nm. It has been generally known that Ps lifetime component in the 15-50 ns range suggested the existence in mesopores while the component less than 7.0 ns reflected the influence of micropores. This study found out that Ps in the matrix film did not escape through interconnected micropores into the vacuum and there was a distribution of pore diameters over this nominal range, which implied that these inherent voids were isolated. The three porous films had additional Ps lifetime due to the mesopores of the films. As the porogen loading increased, the pore sizes gradually increased and so did the Ps formation fraction. Increasing porogen loading led to the increase in the interconnection length, as well. While the porous film with 10% porogen possessed low enough porosity to have closed pores, the film with 20% porogen was almost closed with an interconnection length less than 30nm. The film with 30% porogen exhibited highly interconnected pores on the order of the film thickness.

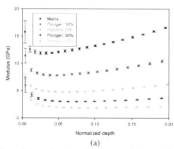
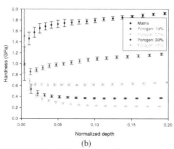

(a)   (b)

Figure 4. (a) Reduced modulus and (b) hardness of porous MTMS:BTMSE=100:50 copolymer films as a function of normalized indented depth.

Figure 4 demonstrates reduced modulus and hardness as a function of normalized indentation depth (indentation depth/film thickness). For current low-k films whose normal thickness was less than 1µm, the indented depth was so small that elastic contact was dominant and the mode of plastic deformation could hardly occur. In the situation, hardness did not provide any information about plastic properties and became proportional to modulus. As expected, modulus decreased with the increase in porogen loading, which was mainly due to the decrease in the film density. The dramatic decrease in modulus at the low porosity film implied that the relationship between modulus and porosity was not linear but it showed power law variation with exponent.

Figure 5 plots the change of residual stress and fracture toughness as a function of porogen loading, which showed a similar trend to the results in modulus. It is certain that the decrease of the residual stress in accord with porogen loading was mainly affected by the decrease in modulus though the stress was also influenced by the thermal expansion coefficient whose effect did not exceed that of the modulus for the porogen range investigated. As porogen loading increased, fracture toughness was monotonically deteriorated and its standard deviation was inclined to increase. Despite of the fact that the MELT method was generally employed to evaluate adhesion

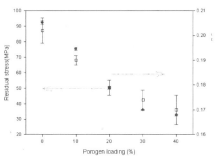

Figure 5 Residual stress and fracture toughness as a function of porogen loading for porous MTMS:BTMSE copolymers.

strength, low-k films were not completely removed from the silicon substrate. In some cases, low-k films were torn away piece by piece from the substrate, which could be ascertained to crack deflection and crack bridging mechanism governed by the presence of pores. In order to verify if the failure was purely related to adhesive fracture, it might be necessary to investigate the failure locus for each sample.

The investigation up to now has focused on the effect of porogen loading on the identical base matrix material. The amount of porogen loading needed to satisfy the target dielectric constant was dependent on the kinds of the matrix materials. The previous study for a series of MTMS-BTMSE copolymers indicated that the addition of BTMSE promoted the condensation reaction and resulted in the increase of crosslinking density [2]. The result suggested that the target dielectric constant of the porous low-k films could be tailored not only by the porogen loading but also by the amount of BTMSE. Table 2 illustrates the thermo-mechanical properties of the porous films when the dielectric constant was fixed to approximately 2.3 by controlling BTMSE and porogen contents simultaneously. The objective dielectric constant was determined with reference to ITRS roadmap in which the dielectric constant suitable for 152 nm pitch might be 2.4 and the value required for 54 nm pitch should be smaller than 2.0 [1].

Table 2 indicates that the necessary amount of porogen loading to satisfy the dielectric objective was proportional to the addition of BTMSE. In spite of considerable differences in the BTMSE contents between MTMS-BTMSE matrix copolymers, the disparity in porogen loading was only less than 5%, which implied that increasing the amount of porogen loading was more effective than controlling BTMSE contents to lower the dielectric constant. For the porous films with the same dielectric constant, the modulus was inclined to increase as the amounts of BTMSE and porogen increased. Not specified here in detail, the decreasing rate of modulus with the increase in porogen content was almost identical over the specified compositional range for a series of porous MTMS-BTMSE polymers. The result explained that the pores were efficiently generated for the copolymers and then modulus of each porous film could be predicted by measuring that of matrix unless pore collapse or film shrinkage happened.

Being different from modulus, fracture strength was gradually lowered with the increase in BTMSE and porogen loading. The previous study showed that fracture toughness of MTMS:BTMSE=100:30 matrix (0.191 MPa·m$^{0.5}$) was much higher than that of MTMS:BTMSE =100:10 matrix (0.166 MPa·m$^{0.5}$) [2]. However, the reversed fracture strength for both the porous films proved that fracture toughness of porous films was highly dependent on the presence of pores rather than the strength of base matrix materials. The simultaneous increase in BTMSE and porogen contents had an advantage in improving modulus and hardness but deteriorated fracture strength. In addition, the increased possibility of generating interconnected pores with the increase in porosity might give rise to the reliability concerns.

Table 2. Physical properties of porous low-k films with different BTMSE and porogen contents.

MTMS:BTMSE	100:10	100:30	100:50	100:70	100:90
Porogen Content (%)	25.0	27.0	30.0	30.0	29.4
Dielectric constant	2.31 ± 0.01	2.35 ± 0.01	2.28 ± 0.01	2.27 ±0.02	2.28 ± 0.01
Reduced modulus (GPa)	2.85 ± 0.01	4.27 ± 0.06	4.59 ± 0.08	5.33 ± 0.14	5.51 ± 0.14
Hardness (GPa)	0.37 ± 0.01	0.59 ± 0.01	0.63 ± 0.01	0.76 ± 0.04	0.72 ± 0.02
K_{IC} (MPa ·m$^{0.5}$)	0.186 ± 0.004	0.181 ± 0.005	0.174 ± 0.005	0.178 ± 0.003	0.162 ± 0.003

CONCLUSION

Compared with MSSQ, MTMS-BTMSE copolymers blended with sacrificial porogen had superior mechanical properties. As the porosity increased, most mechanical properties except residual stress were deteriorated in return for attaining a low dielectric constant. Such a tradeoff in attaining the low dielectric constant and desirable mechanical strength implied that no more pores than necessary to achieve the dielectric objective should be incorporated. In addition to evaluating the physical properties, the pore structure and interconnectivity were characterized by PALS. The result for porous films with composition of MTMS:BTMSE=100:50 indicated that the film with 20% porogen was almost closed but one with 30% porogen could be considered to have fully interconnected pores. This result suggested that it should be attempted to lower the porosity less than 20% to formulate the porous low-k films with homogeneous and nanometer-sized closed pores as well as to avoid the reliability concern caused by percolation. When the dielectric constant was fixed to approximately 2.3 by controlling BTMSE and porogen contents simultaneously, the thermo-mechanical properties of the porous films were also investigated for the comparison purpose. Under the same dielectric constant, the increase in BTMSE and porogen contents led to the improvement in modulus but the deterioration of adhesion fracture toughness obtained.

REFERENCES

1. International Technology Roadmap for Semiconductors, Semiconductor Industry Association, San Jose, CA (2003).
2. B.R. Kim, J.W. Kang, K.Y. Lee, J.M. Son, and M.J. Ko, submitted.
3. H.G. Tompkins, *A User's guide to Ellipsometry*, Academic Press, Boston, 1993.
4. D.W. Gidley, W.E. Frieze, T.L. Dull, J. Sun, A.F. Yee, C.V. Nguyen, and D.Y. Yoon, Applied Physics letters, **76**, p1282, 2000.
5. J.N. Sun, Y.F. Hu, W.E. Freize, and D.W. Gidley, Radiation Physics and Chemistry **68**, p345, 2003.
6. T.L. Dull, W.E. Frieze, D.W. Gidley, J.N. Sun, A.F.Yee, Journal of Physical Chemistry Part B, **105**, p4657, 2001.
7. W.C. Oliver and G. M. Pharr, J. Mat. Res., **7**, p1564, 1992.
8. E.O. Shaffer II, F.J. Mcgarry, and L. Hoang, Polymer Eng. Sci., **36**, p2375, 1996.

Characterizing and Understanding
Thin Film Growth Stresses

Transmission Electron Microscopy Characterization of Microstructure and TiN Precipitation in Low-energy Nitrogen Ion Implanted V-Ti Alloys

M.I.Ortíz[1], J.A. García[2], M.Varela[3], J.P. Rivière[4] R. Rodríguez[2] and C. Ballesteros[3]
[1]Departamento de Tecnología Electrónica, E.T.S.I.T, Universidad Politécnica de Madrid, 28040, Madrid, Spain
[2]AIN-Centro de Ingeniería Avanzada de Superficies, Cordovilla, 31191 Pamplona, Spain,
[3]Departamento de Física, Escuela Politécnica Superior, Universidad Carlos III de Madrid, Avda. Universidad 30, 28911 Leganés, Madrid. Spain, balleste@fis.uc3m.es
[4]Laboratoire de Metallurgie Physique, UMR 6630 CNRS, Universite de Poitiers, Batiment SP2 M1, B.P. 30179, F-86962 Futurscope-Chasseneuil Cedex, France

ABSTRACTS

A detailed structural characterization of low-energy, nitrogen implanted V5at.%Ti alloys is presented. Samples were nitrogen-implanted at 1.2 kV and 1 mA/cm^2, up to a dose of 4×10^{-19} ions/cm^2, at temperatures between 400-575 °C. Alloys were analysed by transmission electron microscopy. Depending on the implantation temperature, the ion beam treatment dramatically changes the microstructure of the material. Partial amorphization, nitride precipitation and dislocations are imaged. A clear correlation between the microstructure of the implanted layer and the reported improvement in the tribological properties has been demonstrated. For implantation at 575°C a nanocomposite layer forms at the sample surface, where the reinforcement particles are TiN precipitates.

INTRODUCTION

Vanadium-based alloys are promising candidates as structural materials for fusion power devices. However these alloys show a poor tribological performance. Recent results have proved that high and low-energy nitrogen implantations are an effective tool to improve the tribological properties of V-Ti alloys[1,3]. Low-energy high-temperature nitrogen ion implantation is an intermediate treatment between ion implantation and plasma nitriding, where ballistic and diffusion processes are combined. The ion beam induces amorphization, solid solution formation and/or second phase precipitation, the last at lower temperatures than with a thermal treatment only. The main advantage of low energy high-current-density ion implantation is the high depth of ion penetration, >1µm, as compared with the conventional high-energy low-current-density implantation, 0.1- 0.3µm. The beam energy, flux and implantation temperature determines the final depth of the implanted layer [4]. Considerable progress has been achieved, but we have not a clear understanding of the microscopic phenomena which occur in the implanted layer. Detailed investigations looking for the relationship between microstructure modifications and new properties are needed to use this technique effectively

In this paper the role of the sample temperature in low-energy nitrogen ion implantation of V5at.%Ti alloys was analyzed. Transmission electron microscopy (TEM) and energy-dispersive x-ray spectroscopy (EDX) have been used to characterize the structural modifications induced by N-implantation. A clear correlation between the crystal structure and the improvement of the measured tribological properties is obtained.

EXPERIMENTAL PROCEDURES

The V-Ti alloys with Ti concentration of 5at. % were produced from 99.9% pure V and 99.5% pure Ti by repeated arc melting in a high-purity He atmosphere. Prior to ion implantation the alloys were solution annealed at 1573 K for 6h in an oil-free vacuum of 10^{-3} Pa or less and the surfaces polished down to a 1μm finish using diamond paste.

The samples were N-implanted by means of an ID 2500 Kaufman source at an accelerating voltage of 1.2 kV and a beam current of 1 mA/cm^2. Samples were maintained at constant temperature of 400 °C and 575 °C. The implanted dose was 4×10^{19} ions/cm^2, and taking into account the special behaviour of the Kaufman sources the implanted species were N$^+$ at 1.2 keV and N$_2^+$ at 0.6 keV. The ionic percentage was about 55-45 % of atomic-molecular ionised species.

Cross-sectional specimens suitable for transmission electron microscopy (TEM) were prepared by standard procedures: mechanical grinding, dimpling and argon ion milling in a liquid-nitrogen-cooled holder with an acceleration voltage of 5 kV and an incidence angle of 8°. The electron diffraction, energy-dispersive x-ray analysis and TEM images were carry out using a Philips Tecnai 20F FEG analytical microscope operating at 200 kV equipped with a dark field high angle annular detector (HAAD) for Z-contrast analysis.

Microhardness measurements were made using a Fischercope H100VP microindenter. A Vickers indenter was used and four maximum loads were tested: 2 mN, 5 mN, 10 mN and 25 mN. Friction measurements were made employing a ball-on-disk tribometer FALEX 320 PC, the test parameters were 50 % of humidity, linear speed of 0.04 m/s and applied loads of 25 g, 50 g, 75g and 100 g. The wear tests were carried out against a 1/8" WC ball and the calculated contact pressures were: 0.4 GPa, 0.5 GPa, 0.6GPa and 0.65 GPa. The wear coefficient after ball-on-disc tests and the removed volume in the wear track was measured by using an optical profilometer WYCO RST 500[5].

RESULTS AND DISCUSSION

Tribological properties

The effects of the N-implantation on the tribological properties of the alloys are summarized in table 1. The hardness, friction and wear coefficients for the unimplanted and the implanted alloys correspond to values measured for the same sample on both, the unimplanted and the implanted sides.

Microindentation tests reveal an effective enhancement of hardness that increases with the implantation temperature. The thickness of the layer active for hardness increase can be easily estimated from the microhardness profiles of the implanted and unimplanted surfaces see figure1. For an implantation temperature of 400 °C the hardness profiles exhibit an effective HU increase until a thickness close to 1μm. The thickness of the layer modified by the ion treatment increases as the implantation temperature increases and can not be estimated using this simple method.

The measured friction and wear coefficients are presented in table 1. In the implanted samples for low contact pressure, ≤0.6 GPa, two stages in the friction coefficient are observed. The initial stage, for few cycles only, having a low friction coefficient of μ=0.1, is followed by a sharp transition to a second stage, rising stage, where the friction coefficient rises to the value measured for the unimplanted alloy. For all the implanted samples the friction coefficient was maintained close to 0.1 and no sharp transitions between low and high friction coefficient were observed. For higher contact pressure ≥ 0.65 GPa it is possible to observe the transition from high to low friction regime except for the sample implanted at 575°C, even after 10^5 cycles

The observed wear decrease, see table 1, is expected after the hardness increase measured in the microindentation tests. No wear was measured for the samples implanted at 575°C when the same load than for samples implanted at lower temperature is used even after 10^5 cycles.

Table 1. Tribological properties. The unimplanted area correspond to the back side of each sample

Implantation Temperature		Universal Hardness (Nmm^{-2}) at final load of 2 mN	Friction Coefficient at contact pressure of		Wear Coeff (m^2/N) at load 1 N	Roughness Ra nm
			0.6 GPa	0.65 GPa		
400 °C	Unimplanted	1900±2000	0.9	0.9	3.0 E-13	25±5
	Implanted	**3000±300**	**0.25 →0.9**	**0.42→0.9**	**1.4 E-13**	**60±10**
575 °C	Unimplanted	1800±200	0.6	0.75	3.5 E-13	32±10
	Implanted	**9000±2000**	**0.15**	**0.25**	---	**90±20**

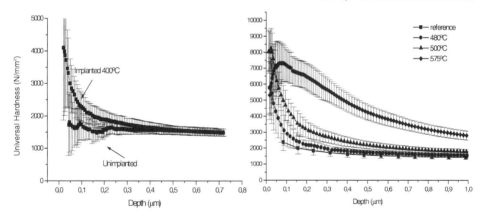

Figure 1. Microhardness profile for the implanted and unimplanted reference surface. Only for the samples implanted at 400°C the thickness of the implanted layer active for the hardness increase can be estimated and was close to 1μm.

TEM results

Important modifications occur in the microstructure of the alloy after implantation, as the TEM analyses reveal. It has been previously found that unimplanted alloys appear free of defects and only isolated needle-shape precipitates, identified as non-stoichiometric titanium carbide were observed[6]. For samples implanted at 400°C, the ion beam treatment changes the microstructure dramatically. Three regions can be distinguished: 1) a surface region, up to 250nm deep, heavily contaminated with carbon and other impurities; 2) a region, from 250 nm to 800 nm deep, consisting of a distribution of nanocrystals embedded in an amorphous matrix, and 3) a completely crystalline region, from 800 nm to 1000 nm deep, where precipitates and dislocations are imaged. An example is shown in the TEM micrograph of figure 2. The size of the nanocrystals in region 2), see fig 2a.2, increases with deep, with a smooth transition from the amorphous to the crystalline region. The three regions observed by TEM, can explain the friction coefficient measurements, with a clear influence of the surface contamination on the initial stage of low friction, before the rising stage.

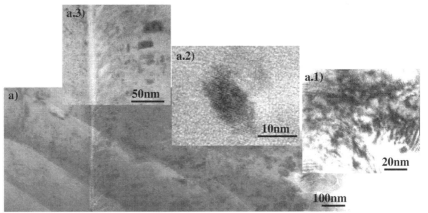

Figure 2. a) Cross-sectional image of sample implanted at 400°C: a.1) Damaged area close to the sample surface. a.2) Nanocrystal embedded in the amorphous matrix from a region 250-800 nm in deep. a.3) Nano-composite layer close to undamaged region of the implanted sample 800-1000 nm in deep. The dark contrast features are precipitates.

The nanocrystals and the precipitates were analyzed by EDX. No X-ray signal can be directly associated to N, due to the overlapping of the K_α line of N with the L lines of Ti and V, and the relatively low intensity expected for the N K_α line. In several spectra the K_α C line was also observed, but in all cases no noticeable differences between the carbon K_α line intensity from the precipitates or the nanocrystals and the surrounding matrix was measured. Carbon is a common impurity in vacuum systems. The X-ray spectra indicate that some of the nanocrystals and all the precipitates analysed are rich in Ti. The size and concentration of the precipitates makes possible to analyse their crystal structure by selected area electron diffraction (SAD). Using the diffraction spots from the unimplanted region as an internal calibration, the patterns were identified as an FCC structure with a lattice parameter of 0.42 nm, essentially the same that the corresponding to TiN. However the differences in lattice parameter do not allow distinguishing between non-stoichiometric titanium carbide and titanium nitride, both with FCC structure. The low number of TiC precipitates observed in the unimplanted alloys, no more than two or three per imaged sample, the clear differences in shape and size and in matrix-precipitate coherence and the fact that in fresh analysed samples no carbon contamination was observed out of the 1) region, close to the sample surface, strongly support that the precipitates are TiN.

Samples implanted at constant temperature of 575°C are polycrystalline and show an inhomogeneous distribution of precipitates. Figure 3a is a typical cross-sectional bright field electron micrograph from a region with precipitates. Precipitates have needle shape and it is possible to observe these until the limit of transparency for the cross-sectional TEM images, close to 9 μm deep. The size of the precipitates is inhomogeneous but the ratio between the length and the maximum width for all the precipitates is close to ten. Strain and defects at the interface precipitate-matrix can be observed.

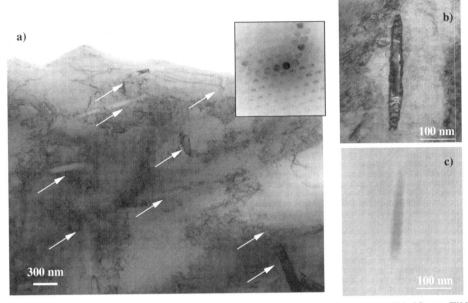

Figure 3. Bright field cross-sectional images of a sample implanted at 575 °C. a) TiN precipitates are indicated by an arrow. Inset: microdiffraction pattern corresponding to a matrix-precipitate interface. b) Detail of a precipitate. c) Corresponding HAAD dark field image, differences in the N concentration along the precipitate can be observed. Atoms with low Z appears darker in the Z-contrast images.

The crystal structure of the precipitates were analysed by electron microdiffraction, an example is shown in the inset of figure 3a. Using the spots of the matrix alloy as an internal calibration, the crystal structure was identified as an FCC structure with a lattice parameter of 0.42 nm. Precipitates are coherent with the matrix, the crystallographic relationship being:

$$(001)_m \parallel (0\bar{1}1)_p \text{ and } (\bar{1}10)_m \parallel (\bar{1}11)_p$$

EDX spectra were obtained under the same conditions than the electron microdiffraction experiments and from the biggest precipitates and in very thin areas of the sample. It was possible to detect N and Ti and can be unambiguously identified as TiN.

HAAD Z-contrast images have been obtained to check differences in N concentration along the precipitates an example is shown in figure 3c. Atoms of low atomic number appear darker in the HAAD Z-contrast images[9]. Changes in the contrast of the dark field Z-contrast in undefaulted areas can be explained as changes in composition.

SUMMARY AND CONCLUSIONS

Tribological properties: microhardness, friction and wear have been improved after low energy N implantation and this improvement increases as the implantation temperature increases. For samples implanted at 400 °C partial amorphization together with the formation of TiN precipitates have been observed, whereas for those samples implanted at 575 °C only precipitation processes have been observed. Assuming ionic displacements for alloy amorphization and thermal diffusion of the ions to form precipitates, both ballistic and diffusive processes operated simultaneously for lower temperature implantation, T=400 °C whereas diffusion was the main activate process for implantation at T=575°C.

Only for samples implanted at 400 °C the active thickness of the implanted layer can be estimated after microhardness tests resulting ~1μm in good agreement with TEM measurements. TEM measurements in samples implanted at 575 °C indicate that the active layer is one order of magnitude bigger, > 9 μm. This is a new evidence for ion diffusion into the material favored by the implantation temperature.

An accurate selection of the implantation temperature determines the tribological performance of the V-Ti alloys and the thickness of the layer modified by ion implantation. The implantation temperature determines the main process activated, ballistic or diffusive.

A direct correlation between the structure of the low energy N-implanted layer and the improvement in the tribological properties is obtained. TiN precipitation appears to be responsible for the improvement in the tribological properties. For implantation at 575°C a nanocomposite layer containing TiN precipitates forms at the sample surface. Further studies are needed to investigate this correlation as a function of the implantation temperature in other metallic alloys.

ACKNOWLEDGMENTS

Research supported by CICYT through the project MAT-99-1012.TEM work has been made at the LABMET of Universidad Carlos III associated to the Red de Laboratorios of the CAM.

REFERENCES

1. M. Varela, J.A. García, R.Rodríguez and C. Ballesteros. *Nanotech. 2003* **3**, 207 (2003)
2. J.A.García, R. Rodríguez, R. Sánchez, R. Martínez, J.P. Rivière, P. Méheust, M. Varela, D. Cáceres, A. Muñoz, I. Vergara, C. Ballesteros. *Vacuum* **67**, 543 (2002).
3. J.A. García, R. Sánchez, R. Martínez, A. Medrano, R. Rico, R. Rodríguez, M. Varela, I. Colera, D. Cáceres, I. Vergara, C. Ballesteros, E. Román, J.L. Segovia. *Surf. Coat. Technol.* **158-159**, 669-673 (2002).
4. D.L. Williamson, J.A. Davis, P.J. Wilbur, *Surf. Coat. Tech.* **103-104,** 178 (1998)
5. SJ Bull, AM Jones, AR McCabe. *Surf Coat Technol* **257-262,** 83 (1996)
6. M. Varela, B. Fernández, A. Muñóz, T. Leguey, R. Pareja, C. Ballesteros. *Philos. Mag. Lett.* **81**, 259 (2001).
7. A. Leyland, A. Matthews, *Wear* **246**, 1 (2000)
8. J. Musil, F. Kunc, H. Zeman, H. Poláková, *Surf. Coat. Tech.* **154**, 304 (1996).
9. S.J. Pennycook . *Advances in Imaging and Electron Physics*. Vol 123 p173 (2002).

Mater. Res. Soc. Symp. Proc. Vol. 875 © 2005 Materials Research Society

Nonlinear Transient Finite Element Analysis of the Relaxation Mechanisms in Strained Silicon Grown on SiGe Virtual Substrate

F. Sahtout Karoui,[1] A. Karoui,[2] G. Rozgonyi[1]

[1]: Materials Science and Engineering Department,
 North Carolina State University, Raleigh, NC 27695-7916, USA.
[2]: Nanoscience and Nanotechnology Research Center, Shaw University

ABSTRACT

Strained-silicon (ε-Si) is essential for future nanoscale MOSFET devices. In this paper we report on the dynamics of strain relaxation in Si/SiGe heterostructures, investigated by transient nonlinear finite element analysis. The contribution to total misfit strain is found largely plastic in the graded SiGe layer and the top of the Si substrate, while it is mainly elastic in the strained Si layer and part of the SiGe constant layer. The calculated lattice parameter for the strained Si layer is about 5.47Å for $Si_{0.8}Ge_{0.2}$ and 5.52 Å for $Si_{0.6}Ge_{0.4}$. Calculated threading dislocation density was about 5.6×10^5 cm^{-2} for x=0.20 and 2.17×10^6 cm^{-2} for x=0.40. A plastic strain rate of 8.4×10^{-3} s^{-1} for $Si_{0.8}Ge_{0.2}$ and 4.1×10^{-2} s^{-1} for $Si_{0.6}Ge_{0.4}$ leading to a density of moving dislocations of ~2.2×10^9 cm^{-2} and ~ 10^{10} cm^{-2}, respectively, have been obtained. The elastic strain in the strained-Si layer appeared to increase with increasing the cooling rate, while plastic work was found to be independent of cooling rates.

INTRODUCTION

Biaxial tensile strain is introduced in Si thin films grown on $Si_{1-x}Ge_x$ "virtual substrate" to enhance carrier transport properties. Indeed, SiGe alloys have larger lattice constant than Si, due to the ~ 4% lattice mismatch between Si and Ge [1]. The reliability of devices made on the strained layer, depends on the stability of the latter as well as on the strain amount, and hence on the misfit dislocation density at the strained-layer interfaces. The ability to consistently predict the level of dislocation and resulting strain relaxation is critical in such devices. Currently, one of the more successful techniques for reducing threading dislocation density in the strained thin Si layer (15 nm in our case) is to grow the strained layer on top of relaxed graded SiGe buffer layer. The self-equilibrated residual stresses, which occur during relaxation in the deposited layers are generally the result of the elastic and/or plastic deformation when inhomogeneously distributed over the volume. They systematically occur during layer growth with variation of material composition. These stresses originate from lattice mismatch of contacting layers, but also from differences in thermal-expansion coefficients and exchange of point defects at the interfaces. For instance during cooling from growth to room temperature, the thermal expansion coefficient α_{Si} of Si varies from 4.9×10^{-6} to 2.57×10^{-6} K^{-1}, and α_{Ge} of Ge varies from 8.55×10^{-6} to 5.9×10^{-6} K^{-1} [2]. When these stresses exceed the yield stress, at growth time, they lead to the onset of plastic deformation.

In this paper, we have investigated the dynamics of strained-Si/$Si_{0.8}Ge_{0.2}$/Si and strained-Si/$Si_{0.6}Ge_{0.4}$/Si heterostructures growth and the stress relaxation mechanisms using a transient and non-linear finite element analysis. Numerical simulations were done using ANSYS package [3]. The elastic model is described by Hook's law, while the material plastic behavior is accounted for using the von Mises yield criterion coupled with isotropic work hardening conditions [3, 4]. Plastic deformation occurs by dislocation generation and interactions. Basically, the larger the number of dislocations produced, the larger their interaction and hence the larger the stresses required for material yielding. Plastic deformation analysis is essential for understanding the relaxation mechanisms as well as to tackle the dislocation generation during the deposition process and to determine their density. The development of elastic and plastic strain accumulating in the structure was undertaken dynamically, to handle the various thermal treatment steps (heating, cooling, deposition, etching/polishing) and different growth rates depending on the layer composition. For this purpose the element 'birth/death' method within finite element treatment was utilized.

From this study we were able to predict for both heterostructures, the residual strains and stresses in the heterostructure, the time dependent plastic relaxation mechanism, as well as the structural properties (relaxation factor, lattice parameters, moving and threading dislocation density) which were compared to experimental data.

ELASTIC-PLASTIC MODEL AND COMPUTATIONAL METHOD

We have considered a 2D plane-strain analysis and symmetric boundary conditions applied on each side, see Fig. 1.. The FEA model comprises about 8000 four-node quadrilateral elements (defined as Plane42 in ANSYS [3]). Both out-of-plane and in-plane distortions can be tracked. The growth process covers up to 40 steps including; deposition, loading and unloading, pre-treatments at high temperatures, and chemical mechanical polishing (CMP). All these steps but the CMP, were assumed to induce a priori plastic deformations. The use of 'birth and death' of elements allowed to simulate dynamically the deposition or removal of sub-layers, replicating thereby the real growth process. Notwithstanding the continuous change of the geometry, the history of the stress is carefully tracked for each element and its surrounding. The model is first generated with all elements that will be used for the simulation. These elements are then activated or deactivated to describe the various steps of the process. The set of simultaneous nonlinear equations generated for one step were solved using the Newton Raphson iterative method on the IBM P690 supercomputer. One step necessitates between few seconds to 30 minutes CPU time to converge, depending on the time-step Δt, the step duration and the temperature.

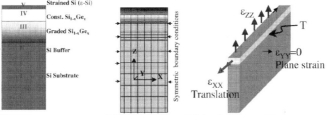

Fig. 1: Si/SiGe/Si Heterostructure (x=0.20 & x=0.40) and Finite element model.

Elastic Model

For simple elastic analysis the total strain follows Hooke's law, and is given by :

$$[\varepsilon] = \left[\varepsilon^{el}\right] = [D]^{-1}[\sigma] \qquad (1)$$

where |D| is the elastic stiffness matrix.
During heating or cooling, differences in thermal expansion coefficients between the layers induce additional strain and stress in the layers, the total strain becomes:

$$[\varepsilon] = [D]^{-1}[\sigma] + \left[\varepsilon^{th}\right] \qquad (2)$$

where ε^{th} defines the thermal induced stress, given as: $\varepsilon^{th} = \alpha.\Delta T$
where α is the thermal expansion coefficient. In the case of plane strain situation and when isotropic elasticity is assumed, the stiffness matrix is simplified to become:

$$[D] = \frac{E}{(1+v)(1-2v)} \begin{bmatrix} 1-v & v & 0 \\ v & 1-v & 0 \\ 0 & 0 & \frac{1}{2}-v \end{bmatrix} \qquad (3)$$

Model Used for Plasticity

Plasticity is a complex phenomenon to model and numerically simulate, especially if the goal is to obtain the relaxation mechanisms through dislocation emission. We consider that the theorem of

410

material plastic deformation, often used for solids at macroscopic scales, and based on stress exceeding a certain threshold expressed as yield stress, can still be applicable at nanoscale.

Plastic deformation occurs via motion of large number of dislocations, as stated by Orowan et al in 1934. Plasticity in SiGe heterostructures occurs mainly by incremental bond breaking ending up by disregistry of atomic planes and dislocation formation. As the plastic deformation proceeds, dislocation density increases dramatically creating threading dislocations that can glide over long distances, see Fig. 2. In our model the plastic behavior is described using von Mises yield criterion coupled with isotropic work hardening conditions [3]. Elastic-plastic analysis is much more delicate than simple elastic analysis. The FEA equation in that case is no longer linear and is replaced by a set of nonlinear equations that are solved iteratively. When elastic and plastic analysis is considered the total strain can be decomposed as follows:

Fig. 2: Dislocation scheme in graded SiGe epitaxial layers.

$$[\varepsilon] = [\varepsilon^{el}] + [\varepsilon^{pl}] + [\varepsilon^{th}] \tag{4}$$

The elastic part of the strain is related to stress by the usual elastic stress-strain equations. For the plastic component ε^{pl}, the stress is no longer proportional to the strain [4]. Plasticity theory provides a more complete description that characterizes simultaneously the elasto-plastic response of the materials which involve three important components: the yield criterion, the flow rule and the hardening rule.

Taking into account that x does not exceed 40% and that at the maximum growth temperature (1050°C) the alloy is still solid (closer to the phase diagram of Si than Ge) we have utilized the yield stress of Si for that alloy. A temperature dependent yield stress (σ_Y) which, range from 4 GPa (room temperature) to 188 MPa (1050°C) has been chosen from Ref. 5 and 6. We applied the von Mises yield criterion given by:

$$F(\sigma) = (\sigma_1 - \sigma_2)^2 + (\sigma_2 - \sigma_3)^2 + (\sigma_3 - \sigma_1)^2 - 2\sigma_0^*(\kappa)^2 = 0 \tag{5}$$

$\sigma_1, \sigma_2, \sigma_3$ are the principal stresses and $\sigma_0^*(\kappa)$ is the actual yield stress and κ is the hardening parameter. We use the Levy-Mises flow rule which was proved a good enough fit for most practical purposes:

$$d\varepsilon^{pl} = d\lambda \frac{\partial F}{\partial \sigma} \tag{6}$$

where $d\lambda$ is the plastic factor. The hardening rule that we chose is the isotropic hardening where the yield surface remains centered about its initial centerline and expands in size as the plastic strains develop. The von Mises stress (deviatoric stress) is given as:

$$\sigma_e = 2G\varepsilon_e \tag{7}$$

where G is the bulk modulus which is equal to: $\dfrac{E}{2(1+v)}$

We applied the Newton Raphson method to solve this set of simultaneous nonlinear equations. Only few sub-layer of elements could be defined for the Si-strained film, because the latter is extremely thin (15 nm) as compared to the specimen thickness (700 μm). The element aspect ratio being a key parameter for computational efficiency. Consequently, in the following section, the magnitude of the bi-axial strain and von Mises stresses in the epitaxial layers are reported. Besides, the directional components were considered to check for the stress and strain signs, details will be published elsewhere.

RESULTS AND DISCUSSION

We found that the elastic component of the strain is maximum in the strained Si layer and is about 0.72% for x = 0.20 in agreement with measured strain of 0.76% [7] and 1.2% for x = 0.40. The corresponding calculated residual von Mises stresses are 730 MPa for $Si_{0.8}Ge_{0.2}$ and 1.2 GPa for $Si_{0.6}Ge_{0.4}$, respectively, see Fig. 3. It is worth noticing that within the elastic theory the calculated residual stress would be equal to 1.28 GPa for x=0.20 and 2.13 GPa for x=0.40, in excess compared to a complete elastic-plastic treatment. Usually the elastic theory is employed to interpret the X-ray diffraction and Raman scattering data. As shown in Fig. 4, the contribution to total misfit strain is mainly elastic in the strained Si layer and part of SiGe constant layer and largely plastic in the graded SiGe layer and the top of the Si substrate. Meaning that the relaxation mechanism occurs from the middle of the SiGe constant layer and propagates to the Si substrate.

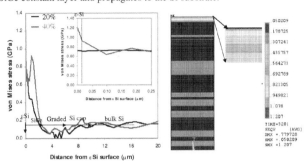

Fig. 3: Calculated von Mises stress (deviatoric) stress at the end of the growth process.

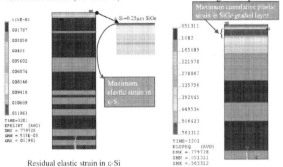

Fig. 4: Distribution of the elastic strain (a) and cumulative plastic strain (b) after growth of $Si_{0.6}Ge_{0.4}$.

The lattice parameter ($a_\parallel$) of the strained Si layer and its relaxation factor (r) where deduced from the misfit strain as follows:

$$\varepsilon_\parallel = (a_\parallel - a_{relax})/a_{relax} \tag{8}$$

$$r = (a_\parallel - a_s)/(a_{relax} - a_s) \tag{9}$$

The threading dislocation density is given by equation (10), as suggested in Ref [8]:

$$\rho_{TD} \approx \frac{8f}{L_{MD}b} \tag{10}$$

where f is the total misfit strain; L_{MD} the length of a misfit dislocation projected in (100) and b the Burgers vector for 60° dislocation, $b = a_{Si}/2\sqrt{2}$ [9]. Indeed, misfit dislocations in SiGe heterostructure are mainly 60° dislocations [10].

The calculated lattice parameter for the strained Si layer from Eq. 6 is 5.47 Å for $Si_{0.8}Ge_{0.2}$ and 5.52 Å for $Si_{0.6}Ge_{0.4}$. This shows that the 15 nm strained Si grown on $Si_{0.8}Ge_{0.2}$ layer is already 4% relaxed and 22% relaxed when grown on $Si_{0.6}Ge_{0.4}$. A simplified approximation for the critical thickness is given by $h_c = b/\varepsilon$ [11] giving a critical thickness of about 20 nm for $Si_{0.8}Ge_{0.2}$ and 10 nm for $Si_{0.8}Ge_{0.4}$. This is in line with the elevated degree of relaxation calculated for the 40% Ge heterostructure. Meanwhile the SiGe graded layer is about 81% relaxed for 20% Ge and 97% for x=0.40. Figure 5 shows the depth profile of the plastic strain accumulating in the structure during the entire growth process. It is high in the graded layer where much of relaxation occurs. It is much higher for the x=0.40 compared to x=0.20 and extends up to 14 μm into Si bulk, in agreement with the high dislocation density observed in that region [12]. The calculated Threading Dislocation (TD) density from Eq. 9 in the as-grown strained Si layer is about 5.6×10^5 cm^{-2} for x = 0.20 and about 2.17×10^6 cm^{-2} for x = 0.40. This variation is in agreement with that of TD density obtained by etching/Normarski optical micrograph and EBIC measurements [13, 8].

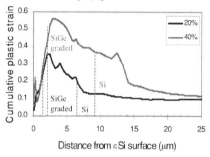

Fig. 5: Depth profile of the cumulative plastic strain.

The time and temperature dependent plastic and elastic deformation behavior was one of the most important considerations in this modeling in addition to the dynamic growth. Figure 6(a) shows the plastic strain accumulated during the growth cycle, for four locations, three in the graded layer and one in the underneath Si buffer layer. The transient analysis shows that plastic flow occurs consistently during SiGe graded layer deposition and $Si_{1-x}Ge_x$ constant layer re-growth. No yielding occurs when the $Si_{1-x}Ge_x$ constant layer is first deposited (i.e. before CMP) and during strained Si layer deposition. The stress state ratio N (N = σ_e/σ_y) for x = 0.40 is shown as a function of time in Fig. 6(b). It shows the sequence of yielding (N > 1) and no-yielding (N < 1) in two regions of the graded layer while the deposition process goes on. As stated before, yielding occurs during the growth of the graded layer and re-growth of the uniform SiGe layer.

One of the key issues is to predict the evolution of the mobile dislocation density ρ_m with strain and stress build-up. Whereas it is experimentally possible to determine the dislocation density, it is impossible to predict which part of this population is mobile, .i.e. which part of dislocations participates in the plastic deformation at a given time. The time dependent cumulative plastic strain enables us to predict the moving dislocation density as a function of x. The Orowan law relates the plastic strain rate to mobile dislocation density, dislocation velocity, and Burgers vector:

$$\frac{d(\varepsilon_{pl})}{dt} = \rho_m b v \qquad (11)$$

We have used the gliding velocity measured by C. G. Tuppen and al [14] that is equal to 10^{-2}cm/s for the growth temperature. We derived from our calculations a plastic strain rate of 8.43×10^{-3} s^{-1} for $Si_{0.8}Ge_{0.2}$ and 4.1×10^{-2} s^{-1} for $Si_{0.6}Ge_{0.4}$, corresponding to a density of mobile dislocation of 2.2×10^9 cm^{-2} for $Si_{0.8}Ge_{0.2}$ and 1×10^{10}cm^{-2} for $Si_{0.6}Ge_{0.4}$. The effect of cooling rate on strain in the strained Si layer has then be studied for $Si_{0.8}Ge_{0.2}$. The residual elastic strain in the strained layer increases with increasing cooling rate from 0.72% for rapid cooling (67°C/s) to 0.52% for slow cooling rate of 1.7°C/s. Plastic work is identical for both cooling rate, therefore the threading dislocation density is expected to be independent of cooling rate. However, it is likely that very slow cooling rate would result in relaxation of the strain in the entire structure via dislocation motion.

413

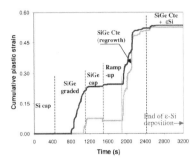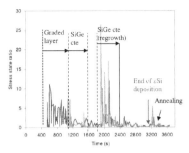

Fig. 6: Time dependent plastic strain (a) and stress state ratio (b) during the growth of $Si_{0.6}Ge_{0.4}$.

CONCLUSION

The dynamic, transient and nonlinear finite element analysis has been used to explore the evolution of elastic and plastic deformations in strained-$Si/Si_{0.8}Ge_{0.2}/Si(001)$ and strained-$Si/Si_{0.6}Ge_{0.4}/Si(001)$ during the growth process. The contribution to total misfit strain is found largely plastic in the graded SiGe layer, and the top of the Si substrate, while it is mainly elastic in the strained Si layer and part of SiGe constant layer. The calculated lattice parameter for the strained Si layer is about 5.47 Å for $Si_{0.8}Ge_{0.2}$ and 5.52 Å for $Si_{0.6}Ge_{0.4}$. The calculated threading dislocation density is about $5.6x10^5$ cm^{-2} for x = 0.20 and about $2.17x10^6$ cm^{-2} for x = 0.40. A plastic strain rate of $8.43x10^{-3}$ s^{-1} for $Si_{0.8}Ge_{0.2}$ and $4.1x10^{-2}$ s^{-1} for $Si_{0.6}Ge_{0.4}$ has been derived leading to a density of moving dislocations of ~$2.2x10^9$ cm^{-2} for x=0.20 and ~ 10^{10}cm^{-2} for x=0.40. The elastic strain in the strained-Si layer appeared to increase with increasing cooling rate, while plastic work was found to be independent of cooling rates. Dynamic and transient non-linear finite element modeling can be a convenient tool to improve the growth process of strained-Si hetero-epitaxial layers for Nanoscale MOSFET Devices.

ACKNOWLEDGMENTS

This work is supported by the Silicon Wafer Engineering and Defect Science center (SiWEDS), under NSF Grant # EEC-9726176.

REFERENCES

1. S C Jaint and W Hayes, Semicond. Sci. Technol., 6, 547-576 (1991).
2. E. Kasper, *Properties of Strained and Relaxed Silicon Germanium*, EMIS Datareviews, **12**, INSPEC (1995).
3. ANSYS Inc., Theory Reference , Release 7.1 Manual, ANSYS Inc.
4. M. A. Crisfield, Non-linear Finite Element Analysis of Solid and Structures, Vol. **2**, John Wiley & Sons, New York, 1995.
5. W. Schroter, H. G. Brian, H. Siethoff, J. Appl. Phys., **54**(4), 1816 (1983).
6. D. Lowney, et al, Semicond. Sci. Technol. **17**, 1081 (2002).
7. S. Nakashima et al., Appl. Phys. Let., 84 (14), 2533 (2004).
8. E.A. Fitzgerald, Y.-H. Xie, M. L. Green, D. Brasen, A. R. Kortan, J. Michel, Y.-J Mii, and B. E. Weir, App. Phys. Lett., **59**(7), 811 (1991).
9. E. Bugiel, P. Zaumseil, Appl. Phys. Lett., **62**(17), 2051 (1993).
10. E. Koppensteiner, A. Schuh, G. Bauert, V. Holy, G. P. Watson, E.A. Fitzgerald, J. Phys. D, **28**, A114 (1995).
11. A. E. Romanov, W. Pompe, S. Mathis, G. E. Beltz, J. S. Speck, J. App. Phys., **85**(1), 182 (1999).
12. M. T. Currie, S. B. Samavedam, T. A. Langdo, C. W. Leitz, E. A. Fitzgerald, Appl. Phys. Let., **72**(14) (1998).
13. S. B. Samavedam, W. J. Taylor, J. M. Grant, J. A. Smith, P. J. Tobin, A. Dip, and A. M. Phillips, J. Vac. Sci. Technol. B, **17**(4), 1424 (1999).
14. C. G. Tuppen and C. J. Gibbings, Thin Solid Films, **183**, 133 (1989).

Thin Film Processing

Mater. Res. Soc. Symp. Proc. Vol. 875 © 2005 Materials Research Society

Passive Layer Formation at Ferroelectric PbTiO$_3$/Pt Interfaces Studied by EELS

S. J. Welz[1], L. F. Fu[1], R. Erni[1], M. Kurasawa[3], P. C. McIntyre[3], and N. D. Browning[1,2]

[1] Department of Chemical Engineering and Materials Science, University of California Davis, One Shields Ave., Davis, CA 95616, USA
[2] National Center for Electron Microscopy, Lawrence Berkeley National Laboratory, One Cyclotron Road, Berkeley, CA 94720, USA
[3] Department of Materials Science and Engineering, Stanford University, Stanford, CA 94305, USA

ABSTRACT

Polarization fatigue with repeated electric cycles in ferroelectric thin films is a major degradation problem in ferroelectric nonvolatile memories. However, the origin of this phenomenon is still not properly understood. The fatigue mechanism of a ferroelectric perovskite in a multilayer ferroelectric PbTiO$_3$ thin film material has been investigated here using scanning transmission electron microscopy (STEM). Z-contrast images of the interfaces show that the ferroelectric PbTiO$_3$ layer has partly decomposed into a single crystal PbTiO$_3$ layer and an amorphous layer. Nanometer-sized precipitates are present near the Pt electrode. Electron energy-loss spectroscopy (EELS) analysis reveals that the amorphous layer is a Ti-rich phase between TiO$_2$ and PbTiO$_3$. The precipitates are determined to be a Pt-Pb rich crystalline phase. It is suggested that the formation of the structure-distorted intermediate layer and precipitates may be associated with the ferroelectric degradation process by acting as a passive layer in a ferroelectric capacitor. In addition, the formation of the Pt-Pb rich precipitates may cause an interruption of the consistent Pt electrode, which may result in failure of the device.

INTRODUCTION

Ferroelectric thin film materials, such as PbTiO$_3$, have been widely studied for potential applications as nonvolatile memory since they offer significant advantages over Si-based devices - such as high dielectric constant, large polarization, and low power consumption [1,2]. However, several material-related problems, such as fatigue (reduced polarization under repeated switching cycles), severely hinder commercial applications in microelectronic devices [3]. Although numerous explanations for fatigue degradation of ferroelectric thin films have been proposed, the mechanism remains undetermined. Recently, significant attention has focused on the possibility that the growth of a non-ferroelectric passive layer near the electrodes may result in device failure [4,5]. This concept suggests that the presence of a low permittivity layer in a ferroelectric capacitor may substantially influence the depolarization field and redistribution of charge carriers [6]. Related studies are mainly based on theoretical assumptions or modeling of passive layers from ferroelectric polarization switching measurements. So far few experimental evidence of the formation of passive layers in a ferroelectric capacitor has been provided [4-7]. Therefore, a comprehensive microstructural characterization of ferroelectric-electrode interface is essential to fully understand the proposed fatigue mechanism.

It is well known that the Z-contrast method in the STEM provides an image free of the phase problem. This makes it a very powerful technique for the intuitive identification of interface structures. Using this image as a map, microanalysis can be performed at the highest spatial resolution [8]. The use of STEM to obtain high spatial resolution electron energy loss spectra (EELS) below the high angle annular dark field (HAADF) detector used for the Z-contrast image is also well established. Through the combination of Z-contrast imaging and EELS analysis, direct information on compositional changes and electronic structure between the interfaces can be obtained on the atomic scale [9]. In this paper, we have utilized these Z-contrast imaging and EELS methods to explore the morphological and chemical change near the interface between the PbTiO$_3$ ferroelectric layer and the Pt electrode.

EXPERIMENTAL PROCEDURE

The ferroelectric films under study consist of a lead titanate (PbTiO$_3$) layer grown epitaxially by metalorganic chemical vapor deposition (MOCVD) on a (001)-oriented strontium titanate (SrTiO$_3$) single crystal. A Pt electrode layer was deposited on the PbTiO$_3$ as a top contact by ultra-low deposition rate sputtering in a UHV base-pressure vacuum system. The sputter target and substrate were configured in an off-axis geometry that produces a nearly thermalized vapor incident on the PbTiO3 film surface. To produce a stoichiometric film surface, the PbTiO$_3$ surface was cleaned by dilute aqueous HNO$_3$ etching prior to Pt deposition.

The TEM cross-section samples are prepared using a unique sample preparation technique (shadow technique) of cleavage, focused ion beam (FIB) milling and Ar ion milling [10]. The FIB milling technique has been widely used to prepare the TEM cross-section samples for semiconductor devices at a precisely pre-selected area on the submicron scale. To reduce the damage layer caused by FIB milling technique, we have developed the conventional FIB technique into this combined sample preparation process. After this process, the samples are analyzed using a 200 kV Schottky field-emission gun FEI Tecnai F20 UT microscope. The microscope operates in STEM mode with a ~0.14 nm probe size and an energy resolution of 0.5 eV. The Z-contrast image is formed by collecting the scattered intensity of illuminated atomic columns on a high-angle detector (35–100 mrad). As the scattering amplitude at high angle is essentially Rutherford scattering, the measured intensity is proportional to the square of average atomic number of the column (i.e. ~Z^2) [8]. The experimental setup of this microscope allows us to use the low-angle scattered electrons that do not contribute to the Z-contrast imaging for EELS. Since the two techniques do not interfere, Z-contrast images can be used to accurately position the electron probe at the desired spot on the sample and to acquire EEL spectra [11].

For the EELS study in this paper, EEL spectra at Ti L$_{2,3}$, O K, Pb M$_{4,5}$ and Pt M$_{4,5}$-edges were recorded using Gatan Imaging Filter (GIF) system. The dispersion of 0.1 eV/channel and acquisition time of 10 s for Ti L$_{2,3}$ and O K-edges were chosen to compare the EEL spectra of different layers or regions. The dispersion of 0.5 eV/channel and acquisition time of 15 s were chosen for Pb M$_{4,5}$ and Pt M$_{4,5}$-edges. Tens of spectra across the interfaces were recorded. Each spectrum has been corrected for dark current and spectrometer gain variations. The background for each spectrum was subtracted by a power-law fitting method before the ionization edge using the Digital Micrograph 3.71 software.

RESULTS AND DISCUSSION

A low magnification Z-contrast image of ferroelectric $PbTiO_3$ thin film specimen prepared by the shadow technique can be seen in Figure 1a. The sample thickness is uniform and is proved transparent to the electron beam over a region of several square micrometers. As supported by EELS, the multi-layer compositions consist of the $SrTiO_3$ substrate, a $PbTiO_3$ single crystal film, a Pt electrode and a top $PbCO_3$ and Pb_3O_4 layer. The area inside the white box is magnified in Figure 1b. The $PbTiO_3$ layer has a uniform epitaxial orientation relationship $[001]_{PbTiO3}$ // $[001]_{SrTiO3}$ on $SrTiO_3$ substrate and no defects are observed at the interface. The $PbTiO_3$ layer should have a thickness of around 30 nm. However, a contrast change along with the loss of crystallinity can be observed close to the Pt electrode, which indicates a different phase formed. This intermediate layer is extended to about 10 nm. The structure is amorphous as confirmed by convergent beam electron diffraction. In addition, precipitates form at the Pt electrode, varying in size from 2 to 5 nm marked by the dashed arrows.

In order to confirm the compositional and structural changes among the crystalline $PbTiO_3$, the intermediate layer, and the precipitates, point analysis of core-loss EELS across the our interfaces has been performed and shown in Figure 2a. The main core loss energy of interest are modifications in the Ti $L_{2,3}$-edges and O K-edge. The Ti $L_{2,3}$-edges originate from dipole-allowed excitations of electrons from the inner $2p_{3/2}$ (L_3-edge) and $2p_{1/2}$ levels (L_2-edge) to the unoccupied 3d band [12]. The O K-edge represents O 1s $\rightarrow$ 2p transitions. By analyzing the Ti $L_{2,3}$ and O K-edges, information about Ti-O-Ti bonding can be obtained. In addition, O K-edge

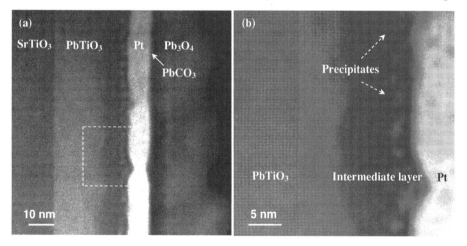

Figure 1. (a) Cross-sectional low magnification Z-contrast image of ferroelectric $PbTiO_3$ thin film sample and (b) High-resolution Z-contrast image acquired from inside the white box in Figure 1a showing the decomposition of the $PbTiO_3$ layer and the formation of precipitate (marked by arrows)

contains valuable information about the hybridization with Ti or Pb sp bands.

In Figure 2a, the EEL spectra from PbTiO$_3$ show splitting of the Ti L$_{2,3}$-edges, which originates from Ti 3d state splitting into t$_{2g}$ and e$_g$ sub-bands in an octahedral crystal field. However, for the intermediate layer, the splitting at the Ti L$_3$-edge is reduced. More remarkably, the splitting at the Ti L$_2$-edge disappears. Since the splitting of Ti L$_{2,3}$-edges reflects the hybridization and ligand field strength of Ti-O atomic interaction, the reduction of splitting in the intermediate layer reflects the weaker bonding force between Ti and O. As a result, the Ti-O octahedral structure loses its symmetry and is distorted in the intermediate layer. EEL spectra of the precipitates show that the ligand field splitting in Ti L$_{2,3}$-edges has minimally reduced in comparison with the signal in the single crystalline PbTiO$_3$, which indicates the existence of a remnant tight-bonding crystal field.

The O K-edge from their corresponding EEL spectra confirms the structural changes. Due to the similar hybridization of O 2p states with Ti 3d in the conduction band, the region from 530 eV to 536 eV maps Ti 3d bands and is split by the crystal-field effect into two sub-bands t$_{2g}$ and e$_g$ (marked peaks a and b in Figure 2a). In consistence with the Ti L$_{2,3}$-edges, the splitting between two sub-bands t$_{2g}$ and e$_g$ is reduced in the spectrum recorded in the intermediate layer. This near edge structure at the intermediate layer confirms the distortion of the linear Ti-O-Ti bond of the perovskite structure, as described by de Groot et al. [13]. Concerning the precipitates, the splitting between two sub-bands t$_{2g}$ and e$_g$ is present in the O K-edge of precipitates in agreement with the splitting in the Ti L$_{2,3}$-edges. The second region of the O K-edge (from 536 eV to 549 eV, marked peak c) maps both Ti 4sp bands and additional bands, which contain counter-ion character [12]. In the case of PbTiO$_3$, they are Pb 6s and 6p bands. The complex

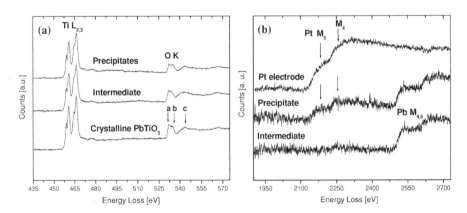

Figure 2. (a) EEL spectra of the Ti L$_{2,3}$-edges and O K-edge acquired from three different regions. The peaks in O K-edge are donated by letters from a to c. (b) EEL spectra of Pb M$_{4,5}$-edges and Pt M$_{4,5}$-edges acquired from the corresponding region in the ferroelectric PbTiO$_3$ thin film sample. The black arrows denote the positions of Pt M$_{4,5}$-edges.

shape observed in this region is more sensitive to long-range order. In comparison with the spectrum of PbTiO$_3$, the intensity of peak c is remarkably reduced at the intermediate layer. The reduced intensity of peak c can be attributed to the distortion of the linear Ti-O-Ti bond of the perovskite structure. Since peak c partially originates from the O-Pb hybridization, the broadened and reduced energy loss near edge structure (ELNES) is possibly due to the preferential reduction of the PbO species or Pb loss in the intermediate layer, as described by Soriano et al. [14].

Using quantitative calculation of several spectra for each layer, we have found that Ti/O ratio in the intermediate layer is increased by 17±2 % relative to primitive PbTiO$_3$. The decomposition of PbTiO$_3$ into a Ti-rich and Pb-poor intermediate layer is consistent with the Z-contrast change in PbTiO$_3$ layer shown in Figure 1b according to their average atomic number. These results are also in agreement with recent angle-resolved photoemission studies of the same sample, which show a Pb-depleted region immediately beneath the Pt electrode layer [15]. The features of EELS and quantitative analysis suggest that the intermediate layer is an intermediate Ti-rich phase between TiO$_2$ and PbTiO$_3$, which has a distorted structure. The composition of precipitates has been determined by the point analysis of EELS from the intermediate layer to Pt electrode using the STEM beam right positioning on them. The strong Ti signal in Figure 2a is a result of the precipitates being embedded in the matrix of the Ti-rich decomposed intermediate layer. However, Figure 2b compares the EEL spectra at Pb M$_{4,5}$-edges and Pt M$_{4,5}$-edges recorded from the corresponding regions. As a result, the precipitates are rich of both Pt and Pb (marked by black arrows). The brighter Z-contrast of precipitates in relation to the intermediate layer in Figure 1b is also consistent with this result since presence of these heavy elements increases the contrast in the image.

A possible explanation for the compositional and microstructural changes in PbTiO$_3$ layer maybe a chemical reaction at the Pt/PbTiO$_3$ interface resulting in Ti or Pb interdiffusion and the decomposition of the PbTiO$_3$. During the decomposition of the PbTiO$_3$ layer, a Ti-rich intermediate layer forms. Pt diffuses from the Pt electrode into the intermediate layer and Pb diffuses from the intermediate layer to the Pt electrode, forming Pt and Pb rich precipitates. This probability is also supported by the occurrence of "mountains" on both sides of the Pt layer (as seen in Figure 1b).

CONCLUSIONS

In conclusion, the multi-layered ferroelectric films after Pt deposition have been studied by means of Z-contrast imaging and core-loss EELS. PbTiO$_3$ layer was identified to decompose into an amorphous intermediate layer and nano-scale precipitates near the Pt electrode. EELS point analysis across the interfaces indicates that the amorphous layer is an intermediate Ti-rich phase between TiO$_2$ and PbTiO$_3$, and the precipitates are Pt and Pb rich crystalline phase. The fact that the PbTiO$_3$ layer decomposes and forms the structure-distorted intermediate layer could be the origin of the inferior reliability of ferroelectric switching typically observed for Pt-electrode ferroelectric capacitors. As a result, the structure-distorted intermediate layer can form a ferroelectrically dead area and act as a passive layer, which interferes with efficient screening of the ferroelectric polarization. The precipitates produced by interface interdiffusion can also play a screening role between ferroelectrics and electrode and influence the depolarization field.

ACKNOWLEDGEMENTS

This work was performed at the National Center for Electron Microscopy, Lawrence Berkeley National Laboratory. It was supported by the U.S. Department of Energy (DoE) under Contract No. DE-AC03-76SF00098, and by National Science Foundation on Grant No. DMR-0205949. Dr. T. Mizoguchi is gratefully acknowledged for helpful discussion of EELS analysis in this work.

REFERENCES

1. J. F. Scott and M. Dawber, *Ferroelectrics* **265**, 119 (2001).
2. T. M. Shaw, S. Trolier-McKinstry, and P. C. McIntyre, *Annu. Rev. Mater. Sci.* **30**, 263 (2000).
3. R. V. Wang, P. C. McIntyre, *J. Appl. Phys.* **94**, 1926 (2003).
4. B. S. Kang, J. G. Yoon, T. W. Noh, T. K. Song, S. Seo, Y. K. Lee, and J. K. Lee, *Appl. Phys. Lett.* **82**, 248 (2003).
5. A. M. Bratkovsky and A. P. Levanyuk, *Phys. Rev. Lett.* **84**, 3177 (2000).
6. S. L. Miller and R. D. Nasby, J. R. Schwank, M. S. Rodgers, and P. V. Dressendorfer, *J. Appl. Phys.* **68**, 6463 (1990).
7. A. K. Tagantsev, M. Landivar, E. Colla and N. Setter, *J. Appl. Phys.* **78**, 2623(1995).
8. N. D. Browning, M. F. Chisholm and S. J. Pennycook, *Nature* **366**, 143 (1993).
9. R. F. Klie, M. Beleggia, Y. Zhu, J. P. Buban and N. D. Browning, *Phys. Rev. B* **68**, 214101 (2003).
10. S. J. Welz, N. D. Browning, prepared paper to be submitted to *Ultramicroscopy* (2005).
11. W. Frogger, B. Schaffer, K. M. Krishnan, and F. Hofer, *Ultramicroscopy* **96**, 481 (2003).
12. G. Duscher, J. P. Buban, N. D. Browning, M. F. Chisholm, and S. J. Pennycook, *Interface Science* **8**, 199 (2000).
13. F. M. F. de Groot, J. Faber, J. J. M. Michiels, M. T. Czyzyk, M. Abbate, and J. C. Fuggle, *Phys. Rev. B* **48**, 2074 (1993).
14. L. Soriano, M. Abbate, J. Vogel, J. C. Fuggle, A.Fernandez, A. R. Gonzalez, M. Sacchi, and J. M. Sanz, *Surf. Sci.* **290**, 427 (1993).
15. M. Kurasawa and P. C. McIntyre, *J. Appl. Phys.,* (2005) (in press).

Mater. Res. Soc. Symp. Proc. Vol. 875 © 2005 Materials Research Society   O14.4

Intrinsic Stress and Alloying Effect in Mo/Ni Superlattices: A Comparison between Ion Beam Sputtering and Thermal Evaporation

A. Debelle, G. Abadias, A. Michel, C. Jaouen, Ph. Guérin, M. Marteau and M. Drouet
Laboratoire de Métallurgie Physique, UMR 6630 CNRS,
Université de Poitiers, SP2MI, Avenue M. et P. Curie, BP 30179
86962 Futuroscope-Chasseneuil, France

ABSTRACT

Epitaxial Mo(110)/Ni(111) superlattices were grown on $(11\overline{2}0)$ single-crystal sapphire substrates, by ion beam sputtering (IBS) and thermal evaporation (TE), in order to investigate the role of deposited energy on the interfacial mixing process observed in Mo sublayers. To separate intermixing and growth stress contributions, a careful and detailed characterization of the stress/strain state of both samples was performed by X-ray Diffraction (XRD). Non-equal biaxial coherency stresses are observed in both samples. For the IBS specimen, an additional source of stress, of hydrostatic type, due to growth-induced point defects, is present, resulting in a triaxial stress state. The use of ion irradiation to achieve a controlled stress relaxation can provide additional data and, as shown elsewhere, allows to obtain the stress-free lattice parameter a_0 solely linked to chemical effects. For the TE sample, a standard biaxial analysis gives a_0. In both samples, the a_0 value is lower than the bulk lattice parameter, due to the presence of intermixed Mo(Ni) layers. However, the intermixing is larger in the sputtered Mo sublayers than in the thermal evaporated ones, putting forward the prime role of energy and/or momentum transfer occurring during energetic bombardment.

INTRODUCTION

The physical properties of many materials when deposited as thin films or multilayers are different from the bulk ones. Large anomalies in the elastic properties (deviations from the continuum elasticity) were reported for some bcc/fcc multilayer systems, such as Cu/Nb or Mo/Ni [1], when the bilayer period is decreased to the nanoscale. For sputtered Mo/Ni superlattices, a drastic softening of the C_{44} elastic shear modulus has been noticed and attributed to a chemical alloying, mainly Ni atoms in Mo layers, observed at the interfaces [2,3], although the bulk phase-diagram exhibits almost zero solubility. However, the driving force for the formation and the stabilization of this metastable alloy remained to be identified; in particular, the possible role of energetic particles involved during the growth must be studied. To address this issue, an accurate characterization of the strain/stress state of the films is required to separate intermixing from growth stress effects [4,5].

For this purpose, epitaxial Mo(110)/Ni(111) multilayers were grown on $(11\overline{2}0)$ single-crystal sapphire substrates by two methods, namely ion beam sputtering and thermal evaporation. The first technique involves energetic particles (sputtered atoms and backscattered neutrals) in the range of several tens of eV (and even more), whereas the deposited energy in the second case is only thermal (<< 1 eV). The resulting microstructure and growth stress level are therefore expected to be significantly different [6,7]. The complete strain/stress state of the Mo/Ni multilayers was determined by X-ray diffraction (XRD) using the "$\sin^2\psi$-method" adapted to epitaxial thin films (i.e. using the ideal directions method). We report in the present paper the results for the Mo sublayers only, for which the stress anisotropy and magnitude of intermixing are large [8]. Differences in stress state and interfacial mixing extent are observed: the role of energetic bombardment is suggested.

EXPERIMENTAL DETAILS

Two Mo/Ni superlattices with similar bilayer period (Λ~7.5 nm) and same thickness (~140 nm), were grown onto (11$\bar{2}$0) oriented single-crystals sapphire substrates using different deposition techniques. The first sample was deposited by ion beam sputtering (IBS) in a ultrahigh vacuum (UHV) Nordiko-3000 system (base pressure of 10^{-8} Torr) at a working Ar pressure of 10^{-4} Torr. Prior to deposition, a 4 nm-thick Mo buffer layer was epitaxially grown onto the sapphire substrate at 650°C [8]. Then, sequential deposition of Ni and Mo layers was performed at room temperature, using a 1.2 kV Ar primary beam and a 80 mA current, yielding a growth rate of 0.6 Å.s^{-1} for Mo and 0.82 Å.s^{-1} for Ni. A bilayer period of Λ~ 7.6 nm was determined by high-angle XRD. The second multilayer was prepared by thermal evaporation (TE) in an UHV Riber chamber, equipped with two electron guns. The power supply was adjusted to obtain growth rates similar to the IBS ones. The base pressure was ~ 4×10^{-8} Torr and never exceeds 4×10^{-7} Torr during deposition. The multilayer was grown at a stabilized temperature of 200°C to insure a good crystalline growth of dense Ni and Mo layers, without introducing significant thermal stress. XRD analysis indicates a bilayer period of Λ ~ 7.1 nm. In both cases, an epitaxial growth is obtained, following the Nishiyama-Wassermann orientation relationship [001] (110) Mo// [1$\bar{1}$0] (111) Ni [8]. The misfit is negative in both principal in-plane directions: 26% for the [1$\bar{1}$0]Ni//[001]Mo directions and 3 % for the [11$\bar{2}$]Ni//[1$\bar{1}$0]Mo ones.

Ion irradiations were performed on the sputtered Mo/Ni multilayer, at room temperature and using 320 kV accelerated Ar ions, in order to induce a growth stress relaxation. The employed fluences, 7.5×10^{13} and 1.1×10^{14} ions/cm^2, correspond to low irradiation doses, 0.1 and 0.15 displacements per atom (dpa) respectively, as calculated using the SRIM computer code [9], so that possible intermixing at the interfaces was limited during irradiation.

The structural properties of the specimen were investigated by high-angle XRD measurements, recorded on a Bruker D5005 diffractometer in Bragg-Brentano geometry, using a Cu radiation. The high-angle XRD spectra were fitted with the Suprex code [10].

The stress state was analyzed in a 4-circle Seifert XRD-3000 diffractometer, equipped with a 1x1 mm^2 point focus and a Ni filter in the direct beam path to absorb the Cu K$_\beta$ radiation. The incident beam was collimated using a 0.5 mm diameter fiber optics, and the exit divergence was limited by two slits (1 mm and 0.5 mm), defining an instrumental resolution of 0.13°.

STRUCTURAL CHARACTERIZATION

High-angle symmetric XRD spectra are shown in Fig.1 for both IBS and TE samples. Both θ-2θ scans exhibit superlattice reflections, which attest of a good crystallinity and coherent stacking of Mo and Ni sublayers. Nevertheless, the smaller peak widths of the TE multilayer indicate a larger coherence length along the growth direction, possibly related to lesser intralayer defects and/or lesser interplanar spacing gradients due to intermixing. The first and the second order (not shown here) spectra were fitted until a consistent solution for both orders was found.

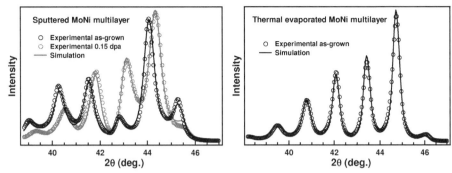

Figure 1: High-angle XRD spectra. Left: experimental (symbols) and refined calculations (lines) for the as-grown (IBS) Mo/Ni superlattice (black) and for the 0.15 dpa irradiated one (grey). Right: experimental (symbols) and refined calculations (lines) for the as-grown (TE) multilayer.

The main parameters used during the fitting process were the out-of-plane interplanar spacings of both layers, d_{Ni} and d_{Mo}, and the number of atomic planes in each elemental sublayer, N_{Ni} and N_{Mo}. The results are reported in Tab.I, and the best-fits of experimental data are displayed in Fig.1. A significant difference in d_{Mo} is found between the two specimen: for the IBS multilayer, d_{Mo} is higher than the bulk distance (2.2253 Å), whereas it is lower in the TE Mo/Ni one. However, no conclusion can be drawn from this unique result since the interatomic spacing depends on the stress state in the layer and is also affected by the chemical effects that are likely to take place at the interfaces. Separating these two contributions requires accurate description and modelling of the strain/stress state of the multilayers. It will be shown that the use of ion irradiation to induce stress relaxation is required to model the complex stress state of IBS multilayers, and to determine the intermixing extent from the stress-free lattice parameter a_0. Regarding the TE sample, a standard biaxial analysis procedure allows obtaining a_0 directly.

Table I: Structural parameters of the IBS and the TE Mo/Ni multilayers obtained from the refinement of the high-angle XRD spectra, achieved with the Suprex code

Mo/Ni multilayer	N_{Ni}	d_{Ni} (Å)	N_{Mo}	d_{Mo} (Å)
as-grown IBS	22.8	2.039	13.1	2.240
0.15 dpa IBS	23.0	2.044	12.9	2.205
as-grown TE	24.7	2.032	9.7	2.218

STRAIN/STRESS STATE AND INTERFACIAL CHEMICAL EFFECTS

Ion beam sputtered Mo/Ni multilayer

The strain/stress state was determined using the so-called "$\sin^2\psi$-method", which relies on the use of the lattice plane spacing d_{hkl} of (hkl) planes as an internal strain gauge. The measured lattice strain $\varepsilon_{\psi,\phi}^{hkl}$ along a particular direction (ψ,ϕ) is given by $\varepsilon_{\psi,\phi}^{hkl} = (a_{\psi,\phi}^{hkl} - a_0)/a_0$, where Ψ is the angle between the surface normal and the normal to (hkl) planes and ϕ is the azimuthal angle; a_0 is the stress-free lattice parameter and $a_{\psi,\phi}^{hkl}$ is the lattice parameter determined from a given {hkl} reflection. In the framework of linear elasticity, the strain is linked to the stress components through the compliances S_{ij} of the material. For isotropic materials, the plot of $a_{\psi,\phi}$ vs $\sin^2\Psi$ is one straight line and the stress value is proportional to the slope. However, the small elastic anisotropy in (110) textured Mo films introduces a weak dependence of $a_{\psi,\phi}$ on the azimuthal ϕ angle. Therefore, a ϕ-reference was defined along the [002] direction, which gives the following ([002], [$1\bar{1}0$], [110]) specimen referential.

In order to appropriately describe the stress state of a sputtered thin film submitted to energetic atomic bombardment during the growth, we developed a stress model originally proposed by Kamminga et al [11]. It was extensively described elsewhere [4,5,12]. Briefly, it depicts the thin film as a free-standing layer, called matrix, into which misfitting inclusions can be introduced. These misfitting particles expand the surrounding matrix, which creates a hydrostatic stress σ_{hyd} in the matrix, but, since this latter must be attached to the substrate, its lateral dimensions are fixed. This generates an isotropic in-plane biaxial fixation stress σ_{fix}. The exact relation between σ_{hyd} and σ_{fix} is complex since it involves the respective elastic properties of the inclusions and of the matrix, as well as their size mismatch; it is however easily conceivable that σ_{fix} is proportional to σ_{hyd}, so that $\sigma_{hyd} = -\beta\sigma_{fix}$, where β is an adjustable parameter. Furthermore, in the present case, an additional stress contribution is expected, owing to the epitaxy. This purely biaxial component, referred as coherency stress, is not equal in all in-plane directions, since the Nishiyama-Wassermann relationship is anisotropic, and consequently introduces a large additional ϕ-dependence. Therefore, it is described with two values, σ_{11}^{coh} and σ_{22}^{coh}, along the [002] and [$1\bar{1}0$] directions, respectively. Finally, the strain observed in the matrix for a given (ψ,ϕ) direction, as obtained from XRD analysis, is expressed as :

$$\varepsilon_{\psi,\phi}^{(110)Mo} = \sigma_{hyd}\left(S_{11}+2S_{12}\right)+\frac{1}{2}\left[\sigma_{fix}+\left(\frac{\sigma_{11}^{coh}+\sigma_{22}^{coh}}{2}\right)\right]\left[\left(J+4S_{12}\right)+\left(S_{44}+J\sin^2\phi\right)\sin^2\Psi\right]$$
$$+\frac{1}{4}\left(\frac{\sigma_{11}^{coh}-\sigma_{22}^{coh}}{2}\right)\left[2J-\left(3J-\left(3J+2S_{44}\right)\cos 2\phi\right)\sin^2\Psi\right] \tag{1}$$

where $J = S_{11}-S_{12}-S_{44}/2$ is the anisotropy factor of the Mo, equal to -0.0013 GPa^{-1}. Accounting for the heteroepitaxial relationship between the fcc and bcc lattices, the term ($\sigma_{11}^{coh}-\sigma_{22}^{coh}$)/2 is expected to be large, and the ϕ-dependence of $a_{\psi,\phi}$ will be particularly visible at $\sin^2\Psi = 1$.

At this stage, it is worth underlining that a complete description of the strain/state of the as-grown Mo sublayers requires the determination of five parameters: a_0, σ_{hyd}, β, σ_{11}^{coh} and σ_{22}^{coh}. An analysis of the sole $\sin^2\Psi$ plots of the as-grown Mo layers cannot provide the whole required information. However, it has been previously shown that ion irradiation can provide additional data, by allowing a controlled stress relaxation of the films. This method has been successfully implemented in the case of pure films [12] as well as in the case of multilayers [4,5].

Fig.2a shows the $\sin^2\Psi$ plots obtained for the Mo sublayers in the as-grown and irradiated Mo/Ni multilayers, for the two investigated ϕ directions (0° and 90°). The slope decrease provides a direct evidence of a biaxial stress relaxation under ion irradiation, but, since the splitting of the two lines observed at $\sin^2\Psi = 1$ remains unchanged, the coherency stress is not released, at least at this low fluence range.

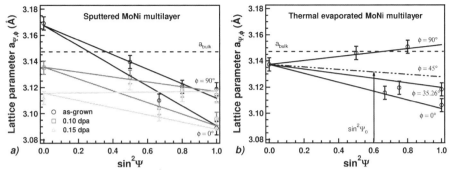

Figure 2: Lattice parameter vs $\sin^2 \Psi$ of Mo sublayers: a) experimental data (symbols) and best-fit lines of the as-grown (black), 0.10 dpa (grey) and 0.15 dpa irradiated (light grey) IBS Mo layers; b) experimental data (symbols), best-fit lines, and average stress direction ($\phi = 45°$) of the as-grown TE Mo layers. The horizontal dotted line corresponds to the bulk Mo lattice parameter.

The fitting of experimental data, using the above-described model, implies too many adjustable fitting parameters. Therefore, following observations in pure Mo films under similar conditions, a complete relaxation of the hydrostatic stress at the highest fluence and a value of $\beta = 1.2$ were assumed. The best fit lines are displayed in Fig.2a. A hydrostatic growth stress component of 4.8 GPa is found in the as-grown layers, akin the value observed in pure Mo films. During irradiation, short-range rearrangements allow the system to evolve towards a closer to equilibrium state; accordingly, the growth stress is relaxed. The stress-free lattice parameter for the as-grown specimen, obtained from this analysis, is $a_0 = 3.118$ Å, which is significantly lower than the bulk Mo lattice parameter (3.147 Å). This indicates that Ni atoms were introduced in the Mo layers during the multilayer growth, since the Ni atomic radius is lower than the Mo one.

Thermal evaporated Mo/Ni multilayer

Fig.2b shows the $\sin^2 \Psi$ plots obtained for Mo sublayers of the as-grown TE Mo/Ni multilayer. For this type of film, only small tensile biaxial growth stress is expected, and the total biaxial stress should be, in the present case, composed of thermal, growth and coherency stresses. Therefore, a standard analysis only, without implementing ion irradiation induced stress relaxation, is required to fully describe the state of stress of the Mo sublayers. Thus, the elastic strain is given by Eq.(1), with σ_{hyd} and σ_{fix} equal to zero. The experimental data were fitted accordingly, and the best fit lines, for the three investigated ϕ directions (0°, 35.26° and 90°), are displayed in Fig.2b. Since the epitaxial misfit is negative in both in-plane principal directions, compressive stresses were expected. However, if the calculated stress in the [002] direction is effectively compressive (–2.3 GPa), the one obtained in [110] is tensile (+ 2.0 GPa). This large tensile stress level cannot be accounted for by growth and thermal stresses. On the one hand, with the present growth conditions, these components are expected to be small, and on the other, in Ni sublayers (not shown here), the maximum biaxial stress is found to be ~ 1 GPa. The origin of this tensile coherency stress could be related to atomic rearrangements of Mo atoms on Ni sublayers during the first growth stages, to minimize the stored elastic energy, the two-fold symmetry of (110) Mo bcc structure being strained to adopt the three-fold symmetry of (111) fcc

Ni planes. The stress-free lattice parameter a_0, obtained from the fitting, is found to be 3.131 A, slightly lower than the bulk Mo value, indicating a small alloying effect. Note that, due to the anisotropy of the coherency stress, the classical stress-free direction $\sin^2 \Psi_0$ (obtained reducing Eq.(1) to zero) depends on σ_{11}^{coh} and σ_{22}^{coh}. Consequently, a_0 cannot be directly graphically determined but requires a fitting of experimental data. This shows again the importance of thoroughly taking into account the stress state, as already demonstrated for the IBS sample.

CONCLUSION

Mo/Ni superlattices were epitaxially grown by ion beam sputtering and thermal evaporation on (1120) single-crystal sapphire substrates. Both specimen exhibit large coherency stresses due to the epitaxy, but only the sputtered multilayer presents a large compressive growth stress, due to energetic bombardment during growth. Therefore, the lattice parameter is expanded, and, since it contains a contribution from growth-induced point defects, a direct evaluation of the interfacial mixing is not possible using a simple biaxial analysis. The proposed triaxial stress model provides, thanks to ion irradiation induced stress relaxation, the stress-free lattice parameter a_0 solely linked to chemical effects, and accordingly an accurate estimation of the interfacial alloying. On the other hand, the thermal evaporated specimen displays biaxial stress only, and the a_0 value can be determined using standard analysis. It is demonstrated that, in both cases, this parameter cannot be obtained without a precise description of the stress state. An interfacial alloying, that is a consequence of a dynamic segregation of Ni atoms in Mo sublayers during growth, is observed in both specimen, which suggests that a lowering of the free interfacial energy promotes exchange mechanisms. This may explain the stabilization of the metastable alloy. Nevertheless, since this mixing is significantly larger in the sputtered multilayer, enhanced surface mobility (energy transfer) and/or collisions effects (momentum transfer) due to the particles involved in the sputtering process may favour its formation.

REFERENCES

[1] I.K. Schuller, A. Fartasch and M. Grimsditch, MRS Bull. XV (10), 33 (1990).
[2] G. Abadias, C. Jaouen, F. Martin, J. Pacaud, Ph. Djemia and F. Ganot, Phys. Rev. B **65**, 212105 (2002).
[3] F. Martin, C. Jaouen, J. Pacaud, G. Abadias, Ph. Djemia and F. Ganot, Phys. Rev. B **71**, 045422 (2005).
[4] A. Debelle, G. Abadias, A. Michel, C. Jaouen, Ph. Guérin and M. Drouet, Mat. Res. Soc. Symp. Proc. Vol. 795, U12.3.1 (2004).
[5] A. Debelle, A. Michel, G. Abadias, and C. Jaouen, Nucl. Instr. Meth. (*under press*).
[6] H. Windischmann, Crit. Rev. Solid State and Materials Sciences **17**, 547 (1992).
[7] R.W. Hoffmann, Thin Solid Films **34**, 185 (1976).
[8] F. Martin, J. Pacaud, G. Abadias, C. Jaouen, Ph. Guerin, Appl. Surf. Sci. **188**, 90 (2002).
[9] J.F. Ziegler, J. P. Biersack, U. Littmark, http://www.srim.org.
[10] E. Fullerton, I. K. Schuller, H. Vanderstraeten, Y. Bruynseraede, Phys. Rev. B **45**, 9292 (1992).
[11] J.-D. Kamminga, Th.H. De Keijser, R. Delhez, E. J. Mittemeijer, J. Appl. Phys. **78**, 832 (1995).
[12] A. Debelle, G. Abadias, A. Michel, C. Jaouen, Appl. Phys. Lett. **84**, 5034 (2004).

Mater. Res. Soc. Symp. Proc. Vol. 875 © 2005 Materials Research Society   O14.5

Kinetic Analysis and Correlation with Residual Stress of the Ni/Si System in Thin Film

F.CACHO[1,2], D.AIME[2,3], F.WACQUANT[2], B.FROMENT[2], C.RIVERO[5], P.GERGAUD[5],
O.THOMAS[5], G.CAILLETAUD[1], H.JAOUEN[2], S.MINORET[4], A.SOUIFI[3]

[1] Centre des Matériaux P.M. FOURT, Ecole des Mines de Paris, B.P. 87, 91003 Evry, France
[2] STMicroelectronics, 850 rue Jean Monet, 38921 Crolles, France
[3] Laboratoire de Physique de la Matière (LPM), INSA Lyon, 69621 Villeurbanne, France
[4] CEA-LETI, 17 rue des Martyrs, 38100 Grenoble, France
[5] TECSEN, UMR CNRS 6122, Univ. Aix-Marseille III, F-13397 Marseille Cedex 20, France
florian.cacho@st.com

ABSTRACT

Reactive diffusion of the Ni/Si system has been studied by annealing nickel thin film on (100) silicon crystal. The measurement of the NiSi sheet resistance as a function of the annealing temperature and the type of annealing (Rapid Thermal Annealing and spike one) has been investigated. A kinetic model based on multiphase diffusion has been developed that fits experimental sheet resistance data. Residual stress in the thin film, measured by a curvature measurement technique, is correlated with the nature of the phases in the film. Finally the viscoplastic mechanical behavior of the Ni_2Si and NiSi phases is analyzed in the case of low and fast thermal ramps.

INTRODUCTION

In microelectronics industry, device performance improvement is due to scale shrinking. According to the International Technology Roadmap Semiconductor, the 65nm node will be released in 2005 and the reduction of contact resistivity in integrated circuit is an important challenge. Silicide thin films have great advantages for the reduction of series resistance in gates and local interconnects of heavily doped contacts[1], but also in Schottky diode for bipolar transistor[2]. In advanced CMOS technology several silicide were used. $TiSi_2$ was commonly used in the past last 20 years, but the low resistivity phase nucleation C54 is limited and for line width less than 0.1μm, the reaction is uncompleted. $CoSi_2$ silicide gives better sheet resistance performance when scaling down dimension; it has therefore replaced $TiSi_2$ for many applications. Nevertheless, NiSi silicide seems to be the best candidate for replacing $CoSi_2$ beyond the 65nm node. There are many advantages in its uses:

- Low consumption of silicon during the formation that avoid spiking of shallow junction

- Low formation temperature for the low resistive NiSi phase (<500°C)

- No line width dependence of the sheet resistance

In MOS technology, silicide is formed by solid-state reaction between metal and silicon in a salicide process (self-aligned silicide process). The phase sequence of the Ni-Si system and kinetics of the reaction were quite extensively studied in the past[3,4,5,6]. Some recent in situ characterization studies[7,8] during silicidation bring information on the formation mechanism and material properties of the silicide film. According to several authors, kinetics[9,10] of Ni_2Si and NiSi are mainly controlled by the diffusion of the most mobile species, the Nickel. In practice a TiN capping layer is deposited on the Nickel film to prevent the diffusion of Oxygen during the silicidation process. Nickel silicide is formed by two annealing steps. Currently, the first step, a Rapid Thermal Annealing, is performed at low temperature, then the unreacted Nickel and the TiN are removed using a selective etch (SE). Then, a second Rapid Thermal Annealing is performed at 450°C to transform the Ni_2Si into NiSi

Since the resistivity of TiN, Ni, Ni_2Si and NiSi materials differs significantly, sheet resistance measurements provide a fast mean to track the phase transformation. The kinetics can be extracted from the analysis of the thin film resistivity as a function of temperature. In this study, we conduct such an analysis for different annealing temperature. A kinetic model based on the diffusion of each phase in the film with temperature is used to fit experimental results of the sheet resistance. Then the correlation between the thickness extracted by this way and the residual stress obtained from curvature measurements at room temperature is discussed. Finally, after the formation of NiSi, the residual stress behavior is analyzed as a function of the cooling down rate.

KINETIC ANALYSIS

A 10nm thick Ni film with a 10nm thick TiN capping layer were deposited on (001) silicon wafers by PVD. Then, two kinds of annealing were carried out before or after the Selective Etch step: a 30s Rapid Thermal Annealing (RTA) or a spike annealing both within the range of temperature 270°-450°C. The temperature is monitored during the annealings using a pyrometer.

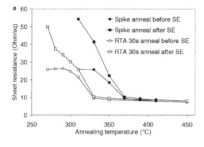
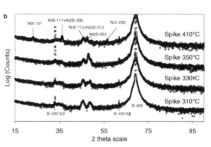

Fig.1a: Sheet resistance vs. annealing temperature for different type of annealing.

Fig.1b: XRD results for the spike annealed samples.

On Fig.1a, the transformation curves are more shifted toward the lower temperature for RTA annealed samples than for the spike annealed ones. At low temperature, the high sheet resistance corresponds to the Ni_2Si phase growth. A significant difference of the sheet resistance behavior for annealed samples before or after the selective etch step (SE) can also be noticed at low temperature, due to an excess of unreacted Nickel. After the formation of NiSi phase in the whole film, the sheet resistance remains constant. For a given high annealing temperature, for example 450°C, the difference in sheet resistance between annealed samples before or after the SE step corresponds to the TiN film removal. Furthermore, XRD analysis conducted on the spike annealed samples (Fig.1b) confirm the presence of Ni_2Si at low temperature and NiSi at higher temperature.

As represented on Fig.2, the growth of all the phases is sequential in thin film, contrary to the bulk case. Ni diffuses in Silicon to form the less resistive and Nickel rich phase, Ni_2Si, until there is no more Ni to react. Then, Ni_2Si is decomposed to form NiSi, releasing Nickel atoms in solid solution (Eq.1). The Nickel atoms diffuse through the NiSi layer and react with Silicon to create a new NiSi layer (Eq.2). Note that Ni_2Si and NiSi are not the only phases in the system. Ni_3Si_2 appears during the sequence but will not be considered here.

$$Ni_2Si \rightarrow NiSi + (Ni)_{ss} \quad (Eq.1)$$
$$Si + (Ni)_{ss} \rightarrow NiSi \quad (Eq.2)$$

Fig.2: Schematic representation of the phase change in Ni/Si system. Subscript ss mean solid solution.

Assuming the planar growth of the different phases, the thin film sheet resistance R_{S1} before SE, or after the selective etch R_{S2}, where the TiN capping layer is removed, can be expressed as:

$$\frac{1}{R_{S1}} = \frac{1}{R_{TiN}} + \frac{1}{R_{Ni}} + \frac{1}{R_{Ni_2Si}} + \frac{1}{R_{NiSi}} \text{ and } \frac{1}{R_{S2}} = \frac{1}{R_{NiSi}} + \frac{1}{R_{Ni_2Si}} \quad (Eq.3)$$

Where R_{TiN}, R_{Ni2Si}, R_{Ni} and R_{NiSi} are respectively the sheet resistance of the TiN, Ni_2Si, Ni and NiSi layers.

MODELING OF KINETICS

The kinetic model must be able to simulate the sheet resistance behavior as a function of the annealing temperature before and after selective etch. In this part, only the spike annealed samples will be studied. The model consists in writing the sheet resistance of each phase as the ratio between the thickness, e, and the resistivity, ρ, of the phase. Assuming that the phase growth is planar, the total sheet resistance is written as:

$$\frac{1}{R_S} = \frac{e_{Ni}}{\rho_{Ni}} + \frac{e_{Ni_2Si}}{\rho_{Ni_2Si}} + \frac{e_{NiSi}}{\rho_{NiSi}} + \frac{e_{TiN}}{\rho_{TiN}} \text{ with } \dot{e}_{Ni_2Si} = \left|\frac{D_{Ni_2Si}}{e_{Ni_2Si}}\right| \text{ and } \dot{e}_{NiSi} = \left|\frac{D_{NiSi}}{e_{NiSi}}\right| \quad (Eq.4)$$

In the case of diffusion limited reaction, as it is the case for spike annealed samples, the thickness of the silicide phases is given by a parabolic law as a function of time. The evolution equations of Ni and Ni$_2$Si layers are composed of a term of growth (Eq.4) and a term of consumption. The diffusion coefficient D is temperature dependent, with an activation energy Q taken from literature[11]:

$$D_{Ni_2Si} = D_{01}e^{\frac{-1.5}{kT}} \text{ and } D_{NiSi} = D_{02}e^{\frac{-1.4}{kT}} \tag{Eq.5}$$

The diffusion prefactors D_{01} and D_{02} are the two unknowns of the model; they will be fitted on the experimental evolution of the sheet resistance as a function of the annealing temperature. The resistivity of each layer, Ni, Ni$_2$Si and NiSi will be determined by a global fitting of the experimental sheet resistance curve, whereas those of the TiN layer will be extracted by comparing samples annealed before or after the SE. The same parameter set is used to optimize the model by a Levenberg-Marquardt algorithm of this system composed of 5 unknown parameters (diffusion prefactors and resistivities) which is the solution of 14 equations (7 annealing temperatures with and without selective etch). After optimization, the results of the resistivity presented in Table1 show a ratio ρ_{Ni2S}/ρ_{NiSi}=1.71 close to those given in open literature ($\approx$1.8). Fig.3 shows good agreement between the sheet resistance simulated curve and the experimental measurement. Then, from this simulation, the layer thicknesses extracted have been plotted in Fig.4 as a function of the annealing temperature, whereas the parameter set is reported in Table1. The different profiles obtained exhibit the Ni/Si sequential formation described previously.

D_{01} (nm²/s)	D_{02} (nm²/s)	ρ_{TiN} ($\mu\Omega$.cm)	ρ_{Ni} ($\mu\Omega$.cm)	ρ_{Ni2Si} ($\mu\Omega$.cm)	ρ_{NiSi} ($\mu\Omega$.cm)
138.10[11]	35.10[11]	90	60	24	14

Table1: Model parameters after optimization: diffusion prefactors and average resistivities.

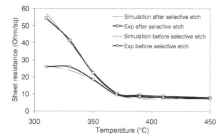

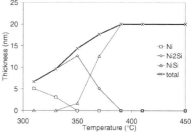

Fig.3: Sheet resistance vs. temperature for spike annealed samples. The kinetic model of multiphase growth is optimized on experimental curve before and after selective etch.

Fig.4: Evolution of the final thickness vs. annealing temperature of the different layers given by the kinetic model.

The evolution of the kinetics is presented for spike annealed samples at 330°C and respectively at 410°C on Fig.5. Temperature is given from pyrometer measurement at each time step whereas the thickness of each layer is computed from the kinetic model. For the 330°C spike annealing, a small amount of Ni$_2$Si is created because the time slot above 300°C is very short, whereas for the 410°C spike annealed sample, Ni is totally consumed to form the Ni$_2$Si phase which is partially consumed to form the NiSi phase. Note that the diffusion and so the phase transformation at this temperature is quickly performed in 10 seconds.

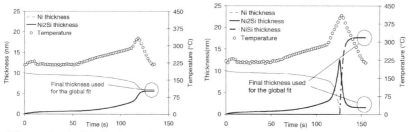

Fig.5: Phase transformation during the spike annealing time, on the left for a spike annealing at 330°C, on the right for a spike annealing at 410°C.

This model is thus able to give the average thickness of Ni$_2$Si and NiSi as a function of the annealing temperature. A same optimization was carried out for RTA annealed samples but not reported here.

CORRELATION WITH RESIDUAL STRESS

After extracting the kinetic of phase change for Ni₂Si and NiSi, a correlation between the residual stress at room temperature in the thin film and the nature of the phases is firstly conducted. Then, the mechanical behavior of Ni₂Si is discussed for explaining the residual tensile stress trend when increasing temperature. Finally, the residual thermal stress is reviewed for different temperature of annealing.

Evolution of residual stress

Several in-situ characterizations of the stress have been done for the Ni/Si system by curvature measurement in the open literature. For most of the authors[12], the first phase Ni₂Si seems to be compressive during its growth. Then, at higher temperature the apparition of the NiSi phase and the consumption of Ni₂Si layer lead at the end of the reaction to a global stress free film.

Curvature measurements have been conducted by a laser scanning technique at room temperature on 300mm wafers after annealing and after TiN removal by the selective etch. The curvature radius of the substrate is related to the force F in the thin film using the Stoney equation:

$$F = \sigma_{film}.e_{film} = \frac{E_s}{1-\upsilon_s}\frac{e_s^2}{6}\left(\frac{1}{R}-\frac{1}{R_0}\right)$$ (Eq.6)

Where R_0 is the initial curvature before metal deposition, R the current curvature, e_{thin} and e_s are respectively, the thickness of the thin film and those of the substrate. E_s and υ_s are respectively the Young's modulus and the Poisson's ratio of Silicon. The stress of the thin film is deduced from the force if the thickness is known. Thanks to the previous kinetic model, the stress is given for RTA and spike anneal (Fig.6 and 7).

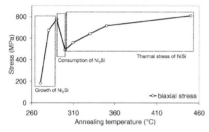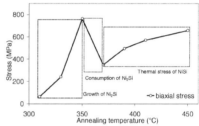

Fig.6. Evolution of residual stress as a function of the temperature of the RTA annealed samples.

Fig.7: Evolution of residual stress as a function of the temperature of the spike annealed samples.

The residual stress behavior has the same trends for the spike and the RTA annealed samples although the spike annealed sample behavior is shifted to higher temperature due to a slower kinetic. At low temperature, the Ni₂Si phase has a tensile residual stress which seems to be opposite to the compressive strain observed during it growth. This apparent contradiction will be explained below. This residual stress increases from 150MPa to 800MPa when the thickness and the temperature become more important. Then, the consumption of Ni₂Si decreases the tensile stress and NiSi grows with a tensile residual stress.

Although the transformation seems to be finished at 390°C for spike annealed films and at 310°C for the RTA annealed ones, the residual stress continues to increase for higher temperature. However, the NiSi trend seems to be linear after 400°C, it means that after this temperature, the behavior is only due to thermal strain. The thin film totally relaxes at the annealing temperature and the residual stress is expressed as $M_{NiSi}.(\alpha_{NiSi}-\alpha_{Si})(T_{annealing}-T_{ambiant})$ with M_{NiSi}, α_{NiSi}, α_{Si} respectively the biaxial modulus and the thermal expansion coefficient of NiSi and Si. This evolution is linear with $T_{annealing}$ as shown in Fig.6 and Fig.7.

Mechanical behaviour of Ni₂Si

In the Ni/Si system, compressive stress occurs during Ni₂Si formation. This is the consequence of a volume change at the growing interface. Some models, like the Zhang and d'Heurle model[13], take into account stress evolution in terms of competing growth and stress relaxation rates. However, the experimental residual Ni₂Si stress as a function of the annealing temperature in Fig.6 and Fig.7, gives a tensile trend when the thickness and the temperature become more important. For the sake of understanding the residual stress behavior in Ni₂Si layer, a thicker film has been processed. In the aim of being sure that the specimen is meanly composed of the rich Ni phase, an XRD and Auger analysis is conducted.

An excess of nickel (145nm) was deposited on a substrate composed with a 120nm thick polysilicon film above an oxide layer 2nm thick. A 330°C RTA annealing during 9min has been performed to form mainly the Ni_2Si phase. Then the excess nickel was removed by Selective Etching. A second annealing at 450°C during 4min was carried out to compare the evolution of the composition and the residual stress of the Ni_2Si film when there is no more Silicon and Nickel to continue the silicidation process.

The Auger profiles on Fig.8 show the existence of the top Ni_2Si layer with a gradual decrease of the Ni/Si ratio to the Ni_3Si_2 phase for the two samples (before and after anneal at 450°C). It means that there is a grain mixture between the two phases with the richest Nickel phase on the top and the richest Silicon phase on the bottom. XRD analysis confirms the presence of Ni_2Si phase on Fig.9. The other peaks observed might be attributed to the $Ni_{31}Si_{12}$ phase. Moreover, there is no significant difference between the specimen annealed at 350°C and the one annealed at 450°C. The two experiments show that the composition and the nature of the phases remain stable when there is no source of silicon and nickel whatever the thermal budget is. Concerning the residual stress, curvature measurement gives respectively a residual tensile stress before and after the 450°C anneal of 1 GPa and 1.9 GPa.

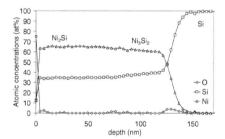
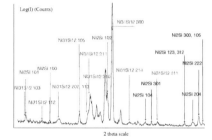

Fig.8: Atomic concentration profiles in the polysilicon layer totally silicided after the second annealing at 450°C. The profile before the post-annealing is exactly the same.

Fig.9: XRD spectrum on the same specimen than Fig.8 after the second annealing at 450°C. The same diffraction peaks are present before the post-annealing.

The residual stress of a silicide film mainly composed of Ni_2Si is tensile whereas it is well known that the Ni_2Si growth induces a high compressive stress during its formation. A simulation which takes into account the temperature and the time of annealing is carried out to explain this opposition.

This behavior can be explained by decomposing the strain responsible for the residual force. When assuming the planar growth of silicide, only the biaxial strain component inducing the curvature of the wafer is monitored. Strain partition locally is a sum of the elastic strain ε^{el}, the viscoplastic strain (highly dependant on temperature) ε^{vi}, the thermal strain ε^{th} and a constant specific to the phase change dilatation ε^{ph}, such as:

$$\varepsilon^{to} = \varepsilon^{el} + \varepsilon^{th} + \varepsilon^{vi} + \varepsilon^{ph} \qquad (Eq.7)$$

Viscoplasticity is modeled by a power law of the Von Mises stress invariant with a threshold R_0. With the convention $\langle a \rangle = \max(a,0)$, the viscoplastic strain rate is written as:

$$\dot{\varepsilon}^{vi} = \left\langle \frac{\sigma - R_0}{K} \right\rangle^n \qquad (Eq.8)$$

The threshold R_0, n and the parameter K of the Norton law are temperature dependent; σ is the biaxial stress in the film. At low temperature the relaxation time is very long and decreases drastically when increasing temperature. K follows an Arrhenius law with an activation energy Q_K: $K=K_0\exp(Q_k/kT)$. The thermal strain rate depends on the thermal expansion coefficient α_{Ni2Si} and on the current temperature rate.

$$\dot{\varepsilon}^{th} = \alpha_{Ni2Si}\dot{T} \qquad (Eq.9)$$

For the simulation, the intrinsic stress of the phase change is assumed to be equal to 1.5GPa. The relaxation coefficient K_0 and Q_k are respectively 1.5eV and 1.3Pa.s and the thermal dilatation is 16ppm/°C. The thermal history of the simulation is composed of an isothermal temperature plateau followed by a cooling down ramp to room temperature fast enough that no relaxation occurs during this step.

Assuming a temperature profile constant up to 80s and then decreasing to room temperature at 90s, the schematic representation of the stress within the thin film during the Ni_2Si formation is simulated and reported in Fig.10 for three isothermal temperatures: 250°C, 300°C and 350°C. As the threshold R_0 is assumed to be null when the relaxation is very fast for the high temperature case at 350°C, the stress decreases drastically to the null value before the cooling down regime. Whereas for the 250°C and 300°C annealings, the relaxation is slower, leading to a residual stress less and less tensile after the cooling down regime.

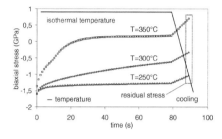
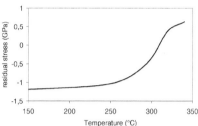

Fig.10: Schematic representation of the stress in the thin film during the Ni_2Si formation for different isothermal annealing of 80 seconds.

Fig.11: Schematic representations of the residual stress of Ni_2Si film vs. annealing temperature for an isothermal annealing time of 80 seconds

The evolution of the residual stress as a function of the isothermal annealing temperature is reported in Fig.11. For the lower temperature (below 250°C), the relaxation time is so long that the residual stress in the film stays constant with the annealing temperature. When the isothermal annealing temperature increases, the relaxation time decreases drastically and the residual stress becomes more and more tensile.

The trend of residual stress of the Ni/Si system described in Fig.6 and 7 can be explained by the combination of the high compressive growth strain which relaxes with temperature and the tensile thermal strain. Note here that the time of annealing is also an important parameter for relaxation and the kinetic of reaction, for shorter time, the evolution curve of Fig.11 is shifted at higher temperature.

Mechanical behaviour of NiSi

The mechanical behavior of NiSi is discussed between a fast and a slow cooling down ramp. Firstly for a fast cooling down ramp, the Fig.12 shows the residual stress of the NiSi film after different successive RTA post-annealing. The formation of NiSi is performed at 450°C. In this experiment, the cooling down rate is very fast and no relaxation is able during this step. Only the time and the temperature of the isothermal step are responsible for the difference of residual stress. During the annealing at 400°C 60s, the NiSi film relaxes, whereas for lower thermal budget the stress stays constant. It means that the viscoplasticity mechanism is important at high temperature (above 400°C) but the 60s annealing time is not long enough for relaxing the stress below 400°C.

When the cooling down is done with slower ramp rate, 2°C/min, as shown in Fig.13, the residual stress is significantly lower than for the fast cooling down case, nearly 430MPa. Two trends are clearly identified:

- For temperature lower than 280°C, the total force is linear versus temperature. The stress in the NiSi film is due to thermal dilatation, the slope of the cooling curve is function of $(\alpha_{Si} - \alpha_{NiSi})E_{NiSi}/(1 - \nu_{NiSi}) = 1,77 MPa/°C$, with respectively α, E and ν the thermal expansion coefficient, the Young's modulus and the Poisson's ratio of NiSi.

- For temperature higher than 280°C, a high temperature dependent viscoplastic strain regime takes place.

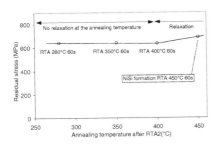

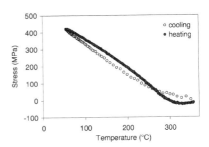

Fig.12: Measurement of residual stress after post- annealing. The formation of NiSi is made at 450°C and the cooling down rate is very fast.

Fig.13: In-situ measurement of NiSi mechanical behavior during annealing at 2°C/min up to 350°C.

The temperature of transition between the viscoplastic and thermo-elastic regime depends on the annealing rate. An annealing rate higher than 2°C/min would give a transition temperature higher than 280°C. However for RTA, the cooling down is similar to a quench, so the time of the isothermal annealing is the key parameter which modify the temperature of transition.

CONCLUSIONS

Integration of the Ni silicidation technique in CMOS technology requires an improvement of the control of the transformation kinetic. RTA and spike anneals at different temperature show the diffusion limited character of the reaction: a fast anneal at high temperature is equal to a long low temperature anneal.

The proposed kinetics model gives the resistivity and the composition of the thin film after annealing. The first phase, Ni_2Si grows by consuming Ni, and then when there is no more metal, NiSi grows and consumes the Ni_2Si in the film.

The behavior of Ni_2Si film is a sum of a high compressive growth strain which relax with temperature and a tensile thermal strain. When the temperature of the Ni_2Si formation increases, the Ni_2Si residual stress becomes more tensile.

Finally the stress behavior of NiSi film was studied. For usual RTA, the annealing time is shorter than the viscoplastic relaxation time. On the contrary for slow ramp rate at 2°C/min, NiSi slowly relaxes above 280°C and its behavior is thermo-elastic below this temperature.

ACKNOWLEDGMENTS

The authors would like to thank G.Rolland for XRD analysis and M.Hopstaken for Auger experiments and are grateful to the CEA-LETI for their support and efficiency.

REFERENCES

1 K.Ng and W.T.Lynch, IEEE Trans. Electron. Devices ED-34 (1987) 503

2 D.B.Scott, W.R.Hunter and H.Shichijo, IEEE Trans. Electron. Devices ED-29 (1982) 651

3 J.Coe and H.Rhoderick, J. Phys. D 9 965 (1976)

4 S.S.Lau, J.W.Mayer, K.N.Tu, J. Appl. Phys.49 4005 (1978)

5 F.M.d'Heurle, C.S.Petrsson, J.E.E.Baglin, S.J.La.Placa and C.Y.Wong, J. Appl. Phys. 55 4208 (1984)

6 P.Sullivan, R.T.Tung and F.Schrey, J. Appl. Phys. 72 478 (1992)

7 C.Lavoie, C.Cabral, F.M.d'Heurle, and J.M.E.Harper, Defect Diffus. Forum 1477 (2001)

8 C.Lavoie, C.Cabral, F.M.d'Heurle, J.L.Jordan-Sweet, J.M.E.Harper, J. Electron. Mater. 31 597 (2002)

9 K.N.Tu, J. Appl. Phys. 48 (1977) 3379

10 J.Angilello, F.M.d'Heurle, S.Peterson and A.Segmüller, J. Vac. Sci. Technol., 17 471 (1980)

11 J.P.Gambino, E.G.Colgan, Materials Chemistry and Physics, pp.99-146, 52 (1998)

12 C.Rivero, Contraintes mécaniques induites par les procédés de la microélectronique : développement des contraintes lors des réactions Co-Si et Ni-Si, Faculté des Sciences et Technique de Saint Jérôme (2005)

13 S.L.Zhang and F.M.d'Heurle, Thin Solid Films 213, 34 (1992)

Mater. Res. Soc. Symp. Proc. Vol. 875 © 2005 Materials Research Society   O14.6

Thermal stress relaxation of plasma enhanced chemical vapour deposition silicon nitride

P. Morin, E. Martinez, F. Wacquant and J. L. Regolini
ST Microelectronics, 850 rue Jean Monnet, F-38926 Crolles Cedex
Pierre.morin@st.com

ABSTRACT

Mechanical and thermal properties of silicon nitride films deposited by different plasma process type have been studied. The initial mechanical stress, thickness and hydrogen content have been evaluated respectively by wafer curvature measurements, ellipsometry and Fourier Transform Infra Red spectrometry. These nitrides presented as-deposited stress values ranging from compressive to tensile. High temperature Rapid Thermal Anneal (RTA) at 1100°C or longer thermal treatments at medium temperature, from 700°C to 850°C were carried out on these materials. The evolution of their properties along the different anneals have been measured and compared to the behaviour of high temperature thermal nitride. One can observe that these stoechiometric plasma nitrides have shifted to an equilibrium tensile stress around 1100-1200 Mpa when submitted to the RTA, independently of their initial stress values. Results are interpreted in terms of H desorption and Si-N bond formation. Chemical reaction $Si_2-N-H + 2 N-H \rightarrow 2 Si-N + NH_3$ appears to be the best candidate to figure out the phenomena.

INTRODUCTION

Silicon nitride is extensively used in CMOS VLSI, as hard mask, sidewall spacers or etch stop layer. Plasma Enhanced Chemical Vapour Deposition (PECVD) processes provide low temperature nitrides films that fit the advanced CMOS front end of line requirements. These nitride films are also used as stressor liners to generate strained silicon channel and enhanced carrier mobility [1] especially because their mechanical stress can be tuned by deposition process parameters, from highly compressive to highly tensile values.

However, these layers can drastically evolve during the subsequent CMOS thermal budgets. In particular, a general trend of PECVD nitride layers is an increase to more tensile stress when annealed [2-4]. These authors suggest that this irreversible stress increase is due to H release and Si-N bonding reorganisation. On the contrary, in [5], it is proposed that the stress change is generated by micro voids shrinkage.

In this paper, we have studied several PECVD nitride films deposited on silicon wafers with different kind of reactors. They were subjected to different anneal, to study the evolution of the material properties under thermal treatment. Dynamic behaviour of stress and hydrogen content are presented and compared.

EXPERIMENTS

PECVD silicon nitride layers were deposited on (100) 200mm silicon wafers, either in a "Cxz" Centura chamber from Applied Materials or in a "Sequel" reactor from Novellus. Both processes use silane amonia chemistry, and samples with thickness ranging between 1300A and 2400A have been prepared. For comparison purpose, stoechiometric Low Pressure CVD

(LPCVD) reference films were processed in furnace, using di-chloro silane and amonia precursors.

The mechanical stresses were evaluated by wafer curvature method using the Stoney formula, with accuracy better than +/- 10 MPa. Because the dilatation nitride and silicon coefficients are not significantly different, thermal stress is low compared to residual stress (below 100 Mpa) and will be negligible in the following. Thickness and hydrogen content are measured by respectively reflectometry and Fourier Transform Infrared spectrometry applying the methodology of Landford and Rand [6].

The PECVD films prepared for this study present as-deposited stress values ranging from compressive to tensile (Figure 1) and refractive index spanning between 1.93 and 2.02, from slightly N-rich to stoechiometric values. LPCVD nitride refractive index and stress respectively equal to 2.0 and 1200 Mpa.

These nitrides were subjected to annealing in Rapid Thermal Anneal (RTA) chamber, at temperature between 700 to 1100°C under N2 atmosphere. The impact of the temperature ramp up and down are considered negligible, since the ramp rate is high (75°C/sec) compared to the annealing time plateau. Mechanical stress, film thickness and hydrogen content measurements are performed at room temperature, before and after annealing periods. In case of long thermal treatment, from 700°C to 850°C, samples were subsequently annealed and cool down to allow the measurements at room temperature.

RESULTS

During a first experiment, samples were annealed at 750°C during 15 minutes. One can observe on Figure 1a that all PECVD films (A-D) shift to a higher stress, more tensile or less compressive, while LPCVD nitride (F) stays at 1200 Mpa.

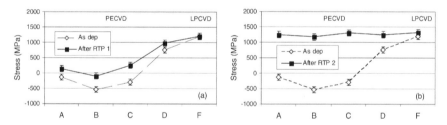

Figure 1. Effect of anneal on stress values; (a) 15' at 750°C and (b) RTP 1 min at 1100°C.

On the other hand, when annealed at 1100°C during 1 minute (Figure 1b), similar PECVD nitride films shift to a final stress around 1200 Mpa. In parallel, LPCVD nitride again stays nearly unchanged at 1200 MPa again. When submitted to an 1100°C extra RTA (not shown here), all the films remain at the same stress values, 1200 Mpa. Other similar PECVD samples (A-D) where subjected to lower temperature anneals, from 700 to 850°C, on longer period. The stress evolutions, measured at room temperature, are plotted on Figure 2. Again, a shift to higher tensile stress has been observed, and saturation is achieved on sample C and D after 1 hour of anneal. In addition, the stress evolution is higher at 850°C than at 700°C.

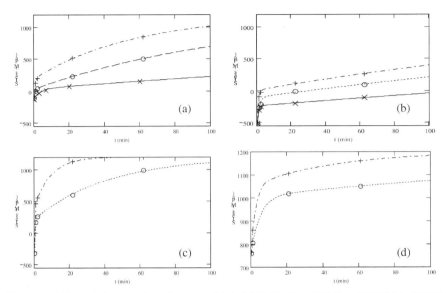

Figure 2. Effect of the anneal on the stress (a) A (b) B (c) C and (d) D at 700°C (x), 770°C (O) and 850°C (+). This curves where fitted with equation (1).

Hydrogen content measured on these samples after 20-22 min of anneal at 700, 770 or 850°C is presented in Figure 3. In all case, N-H bond concentration decreases with anneal while Si-H bond concentration remains unchanged.

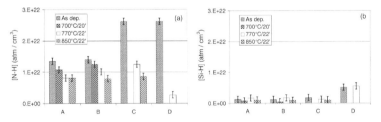

Figure 3. Hydrogen (a) N-H and (b) Si-H content after 22 min of anneal at 700, 770 and 850°C for samples A-D.

DISCUSSION

From these data it is conclude that the PECVD nitride films considered in the study, not far from the stoechiometric composition or slightly N rich, tend to release up to 1200 Mpa, to reach stress similar to that of LPCVD layers.

LPCVD nitrides are deposited at very low deposition rate, with thermal relaxation occurring during the deposition process. The final stress is then very stable, tensile because of the Si-N bond length. On the contrary, as deposited PECVD films are in metastable equilibrium, with specific residual stress, reached through plasma power and ion bombardment. When annealed, the thermal energy allows the relaxation to a more stable equilibrium, similar to that of LPCVD nitride, at 1200 MPa.

Following these remarks, stress values σ plotted on Figure 2 have been fitted with a double relaxation Maxwell relation (1), where τ_1 and τ_2 are material coefficients. σ_{init} and σ_∞ are respectively the initial and final stress values. According to our discussion in the previous paragraph, σ_∞ represents also the equilibrium stress. It is roughly equal for all samples, at 1200 MPa.

$$\sigma(t) = \sigma_0 \exp\left(-\frac{t}{\tau_1}\right) + (\sigma_{init} - \sigma_0 - \sigma_\infty) \exp\left(-\frac{t}{\tau_2}\right) + \sigma_\infty \qquad (1)$$

The generation of compressive PECVD nitride has been extensively studied in the literature. It is often linked to hydrogen incorporation, bonded on nitrogen (N-H). Hasegawa has suggested that the residual stress is decreased by forming a certain proportion of Si-NH-Si structures instead of typical $N-Si_3$ bonds [8]. Due to the N-H bond, only 2 Si-N bonds remain on the nitrogen atom instead of 3, inducing a different Si-N-Si angle and a lower residual stress, compared to the 3 Si-N bonding case. This operation results in the deposition of metastable nitrides with residual low tensile stress or even compressive.

Following these observations, it is assumed that the mechanisms that yield to a stress increase during annealing (up to 1200 MPa) are somewhat inverse to those occurring during the plasma deposition. Consequently, these mechanisms must involve hydrogen bonded on nitrogen. When annealed, N-H bonds are broken and the stress increases by formation of Si-N network. Similarly to the analyses of Hasegawa [8] and Boehme and Lucowsky [9], one can consider the following reactions involving N-H bonds:

Si-Si +N-H → Si-N + Si-H   (2)
Si-H +N-H → Si-N + H_2 (3)
Si_2-N-H + 2 N-H → 2 Si-N + NH_3 (4)

Reactions (2) and (3) result in generation of an additional Si-N while the amount of Si-N bonds remains constant in (4). Reaction (2) requires the presence of Si-Si bonds and significant increase of the Si-H concentration in parallel to decrease of the N-H bonds, without any H desorption. On the other hand, in (3), Si-H decreases at the same rate than N-H. A combination of (2) and (3) implies decrease of N-H without any change in Si-H and increase of Si-N concentration. Figure 4 presents the evolution of the area of the FTIR Si-N peak along the anneal cycles. Any significant increase of Si-N bonds along the anneal is not visible in the figure. As a consequence, (4) is probably the best candidate to explain the decrease of N-H concentration without any change in Si-H and Si-N bond concentration, especially on stoechiometric or slightly N rich films (few Si-Si bonds). In addition, because of one N atom loss (NH_3 gas formation), it gives space for a good reorganisation of the layer compared to the other reactions, in particular to a densification of the Si-N skeleton. However, a process of Si-N bonds formation following (2) or (3) might represents less than 5 % of the total Si-N bonds content and could be difficult to detect by FTIR on thin layers. In reality, it is probably a combination of the three reactions, but reaction (4) is the prominent one to form Si-N bonds on these materials.

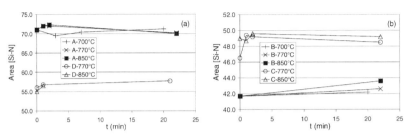

Figure 4. Evolution of the Si-N Area absorption peak (FTIR spectra) along annealing on samples A, D (a) and C, D (b).

Activation energies $E_a(\Delta\sigma)$ of stress relaxation processes are presented in Figure 5 and stays around 1eV or below for all films. They decrease with annealing duration and, as seen before, a double Maxwell model is needed for an accurate Figure 2 data fit. These observations significate that short and long terms behaviour are different.

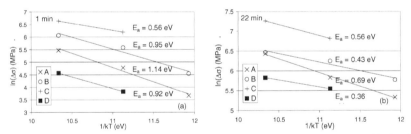

Figure 5. Activation energies of delta stress $E_a(\Delta\sigma)$ calculated at (a) 1 min and (b) 22min, for samples A, B, C, D.

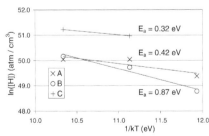

Figure 6. Activation energies of H release after 22 min of anneal, as a function of temperature, for sample A, B, and C.

H desorption is also thermally activated as presented in Figure 6. The activation energies of hydrogen release process E_a (H) obtained from these data values stay below 1 eV. These low activation energies values are consistent with those measured on hydrogen release phenomena by Boehme and Lucowsky [9] and by thermal desorption spectroscopy in [10].

To summarize, thermal stress relaxation and hydrogen desorption present activation energies of same orders of magnitude. This argues positively with the theory of Si-N network reformation by hydrogen release [2-4].

CONCLUSION

Nearly stoechiometric PECVD nitrides, when annealed, tend to become more tensile or less compressive until they reach a final equilibrium around 1200 MPa, which is the high temperature LPCVD nitride stress. In parallel, significant hydrogen degas is observed during these thermal treatments. It is assumed that as deposited PECVD nitrides are in a metastable equilibrium induced by ion bombardment and H incorporation during deposition process. When annealed, the material relax, until the stable stage represented by the LPCVD material. Stress variation and hydrogen release present some activation energies of 1 eV or below, consistent with literature. These values are in line with model of material relaxation involving hydrogen release through N-H bond concentration decrease and stress increase by Si-N bond network formation. Reaction Si_2-N-H + 2 N-H $\rightarrow$ 2 Si-N + NH_3 appears to be the best candidate to figure out the relaxation phenomenon.

REFERENCES

1. F Ootsuka & al., "A Highly Dense, High-Performance 130nm node CMOS Technology for Large Scale System-on-a-Chip Application" IEDM (2000).
2. M. P. Hughey, R.F. Cook "Stress stability of PECVD silicon nitride films during device fabrication" Mat. Res. Soc. Symp. Proc. Vol. 766 (2003).
3. Y. Saito, T. Kagiyama, S. Nakajima "Thermal expansion and atomic structure of amorphous silicon nitride thin films" Jpn. J. Appl. Phys. Vol 42, 1175 (2003).
4. M. P. Hughey, R.F. Cook "Massive stress changes in plasma-enhanced chemical vapor deposited silicon nitride films on thermal cycling" thin solid films 460, 7 (2004).
5. S.-S. Chen, X. Zhang & S.-T. Lin "Intrinsic stress generation and relaxation of plasma-enhanced chemical vapor deposited oxide during deposition and subsequent thermal cycling" Thin Solid Films 434, 190 (2003).
6. W. A. Landford and M. J. Rand, J. Appl. Phys. 49, 2473 (1978).
7. M. P. Hughey and R.F. Cook "Irreversible tensile stress development in PECVD silicon nitride films" Mat. Res. Soc. Symp. Proc. Vol. 795 (2004).
8. Hasegawaa, Y. Amano, T. Inokuma and Y. Kurata "Effects of active hydrogen on the stress relaxation of amorphous SiN_x:H films" J. Appl. Phys. 75, 1493 (1994).
9. C. Boehme and G. Lucovsky "Dissociation reactions of hydrogen in remote plasma-enhanced chemical vapor-deposition silicon nitride" J. Vac. Sci. Technol. A 19(5), 2622 (2001).
10. D. Benoit, P. Morin and J. Regolini "Study of hydrogen desorption from PECVD silicon nitride and induced defect passivation" 207th Electron Chemical Society proceedings (2005).

AUTHOR INDEX

SUBJECT INDEX